Circular Economy and Sustainable Innovation

Pardeep Singh · Shikha Daga · Kiran Yadav ·
Vandana Mishra
Editors

Circular Economy and Sustainable Innovation

Pathways and Applications for a Greener Future, Volume 2

Springer

Editors
Pardeep Singh
Department of Environmental Studies
PGDAV College
University of Delhi
New Delhi, Delhi, India

Shikha Daga
Department of Commerce
PGDAV College
University of Delhi
New Delhi, Delhi, India

Kiran Yadav
Department of Commerce
PGDAV College
University of Delhi
New Delhi, Delhi, India

Vandana Mishra
DU-DDA Biodiversity Parks Programme
Department of Environmental Studies
Centre for Interdisciplinary Studies on
Mountain and Hill Environment (CISHME)
University of Delhi
New Delhi, Delhi, India

ISBN 978-3-032-00436-9 ISBN 978-3-032-00437-6 (eBook)
https://doi.org/10.1007/978-3-032-00437-6

This Springer imprint is published by the registered company Springer Nature Switzerland AG
The registered company address is: Gewerbestrasse 11, 6330 Cham, Switzerland

Preface

The global sustainability crisis demands a transformative shift in how resources are used, products are designed, and systems are managed. In this context, the circular economy—an economic model aimed at minimizing waste and maximizing resource efficiency—has emerged as a powerful alternative to the traditional linear "take-make-dispose" paradigm. However, to translate this concept into actionable strategies, sustainable innovation must play a central role in redefining business models, policymaking, education systems, and technological adoption.

This edited volume, *Circular Economy and Sustainable Innovation: Pathways and Applications for a Greener Future, Volume 2*, brings together interdisciplinary research and case studies that explore the dynamic interplay between circular practices and innovative solutions across sectors and geographies. The book aims to serve as a roadmap for academics, policymakers, industry professionals, and sustainability advocates seeking evidence-based insights and scalable models for driving circular transformation.

Structured in a logical and progressive sequence, the chapters begin with global and regional policy frameworks, offering comparative insights into circular economy implementation across BRICS, WANA, and other regions. The volume then delves into conceptual foundations, including indigenous knowledge systems and regulatory drivers of sustainable innovation. Sector-specific chapters highlight practical applications in agriculture, renewable energy, health care, textiles, fashion, education, and e-waste management—each underscoring the importance of collaboration, technology, and inclusive governance.

The final chapters reflect on the barriers, challenges, and measurement frameworks essential for evaluating circular performance and ensuring accountability. By capturing a diverse range of perspectives and empirical evidence, this book bridges the gap between theory and practice, illuminating the pathways toward a sustainable, circular future.

We extend our deepest gratitude to all contributing authors for their rigorous scholarship and commitment to sustainability. We also thank our editorial and review team, whose expertise and support were instrumental in shaping this volume. It is our sincere hope that this book will inform, inspire, and empower readers to drive systemic change in pursuit of a greener, more resilient world.

New Delhi, India

Pardeep Singh
Shikha Daga
Kiran Yadav
Vandana Mishra

Contents

Policy Approaches for Circular Economy and Green Transition in the BRICS Countries

Anastasia Perminova, Anastasiia Tyukhtina, Mariia Antasheva, and Yulia Milshina

Abstract Nowadays, the world is facing many global issues; therefore, the climate change problem and pollution have become one of the key points of contemporary societies' growing concerns. Such abovementioned environmental crises might lead to terrible consequences for the environment and humanity that are primarily tied to global warming, including sea level rise, droughts, biodiversity loss, etc. These phenomena are caused by greenhouse gases (the so-called "greenhouse effect"), which are produced alongside the generation of electricity, the manufacturing of goods, the use of transport, and the production of industrial waste. These problems are known to be aggravated against the backdrop of rapid economic growth. In this spirit, nowadays countries tend to actively engage in problem-solving activities aimed at ecological protection. BRICS, being one of the key actors shaping the contemporary world agenda, is also recognized for its active SDGs support. However, even though BRICS members are adopting various national climate policies to facilitate the green transition and the circular economy, the level of progress towards these goals varies from country to country. The goal of this paper is to identify the experiences and achievements of the BRICS countries in the implementation of different policy approaches to the circular economy and green transition. To achieve this goal, we evaluated and compared current policies, legislation, and strategies concerning green transition and circular economy initiatives in the BRICS countries; case study helped to identify major examples of SDGs support within the countries. We analyzed the effect these measures have had on shaping the countries' circular economy and

A. Perminova · A. Tyukhtina · M. Antasheva · Y. Milshina (✉)
Institute for Statistical Studies and Economics of Knowledge, HSE University, Moscow, Russia
e-mail: ymilshina@hse.ru

A. Perminova
e-mail: aaperminova1@gmail.com

A. Tyukhtina
e-mail: atyukhtina@hse.ru

M. Antasheva
e-mail: msantasheva@hse.ru

P. Singh et al. (eds.), *Circular Economy and Sustainable Innovation*,
https://doi.org/10.1007/978-3-032-00437-6_1

overall environment protection awareness, the ways these measures were implemented and how they will develop in the future taking into account the increasing power of BRICS in shaping the new world order.

Keywords Climate change · Circular economy · BRICS · Green policy · Green transition · Energy transition

1 Introduction

Climate change has recently become an acute issue in many countries, including BRICS members (Rahman & Turay, 2018); therefore, the topic of a variety of green policy approaches in BRICS has received considerable attention in academic research. Sampene et al. (2021) in their article provided a comprehensive outlook on policies, programs, and plans of BRICS countries (Brazil, Russia, India, China, and South Africa) aimed at promoting low-carbon environment, highlighting the problems connected with implementation of such policies. Besides, the data from the Climate Action Tracker and the corresponding categorization of politics from there were analyzed. Another study (Dudin et al., 2016) scrutinized the green economy of BRICS countries and based on content analysis of empirical and theoretical data formulated problems and perspectives. Kovalev and Porshneva (2021) examined the specifics of the BRICS countries' climate policies at the national and global levels, as well as the countries' positions on green policies within the association and at summits. Meanwhile, the key role of the BRICS countries in dealing with the climate change issue was emphasized.

In addition to considering the BRICS countries as a unified group in green development issues, each country also has its specific green policies, which were described in several studies. For example, Campos de Andrade (2023) in his research investigated the specifics of governing climate change in Brazil, including the discussion of National Policy on Climate Change (PNMC) and other cross-sectoral policies. At the same time, Popova (2023) in her study provided the analysis of the main low-carbon development policies and instruments in Brazil. Some studies conducted research on Russian environmental policies in the context of green transition (Galazova et al., 2023) and Russian climate position as well (Kochtcheeva, 2024). Besides, other studies evaluated the changes in the climate adaptation policy landscape in South Africa, noting the increase in support for climate legislation from 2004 to 2022 (Khavhagali et al., 2023). Al-Sarihi and Mason (2020) assessed challenges and opportunities for climate policy implementation and integration in such oil producing country as UAE, whereas other studies analyzed decarbonization strategies in UEA based on COP26 and COP27 recommendations, stressing the critical importance of implementation of better policy approaches for achieving Paris Agreement targets of the country (Locke et al., 2023).

When it comes to the topic of green transition, China has also received a fair bit of attention in the scholarly literature. For instance, Pingkuo and Huan (2022)

explored factors that drive the green and low-carbon energy transition in China. They found out that incentives from the economic, the behavioral, the technological, the institutional dimension as well as the Dual Circulation Strategy are decisive as far as the current energy transition process in China is concerned. Some scholars explore the same topic from the perspective of digitalization of environmental governance (Mo et al., 2023), while the others study the role of local governments in PRC in the green transition of the country (Xu et al., 2023).

2 Russia

To understand the specifics of climate policy implementation in Russia, it is necessary to look at the country's overall environmental situation. In 2022, Russia was the 4th largest emitter from energy, process emissions, methane, and flaring among other countries in the world, producing 2165.6 million tons of greenhouse gases; in 2023, Russia was still the 4th largest emitter from energy production in the world, but the amount of emissions increased to 2176.1 million tons. Overall, over the ten years from 2013 to 2023, emissions increased by 5.6% (Energy Institute, 2024; Reuters, 2024).

Russia's climate policies can be divided into engagement into international projects and conducting domestic policies aimed at reducing greenhouse gases.

2.1 Russia's Engagement into International Climate Projects

One of the examples of Russia's international engagement into climate projects is the signing of the Paris agreement in 2016. Even though the agreement was not ratified by the Russian government (RBC, 2019), it entered into force in Russia in November 2019 (TASS, 2020), following a decree of the government of the Russian Federation dated the 21st of September 2019 No. 1228 "On the adoption of the Paris Agreement" (Government of the Russian Federation, 2019). Moreover, since 2021, the Russian Federation started participating in the OECD's International Climate Action Program (Government of the Russian Federation, 2021).

2.2 Russia's Domestic Actions on Climate Policy

Concerning Russia's domestic climate policy, the government approved the "Strategy for Low Greenhouse Gas Emissions Socio-Economic Development until 2050" in October 2021. The primary objective of this strategy is to cut net greenhouse gas emissions between 2021 and 2050, ensuring they remain below both Russia's previous levels and those of the European Union. This initiative supports global efforts to limit the rise in average temperatures to well under 2.2 °C. The strategy outlines two potential pathways for Russia's economic development: an inertial scenario and a targeted (intensive) scenario. The inertial scenario maintains the existing economic structure, including the current balance of energy production and consumption. In contrast, the targeted scenario focuses on enhancing Russia's global competitiveness and fostering sustainable economic growth amid the worldwide shift toward cleaner energy. According to projections, under the targeted approach, emissions would fall from 2119 million tons of CO_2 equivalent in 2019 to 1830 million tons by 2050, whereas the inertial scenario would see emissions rise to 2521 million tons over the same period. Given this disparity, the strategy emphasizes a comprehensive, goal-oriented approach across multiple sectors and is favored due to its potential effectiveness (Government of the Russian Federation, 2021).

The following areas are taken into account (Ministry of Economic Development of the Russian Federation, 2023):

Sector/area	Goals
Energy sector	• Reduction of coal's share in electricity generation • Utilization of associated petroleum gas • Improvement of energy efficiency and reduction of energy losses in power grids • Implementation of renewable energy sources
Transport sector	• Electrification and gasification of transport is planned to provide for the population with green transport
Construction sector	• Implementation of stricter requirements for new constructions
Industry sector	• Introduction of regulatory and technical documentation considering emission and energy efficiency indicators • Development of carbon capture, storage, and utilization technologies • Reduction of emissions in carbon-intensive industries (metallurgy, cement, chemistry)
Agricultural sector	• Usage of precision agriculture • Production of biofuels
Forestry and land management	• Flooding of peatlands • Improving the efficiency of forest management, protection, and preservation of forests • Reforestation
Waste management	• Transition to a circular economy • Usage of landfill gas
Inter-sectoral measures	• Introducing fiscal, tax, and budgetary policies to incentivize lower greenhouse gas emissions • Expanding green finance tools, fostering voluntary carbon markets, and engaging in Article 6 mechanisms under the Paris Agreement • Establishing a framework for non-financial public reporting and accountability

In October 2023, Russia approved its Climate Doctrine built upon the principles outlined in the earlier "Strategy for Low Greenhouse Gas Emissions Socio-Economic Development until 2050". This new policy established 2060 as the target year for Russia to reach carbon neutrality (The President of the Russian Federation, 2023).

It should be noted that in 2022 Russia started the process of establishing of voluntary carbon market system as a tool to tackle emissions, however, currently the market is in the pilot stage, as it has only been launched in one region—Sakhalin Oblast (Sitnikov, 2022). Using the cap-and-trade carbon market type (OECD, 2021), The Russian government has designated Sakhalin Oblast as a pilot region for carbon neutrality, aiming to achieve full decarbonization by 2025. However, the testing period for this initiative will extend beyond that target, running from September 1, 2022, to December 31, 2028.[1]

2.3 *Russia Energy Mix*

In 2022, the share of gas in electricity generation in Russia accounted for 45.17%, coal—17.02%, nuclear energy—19.17%, and hydro energy—16.94% (Ember, 2024a, 2024b, 2024c, 2024d). The share of energy received from wind and solar was 0.36% and 0.21%, respectively (Ember, 2024a, 2024b, 2024c, 2024d). In 2023 Russia generated 44.85% of its electricity from gas, 18.45% from nuclear power plants, 17.92% from coal, and 17.05% from hydroelectric power plants (Ember, 2024a, 2024b, 2024c, 2024d). Wind and solar energy accounted for 0.4% and 0.22% respectively (Ember, 2024a, 2024b, 2024c, 2024d). The emission reduction strategy was introduced in the second half of 2021, so it would be logical to look at the first results from 2022 onwards. However, the share of each type of electricity generation source remained almost unchanged in 2022 and 2023, including no noticeable increase in renewable energy sources such as wind, hydro, and solar. Although the share of hydro and nuclear energy is quite large, Russia is more than 50% dependent on fossil fuels such as gas and coal. The evolution of total energy supply in Russia from 2000 to 2022 is presented in Fig. 1.

In summary, as of 2024, Russia has formally adopted a legally binding climate strategy, setting 2060 as the target year for carbon neutrality and committing to substantial emission reductions by 2050. However, recent data from 2023 shows an increase in emissions compared to the previous year, with no major shifts in the energy sector structure. The share of renewable energy sources remained stagnant, while fossil fuels continued to dominate the power generation mix.

---

[1] Cap-and-trade system has a fixed upper limit of emissions and allowances are usually sold at auctions or are given for free if the company meets certain criteria.

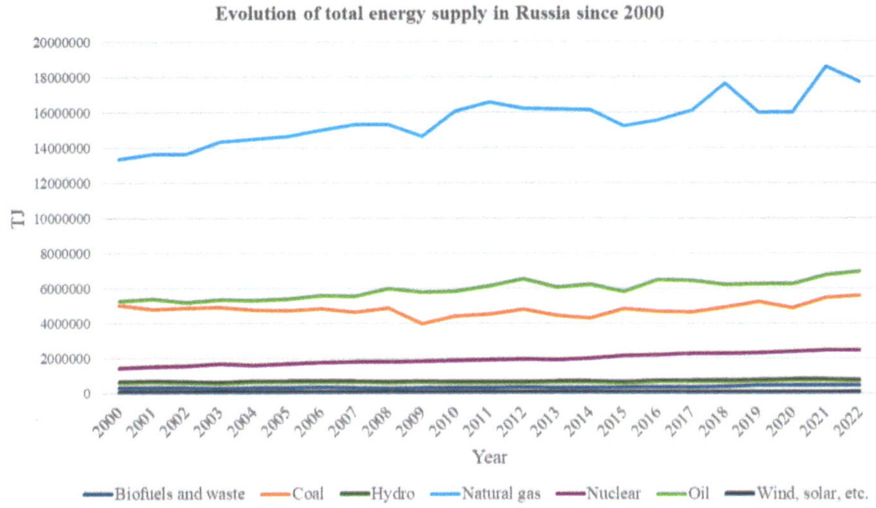

Fig. 1 Evolution of total energy supply in Russia from 2000 to 2022. *Source* IEA Data Services

3 Brazil

The situation with GHG emissions in Brazil is the following: in 2022 Brazil emitted 505.5 million tons of GHG from energy, process emissions, methane, and flaring, while in 2023 the country produced 525 million tons of such GHG emissions. Despite the increase by 2023, over the previous ten years, starting in 2013, emissions decreased by 3.6%. Brazil is also not among the top ten largest polluter countries from emissions from energy, process emissions, methane, and flaring both in 2022 and 2023 (Energy Institute, 2024).

3.1 Deforestation in the Focus of Current Climate Issues in Brazil

One of the main climate issues in Brazil is the deforestation of the Amazon forests. The Amazon Forest is home to about three million plant and animal species and one million indigenous peoples. Approximately 60% of this area is in Brazil. The aggravation of the deforestation issue may lead to the fact that forests will no longer generate enough rainfall to maintain their ecosystem, as well as economically important sectors such as agriculture, hydropower, and water supply. The environmental benefits provided by forest ecosystems throughout Latin America and other regions will likewise experience consequences from these changes. It should be noted that 75% of greenhouse gas emissions in Brazil are connected with land-use change and agriculture, in this regard, it is critically important to stop deforestation and switch to low-carbon farming methods (World Bank Group, 2024a, 2024b); especially regarding the fact that Brazil's agriculture is recognized to be one of the key economic sectors, as the country holds a leading position as an exporter of agricultural products and is known as one of the largest producers of coffee, soybeans, beef and sugar cane (Food Export, 2024).

The deforestation rate in the Brazilian Amazon in 2023 is down almost 50% from the previous year; moreover, Brazil's Environment Ministry announced that this is the lowest deforestation rate in five years (BBC, 2024). Data showed that 5,153,000 km^2 (1 989.6 square miles) of Amazonian territory was deforested in 2023, up from 10,278,000 km^2 in 2022. Brazilian President Luiz Inácio Lula da Silva promised to end deforestation by 2030 when he took office a year ago (BBC, 2024). The president also promised to restore rainforests and prosecute climate criminals during his speech at the COP27 climate summit in 2022 (BBC, 2022). Under his predecessor, Jair Bolsonaro, forest destruction reached its highest level in 12 years (BBC, 2020).

3.2 *Brazil Domestic Climate Policies*

In 2016 Brazil ratified the Paris Agreement and determined its Nationally Determined Contribution (NDC). Since Brazil changed its NDC in December 2021 and April 2022 and established the following plan for climate actions: the level of GHG emissions must be reduced by 37% by 2025, and by 50% by 2030 as compared with the level of emissions of 2005. 2050 is defined as the deadline for achieving carbon neutrality (European Parliament, 2022).

Since Luiz Inácio Lula da Silva became president, Brazil has implemented significant changes to its climate policy framework. A key development occurred in October 2023 when the government resubmitted its Nationally Determined Contributions and subsequently initiated a comprehensive review of the National Climate Policy (PNMC) through the Inter-Ministerial Council on Climate Change. This ongoing review process examines sector-specific mitigation and adaptation measures with planning horizons extending through 2035, while also establishing specific emissions reduction targets for both 2030 and 2035 across various economic sectors. These policy developments mark a substantial shift in Brazil's approach to climate governance under the new administration. These implementations can take about two years.

Moreover, as mentioned earlier, much attention has been paid to combating deforestation, and significant results have been achieved by 2023, but deforestation remains above the 2010 level; in addition, certain forest regions such as Cerrado have become more affected (where deforestation increased by 6% in 2023).

Next, a new strategy, the Ecological Transformation Plan (PTE), is being developed, the main objective of which will be to continue to reduce emissions, as well as to significantly develop the energy sector and make it one of the most critical sectors of the country's economy. Nevertheless, several aspects of climate policy raise questions. For example, the PTE strategy mainly focuses on reducing emissions in the transportation sector with the help of electric motors and low-emission fuels, as well as on the development of renewable sources such as wind and solar energy. However, there are no specific targets for fossil fuel divestment in transportation. Moreover, Brazil also lacks a general timetable for fossil fuel divestment (Climate Action tracker, 2024).

The experience of cooperation between the Brazilian and German governments is also of interest. The Climate Policy Program Brazil (PoMuC), supported by Deutsche Gesellschaft für Internationale Zusammenarbeit (GIZ) GmbH—Brazil, was held from August 2016 to May 2024. The main aspects of the cooperation were to support Brazil in successfully achieving medium- and long-term carbon neutrality targets. The program implemented a national REDD + strategy (National Strategy for Reducing Emissions from Deforestation and Forest Degradation) (Food & Agriculture Organization of the United Nations, 2016) and developed a Federal Payment Program for Environmental Services. The transparency system to track the progress of climate policy was significantly improved. Also, the capacity for greenhouse gas reporting and climate finance was strengthened jointly with the Ministry of Economy of Brazil. The program budget amounted to 11,250,000 € (IKI, 2024).

As for the carbon markets, Brazil is currently in the consultation phase for the launch of the ETS (Emission Trading Scheme). However, currently, many businesses in Brazil are involved in ETS simulation experiment, so the results of this experiment will determine the future of ETS in the country (The Oxford Institute for Energy Studies, 2022). However, the legislative part of this project is already progressing. On the 21st December 2023, the Brazilian Chamber of Deputies approved the new text of Bill 2.148/15, which creates a greenhouse gas emissions trading system in Brazil. The text will now be considered in the Federal Senate and may return to the chamber if changes are made (Global Compliance News, 2024).

3.3 Brazil Energy Mix

Regarding the mix of power generation sources, in 2022, most of Brazil's electricity was generated from renewable energy sources. 63.07% of the energy was produced from hydroelectric power plants, 12.05% from wind, 8.13% from bioenergy, 4.45% from solar energy, and 2.15% from nuclear energy. 6.22% and 2.10% of energy were produced from gas and coal, respectively. By 2023, it can be observed that the mix of resources in power generation has changed as follows: the amount of solar energy increased from 4.45 to 7.25%; and wind energy from 12.05 to 13.45%. It can also be seen that the share of hydropower declined, as it became 60.37%; the use of gas slightly decreased to 5.31%. The other indicators experienced little or no change. Thus, we can say that Brazil produces 91% of its energy from clean sources (if we consider nuclear energy as green energy); therefore, Brazil's energy generation sector is not dependent on fossil fuels (Ember, 2024a, 2024b, 2024c, 2024d).

Despite progress in some areas of Brazilian climate policy, as well as electricity generation, these achievements are not enough to achieve the NDC. It is projected that Brazil's total GHG emissions (excluding land use, land-use change, and forestry sector) will be between 1162 and 1180 Mt in 2030 (Brazil's total emissions accounted for 1170 Mt in 2022). Thus, it is likely that Brazil will not be able to meet either its 2025 or 2030 targets (Climate Action tracker, 2024).

4 South Africa

In 2022, South Africa produced 461.7 million tons of greenhouse gas emissions from energy, process emissions, methane, and flaring, by 2023 the amount rose to 478.1 million tons. In the period from 2013 to 2023, emissions declined by 4.6% (Energy Institute, 2024).

4.1 South Africa Domestic Climate Policies

South Africa ratified the Paris Agreement in 2016 and has adhered to its goals since then (South African Government, 2016). Currently, South Africa has several climate policies and strategies in place. For example, the National Development Plan 2030, which sets out South Africa's NDC and a timeline for emission reductions, according to which the country's total GHG emissions should be reduced to 350–420 Mt by 2030, and by 2050 the country plans to achieve carbon neutrality. To achieve these goals, several policies have been introduced to combat emissions. For example, in 2019, South Africa became the first African country to introduce a carbon tax (Darby, 2019). The tax is based on the principle of levying a tax on fuels based on emission factors and procedures consistent with standards published by the Intergovernmental Panel on Climate and covers about 90% of South Africa's total emissions, but does not cover the following areas: agriculture, forestry, land use and waste. In 2022, the tax rate was US$8 per ton of carbon dioxide emissions, which was significantly lower than in European countries, but the government plans to increase the tax rate over time. So, by 2026 a ton of CO_2 emissions will cost US$20, by 2030 to US$30, and by 2050 to US$120 per ton (Haonan et al., 2023). In addition to the carbon tax, there are other green taxes such as transport fuel tax, engine tax, aviation tax, water tax, etc. South Africa also has a carbon market, called the South Africa Crediting Mechanism, which was launched in 2019 (World Bank Group, 2024a, 2024b). Furthermore, in October 2023, South Africa's National Assembly approved the Climate Change Bill. This document also requires approval from the National Council of Provinces and the President. However, if approved, the bill will have a significant impact on South Africa's climate policy, as it will give formal legislative status to the NDC. In this case, the government will need to set sectoral emission targets and carbon budgets for companies that are significant emitters of GHGs. These companies will also be legally liable for emissions through penalties, including fines (Climate Action Tracker, 2023).

Other countries are also actively assisting South Africa in its energy transition. During the United Nations Climate Change Conference (COP 26), the US government, together with the UK, France, Germany, and the European Union, launched the Just Energy Transition Partnership (JETP) with South Africa. The program aims to help South Africa reduce the use of coal and make a rapid transition to green energy. During the program, donor countries plan to provide South Africa with about US$8.5 billion through various financial instruments, including loans, over a period of 3–5 years to implement various green initiatives. In addition to JEPT funding, South Africa received $45 million for the USAID Power Africa initiative (USAID, 2022).

4.2 South Africa Energy Mix

Regarding the generation of electricity in South Africa, in 2022, coal was the leader among sources, with 84.45%. Then, 4.43% nuclear energy was used, followed by 1.3% hydro energy, 0.17% bioenergy, and 4% solar and wind energy. In 2023, the mix of sources remained almost the same, although there was a slight decrease in coal to 81.57% and nuclear energy to 3.8%. At the same time, solar energy increased to 6.7%, and wind energy increased to 4.96%. The share of other sources remained the same. Overall, South Africa's energy sector relies on fossil fuels for more than 80%, primarily on coal (Ember, 2024a, 2024b, 2024c, 2024d).

However, there are several challenges to South Africa becoming carbon neutral. For example, financing is currently underfunded. South Africa requires an increase in climate finance of at least three to five times the current average of $7.2 billion USD. It is estimated that the country needs an average of $18.37 billion USD per year to meet the net zero target by 2050, and $28.36 billion USD annually to meet the NDC target by 2030. Moreover, in 2023, funding for climate policies came mainly from private investment. The private investment accounted for 86% ($5.99 billion USD) of annual investment, while public investment accounted for only 14% ($954 million USD) (Climate Policy Initiative, 2023; De Aragão Fernandes et al., 2023).

Despite the rapid development of South Africa's climate policies, its climate strategy is likely to fail targets set by the Paris Agreement. Thus, even if South Africa's current climate policies are fully implemented, the global average temperature will exceed 3 °C if all other countries follow the same intensity of climate policy (Climate Action Tracker, 2023).

5 UAE

In the United Arab Emirates, the number of emissions reached 336.3 million tons in 2022 and 340.8 million tons in 2023. Over the decade 2013–2023, the number of emissions grew by 14.1% (Energy Institute, 2024).

5.1 UAE Domestic Climate Policies

In September 2016, the UAE was one of the first Middle Eastern countries to ratify the Paris Agreement and is, therefore, actively implementing climate policies. In 2021, one of the central policies of the UAE was the Government's General Environmental Policy. This policy covered many areas, from cutting greenhouse gases to reducing the loss of biodiversity. For example, the climate change mitigation aspect in support of the country's economic development priorities included the following policy directions: managing the country's greenhouse gases and supporting projects aimed at reducing emissions; developing programs and strategies to increase the adaptability and flexibility of ecosystems, economic sectors, and society in response to climate change; and strengthening the UAE's position in global competitiveness indicators.

It also set specific targets: firstly, to increase the share of green energy sources in the UAE energy mix to 50% by 2050; secondly, to reduce individual and institutional energy consumption by 40% by 2050; and thirdly, to ensure that critical economic sectors are prepared and adaptable to the impacts of climate change by 2030 (UAE Government, 2023).

In 2023, the UAE government set new targets for achieving carbon neutrality by 2050 in its updated "UAE Energy Strategy 2050". The strategy sets out clear actions until 2030 and ambitions to achieve carbon neutrality by 2050. So, the government now plans to triple the number of renewable energy sources (and increase installed clean energy capacity from 14.2–19.8 GW) and invest between AED 150 and AED 200 billion in them by 2030 to meet the country's growing demand for energy because of a rapidly growing economy. In addition, it is scheduled to increase individual and institutional energy consumption efficiency by 42–45%, compared to 2019, and to create 50,000 new green jobs (UAE Government, 2024). Furthermore, in 2023, the UAE launched a voluntary carbon market pilot project under the Dubai Financial Market (DFM). The initiative is aimed at contributing to the goal of achieving net zero by 2050 (DFM, 2024).

5.2 UAE Energy Mix and Strategies Towards Enabling Changes

An Interesting fact about the UAE's generation mix is that until 2020, they produced more than 90% of their energy from gas. However, the situation began to change later, and the mix of energy sources became more diverse. In 2022, for example, 81% of energy was still produced from gas, but the share of nuclear and solar energy increased. In 2023, the share of gas decreased to 72%, while the shares of nuclear and solar power increased, and wind power entered the market with a share of 0.03%. This shows that the UAE is still heavily dependent on fossil fuels, although a slight diversification has happened (Ember, 2024a, 2024b, 2024c, 2024d).

In the current situation, the country will not be able to fulfill its NDC commitments. Projections show that emissions will continue to rise until 2030 and achieving the 1.5 °C compatibility targets will require a 38% reduction in emissions from 2022 levels (Climate Action Tracker, 2024).

6 China

China is among five countries that produce the most Carbon Dioxide (CO_2) (Friedrich et al., 2023a, 2023b). According to the most recent data from the Global Carbon Atlas, China tops the list as the 'worst offender', with the U.S., India, Russia and Japan following it (Global Carbon Atlas, 2023). China included environmental sustainability in its development strategies as early as in the 1980s (Xie, 2020). Apart from this, the country participated in the 1992 UN Conference on Environment and Development held in Rio de Janeiro and issued its own twenty-first century Agenda on sustainable development (Xi, 1999). However, in the twentieth century the economic growth targets outweighed environmental concerns as emission from fossil fuel combustion and industrial processes in China continued to grow.

Participation in the Paris agreement became the pivotal moment for China. It is due to this international treaty on climate change that the country adjusted its economic policy strategies to fulfill its goals of a low-carbon, green transition and sustained economic growth. In 2021, president Xi Jinping addressed the UN General assembly and announced that China would strive to reach carbon dioxide peak by 2030 and carbon neutrality by 2060. This plan is also often referred to as the "30–60 goals" (UN News, 2021).

6.1 China's Policy Strategies Towards Low Carbon Development

Policymakers created an overarching policy framework to guide China's energy transition. Its name is "1 + N", where "1" points to a policy document containing all the guiding principles for all policies towards the country's peaking and neutrality targets (The Working Guidance for Carbon Dioxide Peaking and Carbon Neutrality in Full and Faithful Implementation of the New Development Philosophy) (UNDP, 2021). According to it, China plans to gradually increase the share of non-fossil energy consumption to around 20% by 2025, around 25% by 2030, and over 80% by 2060.

At the same time, the "N" stands for a number of additional policy documents that set the guiding principles for all policies aimed at China's peaking and neutrality targets.

After Xi Jinping delivered speeches in the UN General Assembly, China released two key documents. They are:

- Opinions of the Central Committee of the Communist Party of China and the State Council on Completely, Accurately, and Comprehensively Implementing the New Development Concept to Well Implement the Work of Carbon Peaking and Carbon Neutrality (The Opinion) (De Boer & Fan, 2022);
- Action Plan aimed at Peaking Carbon Emissions before 2030 (the Action Plan) (Department of Resource Conservation and Environmental Protection, 2021).

Taken together, these documents form the cornerstone of China's highest-level policy assessments.

35 policy areas, 10 key targets and 5 principles are determined in The Opinion as a direction for action. At the same time, The Opinion does not contain any action-oriented policies or situation analysis. Instead, the objective of this top-policy document it to show China's eagerness to move forward in green, low-carbon transition to reach 30–60 goals. On the contrary, The Action Plan (AP) includes China's concrete efforts aimed at reducing carbon emissions. It outlines the top 10 areas for action to reach peak—and afterwards reduce—carbon emission levels. Furthermore, the *Action Plan* contains 43 tasks in diverse sectors, such as, for example, industrial development, energy transition, urban and rural construction.

6.2 Key Action Areas

6.2.1 Energy

The largest source of emission in China is energy use (IEA, 2021). Three strategic frameworks underpin the country's energy transition. Through demand-side governance, China simultaneously targets energy conservation and electrification. At the same time, on the supply side, China is pursuing a dual strategy of decreasing fossil fuel consumption while developing renewables.

Currently, China is still heavily dependent on coal, the fossil fuel with the largest greenhouse gas emissions per unit of energy. However, in the coming years, the country is planning to continue its shift from maintaining the energy system based on fossil fuels to creating a system that is much more reliant on renewables.

China's plan of energy transition is presented in Fig. 2.

As seen in Fig. 2, although at present fossil fuels supply 80% of China's energy today, the country's strategy envisions coal, oil, and gas contributing just 14% by 2060. At the same time, China plans to significantly increase its renewables capacity.

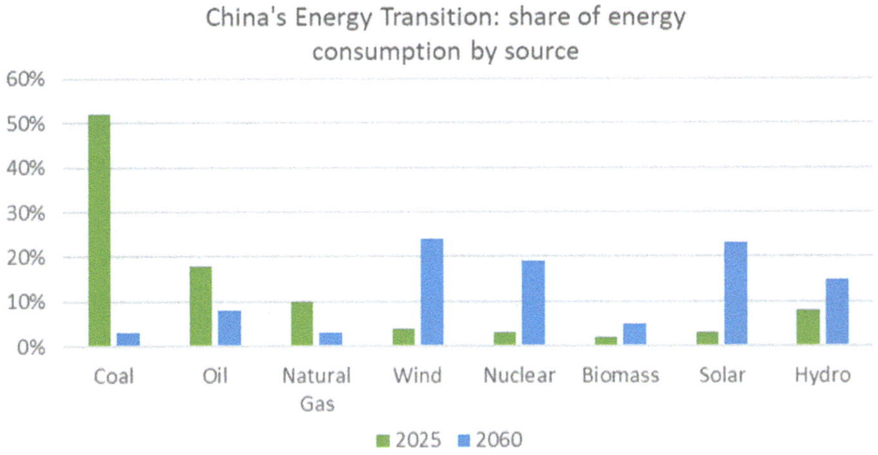

Fig. 2 China's ENERGY transition. *Source* Tsinghua University Institute of Energy, Environment and Economy; U.S. EIA

6.2.2 Transportation System

The shift to clean energy is essential for developing a sustainable, low-carbon transportation system in China (Wang et al., 2022). The emphasis is on developing a more systematic and efficient transportation system while increasing the use of renewable energy within it.

6.2.3 Industrial Sector

The Green Development Plan in Industrial Sector (Climate Cooperation China on behalf of the International Climate Initiative, 2021) states that China is set to develop a comprehensive decarbonization roadmap targeting its four most emissions-intensive sectors: petrochemicals, steel, nonferrous metals, and building materials.

6.2.4 Urban Area

As stated in the policy document on advancing green development in urban and rural construction (issued by the China's State Council) (The General Office of the CPC Central Committee & the General Office of the State Council, 2021), Critical strategies for reshaping urban–rural development structures encompass promoting green building practices, fostering sustainable lifestyles, and reinforcing systemic infrastructure.

Overall, China is making active steps to achieve a low-carbon, green transition and sustained economic growth. Even though the country is still heavily reliant on coal, the government's initiatives on intensifying the use of renewables establish a route for a new energy system.

7 India

7.1 India Takes Actions: Country's Energy Sector Development

Despite active spread of renewable energy sources on the energy mix of the country, India is still known to be one of the world serious carbon dioxide generators, following China and the USA in the World's Top Greenhouse Gas Emitters (Friedrich et al., 2023a, 2023b) and becoming world's third largest CO_2 producer. In this spirit, in 2022 the number of emissions reached 2865 million tons and 3121.5 million tons in 2023 (Energy Institute, 2024).

India has been a significant participant in the Paris Agreement, signed the agreement in April 2016 and ratified it in October 2016 (Press Information Bureau, Government of India, Ministry of Environment, Forest and Climate Change, 2016). Therefore, India seeks to harmonize rapid economic growth with ecological preservation, pursuing its climate targets by implementing clean energy projects and at the same time acknowledging the distinct obstacles of its fast-paced development. India has committed to reducing its carbon intensity (CO_2 emissions per unit of GDP) by 33–35% by 2030 compared to 2005 levels (Press Information Bureau, Government of India, Ministry of Environment, Forest and Climate Change, 2023). Moreover, the second of the country's quantifiable targets aiming fighting with the climate change is to achieve 40% of its energy capacity from non-fossil fuel sources by 2030 (Press Information Bureau, Government of India, Ministry of Environment, Forest and Climate Change, 2023) including investments in renewable energy such as solar and wind power. This is a critical part of India's nationally determined contributions (NDCs).

7.2 Legal Framework of Dealing with Climate Issues: India's Long-Term Low-Carbon Development Strategy

One of India's main official initiatives targeting the climate change is "India's Long-Term Low-Carbon Development Strategy" issued by the country's Ministry of Environment, Forest and Climate Change in 2008 (Ministry of Environment, Forest and Climate Change, Government of India, 2022). A complex strategy aims to synchronize India's domestic climate targets with the global goals outlined in the Paris Agreement. The main target that is set within the strategy is achieving net-zero emissions by 2070 alongside with prioritizing the development of renewable energy

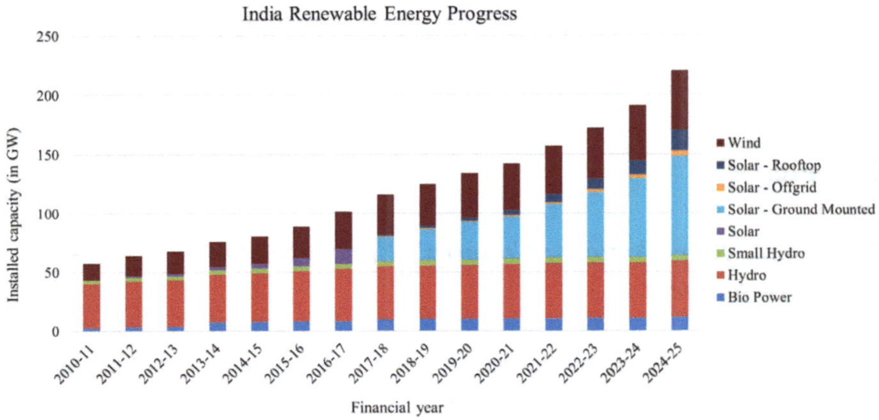

Fig. 3 India renewable energy progress. *Source* NITI Aayog

initiatives in the country and energy efficiency; adopting this strategy also highlights India as a state promoting the principle of"common but differentiated responsibilities and respective capabilities" (CBDR-RC) (Ministry of Environment, Forest and Climate Change, Government of India, 2022). This principle, stated in the United Nations Framework Convention on Climate Change (UNFCCC) of Earth Summit in Rio de Janeiro in 1992, means that even though countries all over the globe need to address the urgent climate change issue, their responsibilities to tackle this issue differ: developed countries having more responsibility than the developing ones—thus, the correlation between the historical process of industrialization and climate change is being admitted. The progress under the initiative regarding the intensification of alternative energy projects development is presented in Fig. 3.

As seen in Fig. 3, India renewable energy mix is becoming more diversified. Due to the government's active initiatives on intensifying the use of alternative energy sources, India has rapidly transformed its green energy sector, reaching 220.1 GW of renewable capacity (as of 2024).

7.3 Initiatives Within the National Action Plan on Climate Change

Next, another important initiative launched by the Indian Government is the National Action Plan on Climate Change (NAPCC) issued in 2008 and revised for several times (Prime Minister's Council on Climate Change, 2008). The plan aimed to counter the climate change issues and its already existing and potential mitigation. Eight key missions targeting different aspects of the environmental concerns strategies that are considered to be particularly crucial for the country were implemented: National Solar Mission, National Water Mission, National Mission for a Green India, National

Mission on Sustainable Habitat, National Mission for Sustaining the Himalayan Eco-system, National Mission for Sustainable Agriculture, National Mission for Enhanced Energy Efficiency, and National Mission on Strategic Knowledge for Climate Change.

Each of the initiatives within NAPCC targets certain ecological issues relevant to the country. Thus, the National Mission for Enhanced Energy Efficiency focuses on improving energy efficiency across industries and introduces market-based mechanisms such as the Perform, Achieve, and Trade (PAT) scheme to incentivize efficiency improvements. Through the National Water Mission, the government aims to enhance water use efficiency by 20%, minimize wastage, and promote equitable distribution of water resources across regions. So, all these missions and the rest ones imply a "green" approach towards using the country's resources and serve as the solid basis for India's sustainable future. The NAPCC underscores a collaborative and participatory approach, brings together government bodies, private enterprises, academia, and civil society. It also emphasizes the importance of raising public awareness and fostering behavioral change to support climate action.

Speaking of the public awareness, state-level initiatives are also driving forces for local communities to be aware and to participate in "sustainable" activities: for example, the Sikkim Organic Mission made the state of Sikkim to be renowned for becoming India's first fully organic state (Government of Sikkim, 2012); the Bhoochetana Program initiated by the Government of Karnataka in 2009, serves as a pioneering agricultural development program aimed at enhancing soil health and improving crop productivity in a sustainable manner (Wani et al., 2015), etc. Thus, by leveraging local resources and knowledge, states are playing a critical role in advancing sustainability and contributing to India's broader climate goals.

7.3.1 Advances in Solar Energy

Among the other NAPCC missions, Jawaharlal Nehru National Solar Mission (launched in 2011) has shown itself to be one the most resonant ones: the mission aimed to achieve grid parity by 2022 and parity with coal-based thermal power by 2030, as well as organizing major R&D activities dedicated to solar energy (Department of Science and Technology, Government of India, 2025). The primary target to generate 20 GW of installed solar capacity by 2022 was achieved in 2018, 4 years ahead of schedule. The effectiveness of India solar power policy is also highlighted by the fact that since the mission implementation India has become the world's third largest solar generator. In Indian Net Zero scenarios, it is the solar energy that is projected to be the main constituent of electricity generation capacity under long-time Net Zero scenarios (Garg et al., 2024).

8 Conclusion

India energy policy strategically balances between rapid renewable energy expansion and pragmatic, "traditional" reliance on fossil fuels to meet its constantly growing energy demand and development needs. The country has emerged as one of the global leaders in solar power, with installed capacity skyrocketing from under 20 GW in 2018 to over 70 GW by 2023, while still deriving about 70% of its electricity from coal. This dual-track approach reflects India's unique challenge of lifting millions of its citizens out of energy shortage while transitioning to cleaner sources. Looking ahead, India's focus on green hydrogen, battery storage, and domestic manufacturing will be crucial for achieving its 500 GW renewable target by 2030 while ensuring energy security.

Acknowledgements The article is based on the study funded by the Basic Research Program of the HSE University.

Annex

Country	Emmissions from energy, process emissions, methane and flaring, 2023 (Mt)	Energy generation mix, 2023	Carbon regulation (carbon market/carbon tax)	Key documents/ climate strategies	Timeline for achieving carbon neutrality
Russia	2176.1	44.85% gas 18.45% nuclear power 17.92% coal 17.05% hydro power 1.04% other fossil 0.4% wind 0.22% solar energy	Pilot voluntary carbon market in Sakhalin Oblast (since 2022)	"Strategy for Low Greenhouse Gas Emissions Socio-Economic Development of the Russian Federation until 2050" Decree of the President of the Russian Federation dated 10.26.2023 No 812 "On approval of the climate doctrine of the Russian Federation"	Under the target scenario, greenhouse gas emissions should decrease between 2019 and 2050 from 2119 million tons of GHG equivalent to 1830 million tons 2060—achiving carbon neutrality

(continued)

(continued)

Country	Emmissions from energy, process emissions, methane and flaring, 2023 (Mt)	Energy generation mix, 2023	Carbon regulation (carbon market/carbon tax)	Key documents/ climate strategies	Timeline for achieving carbon neutrality
Brazil	525	60.37% hydro power 13.45% wind 7.25% solar energy 7.86% bioenergy 5.31% gas 2.04% nuclear power 2% coal 1.71% other fossil	In the consultation phase for the launch of the ETS	National Climate Policy (PNMC) The Ecological Transformation Plan (PTE) The Climate Policy Program Brazil (PoMuC) National REDD + strategy	The level of GHG emissions must be reduced by 37% by 2025, and by 50% by 2030 as compared with level of emissions' of 2005 2050 is defined as the deadline for achieving carbon neutrality
South Africa	478.1	81.57% coal 6.67% solar energy 4.96% wind 3.81% nuclear power 2.09% other fossil 0.72% hydro power 0.18% bioenergy	Carbon tax (since 2019) Carbon market called "South Africa Crediting Mechanism" (since 2019)	National Development Plan 2030 Climate Change Bill The Just Energy Transition Partnership (JETP)	The country's total GHG emissions must be reduced to 350–420 Mt by 2030 South Africa plans to achieve carbon neutrality by 2050
UAE	340.8	72.07% gas 19.55% nuclear power 8.33% solar energy 0.02% bioenergy	Voluntary carbon market pilot project under the Dubai Financial Market (since 2023)	Government's General Environmental Policy "UAE Energy Strategy 2050"	Achiving net zero by 2050

(continued)

(continued)

Country	Emmissions from energy, process emissions, methane and flaring, 2023 (Mt)	Energy generation mix, 2023	Carbon regulation (carbon market/carbon tax)	Key documents/ climate strategies	Timeline for achieving carbon neutrality
India	3121.5	57.66% coal 29.45% crude oil 8.88% gas 1.67% hydro power 1.46% nuclear power 0.88% renewables	Planned compliance carbon market under the Carbon Credit Trading Scheme (CCTS)	National Action Plan on Climate Change (NAPCC) India's Long-Term Low-Carbon Development Strategy	Achiving net zero by 2070
China	12,603.5	60.5% coal 13.2% hydro power 9.4% wind 6.2% solar energy 4.6% nuclear power 3.3% gas 2.2% bioenergy 0.7% other fossil	China National ETS (launched in 2021) The China GHG Voluntary Emission Reduction Program (Implemented in 2014)	1 + N Opinions of the Central Committee of the Communist Party of China and the State Council on Completely, Accurately, and Comprehensively Implementing the New Development Concept to Well Implement the Work of Carbon Peaking and Carbon Neutrality (The Opinion) Action Plan for Peaking Carbon Emissions before 2030 (the Action Plan)	The country aims to peak carbon dioxide emissions before 2030 and realize carbon neutrality before 2060

References

Al-Sarihi, A., & Mason, M. (2020). Challenges and opportunities for climate policy integration in oil-producing countries: The case of the UAE and Oman. *Climate Policy, 20*(10), 1226–1241. https://doi.org/10.1080/14693062.2020.1781036

BBC. (2020). *Brazil's Amazon: Deforestation 'surges to 12-year high'.* https://www.bbc.com/news/world-latin-america-55130304. Accessed: 02 May 2025

BBC. (2022). *COP27: Brazil is back on the world stage; Lula tells climate summit.* https://www.bbc.com/news/science-environment-63625698. Accessed: 02 May 2025

BBC. (2024). *Amazon rainforest: Deforestation rate halved in 2023.* https://www.bbc.com/news/world-latin-america-67962297. Accessed: 02May 2025

Campos de Andrade, A. L. (2023). *Governing climate change in Brazil: a governance gap analysis of selected cases* (Doctoral dissertation, Universität Potsdam). https://publishup.uni-potsdam.de/58733

Climate Cooperation China on behalf of the International Climate Initiative (IKI). (2021, December 6). *China's five-year green development plan for industrial sectors.* https://climatecooperation.cn/climate/chinas-five-year-green-development-plan-for-industrial-sectors/. Accessed: 19 May 2025

Climate Action Tracker. (2023) *South Africa.* https://climateactiontracker.org/countries/south-africa/. Accessed: 19 May 2025

Climate Policy Initiative. (2023). *The South African climate finance landscape 2023.* https://www.climatepolicyinitiative.org/publication/the-south-african-climate-finance-landscape-2023/. Accessed: 19 May 2025

Climate Action Tracker. (2024) *UAE.* https://climateactiontracker.org/countries/uae/. Accessed: 19 May 2025

Climate Action tracker. (2024). *Brazil.* https://climateactiontracker.org/countries/brazil/policies-action/. Accessed: 02May 2025

Darby, M. (2019). South Africa set to introduce long-awaited carbon tax in June. *Climate Home News.* https://www.climatechangenews.com/2019/02/20/south-africa-set-introduce-long-awaited-carbon-tax-june/. Accessed: 19 May 2025

De Boer, D. & Fan, D. (2022, March 11). Impressive Progress in China's 1+N Policy Framework. *China Council for International Cooperation on Environment and Development.* https://cciced.eco/climate-governance/how-is-progress-in-chinas-1n-policy-framework/. Accessed: 19May 2025

De Aragão Fernandes, P., Gwebu, L., Johansson, L., Meattle, C., Radmore, J. V., & Solomon, C. (2023). *South African climate finance landscape 2023.* Presidential Climate Commission, South Africa.

Department of Resource Conservation and Environmental Protection. (2021, October 27). Action Plan for Carbon Dioxide Peaking Before 2030. *National Development and Reform Commission (NDRC).* https://en.ndrc.gov.cn/policies/202110/t20211027_1301020.html. Accessed: 19 May 2025

Department of Science and Technology, Government of India. (2025). *Jawaharlal Nehru National Solar Mission.* India Science, Technology and Innovation Portal. https://www.indiascienceandtechnology.gov.in/st-visions/national-mission/jawaharlal-nehru-national-solar-mission-jnnsm. Accessed: 05 May 2025

DFM. (2024). *Voluntary Carbon Credit Market.* https://www.dfm.ae/investing/products/carbon-credits. Accessed: 19 May 2025

Dudin, M. N., Artemieva, J. A., Bezbah, V. V. E., & Kirsanov, A. N. (2016). Problems and perspectives of BRICS countries transfer to "green economy" and law-carbon energy industry. *International Journal of Energy Economics and Policy, 6*(4), 714–720. https://dergipark.org.tr/en/pub/ijeeep/issue/31919/351132

Ember. (2024). *South Africa.* https://ember-climate.org/countries-and-regions/countries/south-africa/. Accessed: 19 May 2025

Ember. (2024). *Brazil.* https://ember-climate.org/countries-and-regions/countries/brazil/. Accessed: 02May 2025

Ember. (2024). *Russia electricity generation by source.* https://ember-climate.org/countries-and-regions/countries/russia/. Accessed: 18 April 2025

Ember. (2024). *UAE.* https://ember-climate.org/countries-and-regions/countries/united-arab-emirates/. Accessed: 19 May 2025

Energy Institute. (2024). *Statistical review of world energy,* 73rd edn. https://www.energyinst.org/statistical-review

European Parliament. (2022). Briefing. In *Brazil's climate change policies. State of play ahead of COP27.*

Food and Agriculture Organization of the United Nations. (2016). *FAOLEX Database.* https://www.fao.org/faolex/results/details/en/c/LEX-FAOC172762/. Accessed: 02 May 2025

Food export. (2024). *Brazil Country Profile.* https://www.foodexport.org/export-insights/market-country-profiles/brazil/#:~:text=Brazil%20is%20a%20global%20leader,technology%20and%20vast%20arable%20land. Accessed: 02 May 2025

Friedrich, J., Ge, M., Pickens, A. & Vigna, L. (2023). This interactive chart shows changes in the world's top 10 emitters. *World Resources Institute.* https://www.wri.org/insights/interactive-chart-shows-changes-worlds-top-10-emitters. Accessed: 19 May 2025

Friedrich, J., Ge, M., Pickens, A., & Vigna, L. (2023). This interactive chart shows changes in the world's top 10 emitters. *World Resources Institute.* https://www.wri.org/insights/interactive-chart-shows-changes-worlds-top-10-emitters. Accessed: 05 May 2025

Galazova, S., Sitokhova, T., & Dzokaeva, Z. (2023). Russia's new environmental policy in the context of "green" transformation. In *BIO Web of Conferences* (Vol. 63, p. 04006). EDP Sciences. https://doi.org/10.1051/bioconf/20236304006

Garg, A., Patange, O., Vishwanathan, S. S., Nag, T., Singh, U., & Avashia, V. (2024). *Synchronizing energy transitions toward possible Net Zero for India: Affordable and clean energy for all.* A report prepared for Office of the Principle Scientific Advisor (PSA) to Government of India and Nuclear Power Corporation of India Limited (NPCIL). 224 p.

Global Carbon Atlas. (2023). *Fossil fuels emissions.* https://globalcarbonatlas.org/emissions/carbon-emissions/. Accessed: 19 May 2025

Global Compliance News. (2024). *Brazil: What you need to know about the proposed Brazilian carbon market.* https://www.globalcompliancenews.com/2024/02/03/https-insightplus-bakermckenzie-com-bm-energy-mining-infrastructure_1-brazil-what-you-need-to-know-about-the-proposed-brazilian-carbon-market_01102024/. Accessed: 02 May 2025

Government of Sikkim. (2012). *Sikkim Organic Mission.* https://sikkim.gov.in/Mission/Mission-info/1?Mission=Sikkim%20Organic%20Mission. Accessed: 05 May 2025

Government of the Russian Federation. (2019). *Decree of the government of the Russian Federation dated the 21st of September 2019 No. 1228 "On the adoption of the Paris Agreement".*

Government of the Russian Federation. (2021). *Strategy of socio-economic development of the Russian Federation with a low level of greenhouse gas emissions until 2050.*

Haonan, Q., Suphachalasai, S., Thube, S., & Walker, S. (2023). South Africa carbon pricing and climate mitigation policy. In *IMF Selected Issues Paper* (SIP/2023/040). Washington, D.C.: International Monetary Fund. https://www.imf.org/en/Publications/selected-issues-papers/Issues/2023/06/26/South-Africa-Carbon-Pricing-and-Climate-Mitigation-Policy-South-Africa-535220. Accessed: 26 May 2025

IEA. (2021, September). *An energy sector roadmap to carbon neutrality in China.* https://www.iea.org/reports/an-energy-sector-roadmap-to-carbon-neutrality-in-china/executive-summary. Accessed: 19 May 2025

IKI. (2024). *Climate Policy Programme Brazil (PoMuC).* https://www.international-climate-initiative.com/en/project/climate-policy-programme-brazil-pomuc-16-i-205-bra-g-programm-klimapolitik/. Accessed: 02 May 2025

Khavhagali, V., Reckien, D., Biesbroek, R., Mantlana, B., & Pfeffer, K. (2023). Understanding the climate change adaptation policy landscape in South Africa. *Climate Policy, 24*(4), 458–472. https://doi.org/10.1080/14693062.2023.2268576

Kochtcheeva, L. V. (2024). Foreign policy, national interests, and environmental positioning: Russia's post Paris climate change actions, discourse, and engagement. In *Sustainable development, regional governance, and international organizations* (pp. 131–143). Routledge. https://doi.org/10.1080/10758216.2021.1968912

Kovalev, Y. Y., & Porshneva, O. S. (2021). BRICS countries in international climate policy. *Vestnik RUDN. International Relations, 21*(1), 64–78. https://doi.org/10.22363/2313-0660-2021-21-1-64-78

Locke, J., Dsilva, J., & Zarmukhambetova, S. (2023). Decarbonization strategies in the UAE built environment: An evidence-based analysis using COP26 and COP27 recommendations. *Sustainability, 15*(15), 11603. https://doi.org/10.3390/su151511603

Ministry of Environment, Forest and Climate Change, Government of India. (2022). *India's long-term low-carbon development strategy.*

Ministry of Economic Development of the Russian Federation. (2023). *Carbon regulation in Russia.*

Mo, R., Lin, Y., Lin, B. (2023). Does digitalization support green transition in Chinese cities? Perspective from Metcalfe's Law. *Journal of Cleaner Production, 425.*https://doi.org/10.1016/j.jclepro.2023.138769

OECD. (2021). *Emissions trading systems.* https://www.oecd.org/environment/tools-evaluation/emissiontradingsystems.htm. Accessed: 26 May 2025

Pingkuo, L., & Huan, P. (2022). What drives the green and low-carbon energy transition in China?: An empirical analysis based on a novel framework. *Energy, Part E.* https://doi.org/10.1016/j.energy.2021.122450

Popova, I. (2023). Key low-carbon development policies and instruments in Brazil. *International Organisations Research Journal, 18*(4), 34–61. https://doi.org/10.17323/1996-7845-2023-04-02

Press Information Bureau, Government of India, Ministry of Environment, Forest and Climate Change. (2016). *Cabinet approves ratification of the Paris Agreement.* https://pib.gov.in/newsite/PrintRelease.aspx?relid=151205#:~:text=Cabinet%20approves%20ratification%20of%20the,the%20day%20of%20Gandhi%20Jayanti. Accessed: 05 May 2025

Press Information Bureau, Government of India, Ministry of Environment, Forest and Climate Change. (2023). *India achieves two targets of Nationally Determined Contribution well ahead of the time.* https://pib.gov.in/PressReleaseIframePage.aspx?PRID=1987752. Accessed: 05 May 2025

Prime Minister's Council on Climate Change. (2008). *National action plan on climate change.*

Rahman, M. N., & Turay, A. M. (2018). Climate change issues in BRICS countries. *Management and Economics Research Journal, 4,* 174–183. https://doi.org/10.18639/MERJ.2018.04.678933

RBC. (2019). *Russia has joined the Paris Climate Agreement.* https://www.rbc.ru/politics/23/09/2019/5d88a9089a79475f76930863. Accessed: 09 April 2025 (In Russian)

Reuters. (2024). *Top 10 country emitters discharged record amount of CO_2 in 2023.* https://www.reuters.com/markets/commodities/top-10-country-emitters-discharged-record-amount-co2-2023-2024-06-21/. Accessed: 09 April 2025

Sampene, A., Li, C., Agyeman, F., Brenya, R. (2021). Analysis of the BRICS countries' pathways towards a low-carbon environment. *BRICS Journal of Economics, 2*(4), 77–102. https://doi.org/10.38050/2712-7508-2021-4-4

Sitnikov, S. L. (2022). Carbon regulation in Russia: origins and peculiarities. *The Eurasian Scientific Journal, 14*(6). (In Russian, abstract in English)

South African Government. (2016). *International relations and cooperation on Paris agreement on climate change.* https://www.gov.za/speeches/paris-agreement-climate-change-2-nov-2016-0000#:~:text=South%20Africa%20has%20ratified%20the,Secretary%2DGeneral%20in%20New%20York. Accessed: 02 May 2025

TASS. (2020). *Russia in the Paris Climate Agreement. What will this give to the environment and businesses?* https://tass.ru/obschestvo/10230505. Accessed: 09 April 2025 (In Russian)

The General Office of the CPC Central Committee and the General Office of the State Council. (2021, October 21). *Opinions on promoting green development of urban and rural construction.* https://www.gov.cn/zhengce/2021-10/21/content_5644083.htm (In Chinese)

The Oxford Institute for Energy Studies. (2022). *The Evolution of carbon markets and their role in climate mitigation and sustainable development.* 132.

The President of the Russian Federation. (2023). *Decree of the President of the Russian Federation dated 10.26.2023 No 812 "On approval of the climate doctrine of the Russian Federation".*

UAE Government. (2023). *UAE Environment Policy.* https://u.ae/en/about-the-uae/strategies-initiatives-and-awards/policies/environment-and-energy/uae-environment-policy. Accessed: 19 May 2025

UAE Government. (2024). *UAE energy strategy 2030.* https://u.ae/en/about-the-uae/strategies-initiatives-and-awards/strategies-plans-and-visions/environment-and-energy/uae-energy-strategy-2050. Accessed: 19 May 2025

UN News. (2021, September 21). *China headed towards carbon neutrality by 2060; President Xi Jinping vows to halt new coal plants abroad.* https://news.un.org/en/story/2021/09/1100642. Accessed: 19 May 2025

UNDP. (2021, December 21). *Issue brief: China's climate policy documents – 1+N and updated NDC.* https://www.undp.org/china/publications/issue-brief-chinas-climate-policy-documents-1n-and-updated-ndc. Accessed: 19 May 2025

USAID. (2022). *South Africa. Climate change fact sheet.*

Wang, H., Li J., & Wang, M. (2022, December 10). China works vigorously to advance green transportation. *People's Daily Online.* http://en.people.cn/n3/2022/1210/c90000-10182407.html. Accessed: 19 May 2025

Wani, S. P, Anantha, K. H., Sudi, R., Krishnappa, K., Dharmarajan, & B. K. (2015). *Bhoochetana: A compendium of success stories. Research report IDC-3.* Patancheru 502 324. Telangana, India. https://oar.icrisat.org/9268/

World Bank Group. (2024). *The world bank in Brazil.* https://www.worldbank.org/en/country/brazil/overview. Accessed: 18 Apr 2024

World Bank Group. (2024). *State and trends of carbon pricing dashboard.* https://carbonpricingdashboard.worldbank.org/credits/factsheets?mechanism=CR_GOV_ZAF. Accessed: 19 May 2025

Xi, W. (1999). Implementing the Rio declaration and Agenda 21. *Asia Pacific Journal of Environmental Law, 3*(1), 9–32.

Xie, Z. (2020). China's historical evolution of environmental protection along with the forty years' reform and opening-up. *Environmental Science and Ecotechnology, 1*, 1–10. https://doi.org/10.1016/j.ese.2019.100001

Xu, Y., Ge, W., Liu, G., Su, X., Zhu, J., Yang, C., Yang, X., & Ran, Q. (2023). The impact of local government competition and green technology innovation on economic low-carbon transition: new insights from China. *Environmental Science and Pollution Research, 30*, 23714–23735.https://doi.org/10.1007/s11356-022-23857-1

Circular Economy in the West Asia & North African (WANA) Countries: Challenges and Opportunities

Gargee Sahoo and Pratibha Agarwal

Abstract As an environmentally friendly substitute for the conventional linear economic paradigm, the idea of a circular economy has acquired considerable traction. Sharing, leasing, repairing, refurbishing, and recycling current resources and products for a longer period of time is at the heart of the Circular Economy (CE), also known as circularity. Efforts to attain social, environmental, and economic sustainability are driving the global promotion of CE as a policy. The WANA region, characterized by its rich natural resources and rapid economic development, faces unique opportunities and challenges in adopting CE practices. The ongoing economic landscape of this region puts emphasis on region's heavy reliance on resource extraction which has clear cut implications for sustainability. Transitioning to a circular economy model can mitigate environmental degradation, reduce scarcity of available resources, and foster economic resilience. In this chapter an analysis is made of some important industries for instance water management, construction, manufacturing, energy to illustrate the potential benefits of circularity in these economies. While the WANA region faces challenges such as political instability and varying levels of sectoral technological advancement, there exists a growing awareness and commitment among governments, business establishment, and civil society to embrace sustainable practices. Case studies from WANA countries covering Algeria, Morocco, Tunisia, Lebanon, Egypt, Turkey and the Gulf Cooperation Council (GCC) countries etc. highlight pioneering initiatives and policies aimed at promoting the seven basic principles of CE. Those are management of waste, energy, industry, agriculture, water, tourism, and hospitality, renewable energy, infrastructure, and sustainable urban development projects. The information presented in this chapter is on current scenario of development and usage of natural resources in the WANA region; what a shift to the CE model entails for the region; which possibilities a transformation can bring open; the current progress towards a CE model; current challenges and the necessary enabling conditions. While the possibilities within a

G. Sahoo (✉)
School of International Studies, Jawaharlal Nehru University, New Delhi, India
e-mail: gargeesahoo1992@gmail.com

P. Agarwal
Department of Economics, PGDAV College, University of Delhi, New Delhi, India
e-mail: pratibha.agarwal@pgdav.du.ac.in

P. Singh et al. (eds.), *Circular Economy and Sustainable Innovation*,
https://doi.org/10.1007/978-3-032-00437-6_2

CE are immense, there are global and regional gaps in both a shared understanding of the concept. By embracing the CE practices, WANA region can achieve the objective of attaining a higher level of socio-economic wellbeing that is accompanied with an improved environmental sustainability outcome.

Keywords Circular economy · WANA · Energy policy · Waste management · Environmental sustainability · Challenges

1 Introduction

Circular economy (CE) is recognized globally as a model of economic development based on more efficient utilization of resources across the production and consumption value chain (Corona et al., 2019). The concept of the circular economy thus represents a paradigm shift from the traditional linear economic models to more sustainable practices that emphasize resource efficiency, waste reduction, and value preservation (Ferdous, 2021). By repurposing previously used materials into other forms of revenue, this zero-waste strategy aims to increase the useful life and monetary worth of resources and promote long-term economic growth. It has a potential to lead holistic transformation of society with systemic changes aimed at achieving better environmental quality and resilience, social equity, and economic prosperity. According to Circular Economy Gap Report 2024, by adopting a CE model it is possible to fulfil the needs and wants of the society with just 70% of the current usage of materials. A shift to CE can facilitate increased interregional cooperation. It can promote inclusive development and making sure that everyone has a good life. At its core, it is a sector-and supply-chain-agnostic economic model that plots out the production, use, and design of goods and services with an eye on reducing their negative effects on the environment while simultaneously opening the doors to yet underexplored markets (Fig. 1).

As such West Asia and North Africa (WANA) has no standardised definition. The term WANA, typically also known as MENA (Middle East and North Africa), as a regional identity, has kept evolving over the decades. The region is defined as consisting of different territories by different organisations. In this chapter, the grouping of the countries in the WANA region is as specified by the European External Action Service (EEAS). The West Asia and North Africa region includes 22 countries exhibiting varied political, economic, and social characteristics. Egypt, Algeria, and the Gulf Cooperation Council (GCC) countries of Kuwait, Oman, Qatar, Saudi Arabia, Iran, and the United Arab Emirates (UAE), Bahrain have historically been countries that are major producers of oil and gas. Egypt, Morocco, Jordan, and Tunisia have garnered substantial investment and income from mineral resources, including bauxite, cobalt, diamonds, gold, lithium, phosphates, potash, rhodium, silver, iron ore, zinc, and platinum-group metals (Olawuyi, 2018).

This region is strategically significant due to its substantial share of the reservoir of oil in the globe. WANA region contains the world's leading exporters of oil,

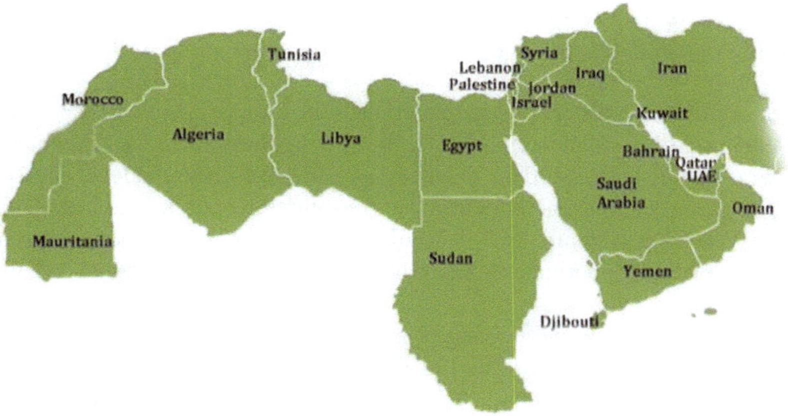

Fig. 1 WANA countries

solid minerals, and natural gas. The WANA region contains the largest proven oil reserves globally, accounting for approximately 59%, along with around 45% of the world's proven natural gas reserves and 30% of global mineral reserves (Aoun, 2013). Although oil constitutes the primary source of income in numerous countries within this region, yet others are endeavouring to diversify their revenue streams. The WANA region faces significant environmental challenges, including scarcity of water, high waste generation, and resource depletion. The possibility of shifting to a circular economy model suggests a promising approach to address these issues.

The circular economy aims to complete product life cycles by enhancing resource efficiency, recycling, and reusing materials. The traditional linear economy refers to a 'take, make, dispose' model. It is an economic approach designed towards waste elimination and continual usage of resources by offering a new paradigm on a comprehensive view of products and processes. A production system that adopts circular economy practices can reduce resource dependencies and gain competitiveness. It also brings in substantial environmental benefits with associated significant reduction in congestion and pollution. While resource efficiency is maximised, the consumption of finite resources is minimised (Corona et al., 2019).

Circular Economy is a cross-sectoral approach to development where circular systems are designed to favour solutions that upgrade ecosystems and retain the value of resources within the system for as long as possible (Geissdoerfer et al., 2016). Renewable and finite resource management is guided by the circular model, which includes the restorative and regenerative practices of rethink, redesign, reduce, refuse, reuse, repurpose, recycle, and recover, to achieve a sustainable use of natural resources while simultaneously advancing social equity and economic prosperity. According to the European Parliament News (2022), CE is a model of production and consumption, which involves sharing, leasing, reusing, repairing, refurbishing, and recycling existing materials and products as long as possible. Further, the EU's definition, as complemented by the Ellen Macarthur Foundation (2014) states that

CE is founded on three principles that are driven by design: To eliminate waste and pollution, regenerate nature while circulating materials and goods at their highest value.

A shift toward renewable energy and materials is the foundation of CE, according to the Ellen MacArthur Foundation (2014). By moving away from the take-make-waste linear paradigm and toward the reduce-reuse-recycle circular model, it stresses the need to separate economic activity from the consumption of finite resources. This transition represents a systemic shift that will build long-term resilience, generate business and economic opportunities, and provide environmental and social benefits (Dincer, 2000).

Concepts such as green entrepreneurship and circular design are still at very early stages in the target countries and have a negligible impact when compared to primary focus areas (Morseletto, 2020). Both concepts serve as references and guides for assessing the Circular Economy in the WANA region, particularly within the specific countries. The region's stage of development can be characterized as explorative, considering the principles of circular economy. With few exceptions, the majority of initiatives and research focus on mitigating existing practices, alongside areas including management of coastlines and efficiency in energy use.

Management of waste and depletion of resources are issues that have been overlooked in the WANA region. In the last ten years, the region has experienced persistent instability due to geopolitical and intra-religious conflicts, resource scarcity, and halted economic and governance reforms, leading to institutional, political, and economic division. The region is currently facing several significant issues, including high unemployment rates, particularly among youth, ineffective public services, costly energy subsidies, and declining currency valuations in various countries, such as Lebanon and Palestine. The project, Reduction of Unintentional POPs through Waste Management in a Circular Economy, was run by Lebanon to improve management of waste disposal sites in the waste sector (World Bank, 2021).

The impact of climate change serves as a multiplier of conflict in the region. WANA has historically deprioritized the adoption of sustainable practices and principles of circular economy by categorizing them as non-urgent or non-essential during periods of significant economic crisis. A business case study from a 2020 project by Acted Lebanon, an NGO, has demonstrated that the ongoing rapid economic decline in WANA has created several unique opportunities for advancing circular economy and resource management, potentially catalyzing economic recovery. The implementation of circular approaches can stimulate job creation, enhance value extraction, boost productivity, reduce government expenditures, and improve the accessibility of public services in the region.

WANA's interpretation of CE primarily puts focus on waste management, renewable energy, and water resource management. Secondary focus areas further include management of the coast, innovation of circular design, and the green economy. The transition of four systems: food systems; manufactured good and consumables; built environment; and mobility and transport, could reduce the global material extraction by 34%, limit the emissions of Green House Gases (GHG) enough to keep global temperate rise to two degrees, and reduce the aerosol loading, freshwater use, ocean acidification, phosphorus cycle, nitrogen cycle, and land system change.

2 Historical Background of Circular Economy in WANA

The linear economy, characterized by the extraction of raw materials, their processing into products, and subsequent disposal, is sometimes referred to as a 'take - make - use - dispose' model. According to the Ellen MacArthur Foundation (2019), the circular economy aims to redefine growth and to build economic, natural, and social capital by gradually decoupling economic activity from the consumption of finite resources.

Since the beginning of the twentieth century, there has been a discernible and exponential increase in the levels of material extraction and consumption around the globe. According to estimates, the quantity of raw materials mined in 2018 was approximately 92 billion tons. Based on the existing trajectory, it is anticipated that this figure will climb to 177 billion tonnes by the year 2050. This is mostly because the linear economy model of production and consumption that is currently being utilized in the global economy is the primary cause of this development. According to estimates, 91.4% of all materials are only utilized once in their lifetime.

Most of the WANA region has historically had low GHG emissions, but this is changing as countries are developing, followed by lifestyle changes and an increased dependency on fossil fuels. Historically, most of the waste entering the circular economy in WANA countries has been collected by the informal sector. For example, in Jordan, an estimated 10% of municipal waste quantities (Aljaradin et al., 2015) or in Lebanon, up to 20% of Beirut's municipal waste stream is collected as recyclables by the informal sector. An accelerating trend has been noted in both the Material Footprint and the Domestic Material Consumption per capita in the WANA region, implying a change of lifestyles connected with a rapid growth in the consumption of biomass, fossil fuels, and minerals.

Following the approval of the 17 Sustainable Development Goals (SDGs) in 22 WANA countries under the 2030 Agenda for Sustainable Development, nearly all nations in the WANA area have formulated Nationally Determined Contributions (NDCs) outlining their contributions to the implementation of the Paris Agreement. The GHG emissions vary greatly among the regions, but as in the rest of the world, a high development standard is closely linked with high emissions. Further, the WANA region is now hosting the greatest inequality of GHG emissions in the world, where the wealthiest 10% have 15 times higher emissions than the lower 50% in the same country on average. There is now a greater inequality of GHGs within countries than between countries. The WANA region shares several challenges connected to its local natural resources ranked as the most water scarce region in the world; habiting large arid land masses leading to a high dependency on food import; and its geographic location being in a 'climate change hotspot' where the climatic changes are exacerbating the already fragile ecosystems, leading to increased vulnerabilities of populations and the creation of new vulnerable groups (Chizaryfard, 2021).

However, the SDGs together with several of the most urgent declarations are far away from being reached, and at the current speed, no goal will be reached by 2030, and the WANA region and its populations will be left behind while other regions progress and reap the advantages of being frontrunners.

3 Circular Economy Initiatives in WANA

The central circular policies being discussed in the present context are focused on controlled dumping, promotion of the hierarchical steps of waste reduction, reuse of products, recycling with different kinds of sorting, recovery of energy, and sanitization of landfills. New dimensions are also being mentioned in policies with reference to a culture of shared responsibility of protecting the environment and public health, and a decentralization of responsibilities, but with a chartered role of central authorities to ensure efficient resource allocation.

At the global level, in the year 2000, plans and policies of a CE were created and adopted in the emergent economy of China. In 2015, the European Commission followed and approved a plan moving towards a CE in the EU. Even though extensive steps in this direction have been carried out, the global economy as of 2023 was only 7.2% circular, while counting the amount of materials that are cycled back into the economy after their primary use. This points to a decrease in circularity compared to just a few years ago, highlighting an increased dependency on virgin materials and a strong incentive to bring all regions in the world together and start carrying out goal-centred action plans. Except in China, the CE has previously gained little attention in low and middle-income countries.

In the last few years, WANA countries have started to question the linear economy and to investigate alternative development trajectories. The transformation from linear to circular economy in the WANA region is progressively becoming a goal while simultaneously working on achieving Agenda 2030. The mapping found that CE and circular processes and solutions are gaining increased attention in the WANA region, and that the ambitions are varied, but the region shows great interest in a new way of natural resource use, and a high level of potential and willingness for change. Almost all WANA countries have produced and submitted Nationally Determined Contribution (NDC), and the majority include either circular processes or circular solutions to reach mitigation goals towards the Paris Agreement and adaptation to climate change. The circular national policies existing in the WANA region are either directly linked to a CE or aimed at tackling connected issues, such as SDG12 (especially responsible consumption and production), energy efficiency in industries, or the reduction of waste creation at the first source.

A recycling company in Saudi Arabia, the largest in West Asia and North Africa, processes 3 million tons of waste each year. This plan seeks to recycle 95% of its garbage, contributing US$31.99 billion to GDP, generating 100,000 employments, and establishing itself as a global leader in sustainability by 2040 through sophisticated waste management practices (Alsaud et al., 2025). In Saudi Arabia, Clorox's facilities in Dammam and Jeddah have attained a 100% zero waste to landfill status. Furthermore, KAUST startup Edama Organic Solutions has inaugurated the Kingdom's inaugural organic waste recycling facility at the KAUST Research (Almadhi et al., 2023).

Oman is initiating its inaugural Waste-to-Energy (WTE) project, anticipated to reduce landfill carbon emissions by 50 million tons over a span of 35 years. Oman's e-waste recycling initiative addresses the escalating problem of electronic waste while generating new economic prospects, reinforcing the country's status as a frontrunner in sustainable waste management and resource efficiency (Ibtisam Sulaiman Al Hosni, 2020). In the State of Qatar, a trash Treatment Center located in the Masa'ada area recycles household trash to produce electricity. The Ministry of Municipality of Qatar intends to construct an engineered landfill at Al Khor, conforming to the highest international standards, and to operate a materials recycling facility.

Several green sukuks have been successfully issued in the WANA region to promote green finance and the circular economy. These comprise one issued by Majid Al Futtaim valued at USD 1.2 billion and another by Saudi Electricity Company valued at USD 1.3 billion. Furthermore, the Islamic Development Bank and Egypt released green sukuks throughout the region in 2021. Formulating strategies, programs, and initiatives that incorporate circular economy ideas into national frameworks; promoting the maturation of nascent markets; and enhancing the capacity of international stakeholders to access the market.

In 2021, the UAE initiated a comprehensive Circular Economy program and formed a Circular Economy council of leaders from federal, local, and business sectors. The council intends to execute the CE policy via national plans and legislation that incorporate and oversee sustainability criteria, utilizing the following processes as a framework. Furthermore, the report by Fadi Al-Shihabi (2025) indicates that the implementation of the CE model in the cities of the UAE could yield savings of up to 28 billion dollars (102.8 billion dirhams) from 2020 to 2030, distributed as 7.2 billion dollars in the urban environment, 11 billion dollars in transportation systems, and 9.8 billion dollars in housing. The UAE has built the world's largest waste-to-energy facility, which will render Dubai the most sustainable and intelligent city by processing 1900 tons of residential garbage annually, while the resultant gases will be managed in the most environmentally safe and effective manner.

Lebanon has stated that the goal is a green economy, described as a sustainable development model with high human development and reduced ecological impact. The United Arab Emirates has stated that a CE is the goal for their waste sector, while Kuwait and Saudi Arabia are including a Circular Carbon Economy (CCE) as the strategy to reach their mitigation targets (Almulhim et al., 2023). Algeria has demonstrated advancement in its circular economy initiatives, emphasizing the development of sustainable waste management methods and renewable energy solutions.

3.1 Flagship Projects

The **United Arab Emirates** has developed a 'Circular Economy Policy 2021–2031' where they state sustainability of manufacturing, nature-based solutions, green transportation, and ethical food consumption and production as prioritized sectors. It is focused on the possibilities of transition to CE, including a strategy of steps on how to transition and connecting the ambitions with vast investment opportunities.

Saudi Arabia's documents with a dedication to circularity are also frontrunners. They are focusing on the energy sector through the Circular Carbon Economy (CCE) and have chartered a 'Circular Carbon Economy National Program' (Almulhim, 2024). The ambition is emissions reduction through efficiency measures, storage, and capture of CO_2 and other such energy-focused measures. The CE concept is applied to the food sector in Saudi Arabia by identifying relevant issues and related CE strategies (Shehri et al., 2023).

Jordan, with a focus on single sector as of now, i.e., the waste sector, has chartered a plan for 'Climate and Resource Protection through Circular Economy in Jordan 2017–2021'. This document, even though not cross-sectoral, is holistic as it sets out the necessary conditions to implement a CE in the Greater Amman Municipality (Al-Masri et al., 2020).

A new project, *REUSEMED,* is being implemented in Jordan through a partnership between the Ministry of Local Administration and the New Deir Allaa Municipality. The project is EU-funded to strengthen the culture of reusing products and create networks to facilitate this. With a focus on the reuse of compost, clothes, books, furniture, and household appliances, the project is promoting both regenerative and restorative solutions. In 2022, they organized the first reuse festival, raising the local public's awareness of reuse and inviting them to participate in reuse activities. Other achievements include the setting up of reuse centres where people can turn in old products which are then resold for a minimal price, a repair café, and a compost factory (Hemidat, 2019).

The project is within the theme of waste management, but instead of the usual focus on recycling and incineration of waste, the more resource-efficient processes of reuse are used (OECD, 2021). The goal is to target the increasing waste production by taking care of already produced products to decrease waste and the need for new production. The project is promoting changed consumption patterns by enhancing the possibilities with used products, raising awareness of waste management, capacity building of reuse, repair, and composting practices, and creating new green jobs. Another future goal is to involve the private sector to increase the medium- and long-term revenues of the New Deir Allaa municipality.

Rapid population growth, industrialization, urbanization, and GDP growth have led to recent economic prosperity in **Saudi Arabia**, but with the drawback of an extensive use of natural resources. The state suffers from insufficient solid waste handling with landfills that have reached their brink, leaking out toxins affecting water, land, and air quality, emitting GHGs, and contributing to bad health. It was estimated that in 2021, the environmental degradation by solid waste reached 1.3 billion dollars in Saudi Arabia (Ghanem & Alamri, 2023).

Saudi Arabia has now put waste management of its 53 million tons of waste as a top priority for the state and produced a strong framework document and a waste management law, to switch to a CE. The National Centre for Waste Management has set the ambitious 2030 goal of diverting 82% of its waste from landfills, mainly through recycling, composting, and incineration, and to reach a 100% handling of new waste through either recycling or conversion into fuel and energy. They have managed to improve their industrial waste handling considerably through strong governance and best practices. Smart waste containers, practical waste management plans, waste valorisation plants, public–private partnerships, and new private investors have been a part of the success in this direction.

Morocco has a high food self-sufficiency and agri-food production accounts for 33% of all employment. But the fertility of the land and profit in the sector is decreasing with changed market conditions and a lack of natural resource availability. The agriculture sector is also responsible for a considerable amount of the world's GHG emissions, partly due to an inefficient use of natural resources through uncontrolled methane production from waste, extensive use of inputs such as chemical fertilizers, and the use of fossil fuels in the production process (El Hafdaoui et al., 2024).

In a collaboration between the company Biodôme du Maroc and the EU-funded SwitchMed Program, a new solution for farmers in Morocco has been developed to help them include circular practices to decrease their emissions and increase their profits and resilience. The solution is based on a small-scale local tank which produces biogas and fertilizer through methane production from organic waste, animal waste, and wastewater. Instead of leaving the waste untreated as such on the farms, which allows leakage of GHGs, the biogas is used to supplement the conventional use of oil to run irrigation systems and other energy-demanding systems on the farm (Mana et al., 2021). Such locally produced organic fertilizers are replacing market-bought chemical fertilizers. The project has also brought the possibility for the farmers to sell organic fertilizers, and for new employment opportunities, as the solution is meant to be self-sustained by local farmers. The solution is being prepared for expansion on a national level and to educate all relevant stakeholders to form a public–private partnership for sustainable agricultural circular solutions (El Hafdaoui et al., 2024).

Egypt has the largest population in the Arab region, with one of the highest population concentrations in the world around the Nile River. The urban population is rapidly increasing and is expected to grow from 43% (2022) to 75% in 2050. Cairo suffers from high air pollution, mainly due to transportation, industry, and the open burning of waste, and is experiencing 10–100 times the recommended international acceptable levels of dangerous air particles. Air pollution is mainly a health hazard and can lead to extreme temperatures, shortness of breath, disease, and premature death (Suiko et al., 2024).

With a focus on public transport, Egypt has put up ambitious plans to lower its air pollution by 50% by 2050 and has several ongoing projects, with new projects being proposed for the near future. The Cairo Metro system, with the goal of 'low carbon mass transit' has been expanded and now has three lines expected to serve 2 million people within the near future (Suiko et al., 2024). A private sector company has also introduced 200 buses providing a high-quality bus service, targeting and incentivizing car drivers to shift to a more sustainable mobility option. Further measures aim to plan a transition from the fuel used to natural gas in public buses, and public encouragement of bicycling through new infrastructure development.

Public transportation is a circular solution with a short loop, with multiple positive outcomes. A strong public transport system is a part of a shared economy, where people can buy a service instead of a product when it is needed. This not only leads to a decrease in usage of natural resources in production since fewer vehicles are needed, but also to an overall decrease in use of natural resources in consumption since public transport is a shared service. Benefits include lower emissions in production, lower emissions in consumption, lower air pollution, increased biodiversity, safer and healthier cities to live in, and other socio-economic benefits such as a higher degree of mobility inclusion, strong investment opportunities, and returns for the green entrepreneurs.

3.2 Vision Documents

Many states are working on their governance structures to make their ambitions of a CE go from paper to reality. Several countries have produced documents with their Visions for 2030, 2035, 2040, or for an undefined time. Some examples of the most progressive vision of circularity and how to get there can be found in documents produced by Oman, Saudi Arabia, Djibouti, Yemen, and Iraq.

Initiated in January 2021, the goal statement of '**Oman** Vision 2040' is to ensure sustainable use of natural resources to support the economy, which will be achieved partly through adopting a green economy and partly through transition to CE. The goal progress is measured through indicators in environmental performance, water index, domestic material consumption, etc. This document is cross-sectoral and captures the shared vision while at the same time sets a mission of transition to CE.

In 2016, the Council for Economic and Development Affairs in **Saudi Arabia** chartered a 'Saudi Arabia Vision 2030'. This vision includes the practices of recycling and decreasing water consumption through efficiency measures, to preserve the environment and the natural resources. The document further contains a whole section about the importance of the development of the council to achieve efficient governance structures to implement the vision, with the creation of an office that will solely work on the coordination of different government programs.

In the 'National Vision for the Modern **Yemin**i State' announced in 2019, the government of Yemen is focusing on improving infrastructure for waste management, targeting both its treatment and recycling. The Vision includes the key circular indicators of 'percentage of treated and cycled waste' and 'utilization of industrial technology to recycle waste and transform it into an essential part of development'.

The aim and ambition are to develop a more sustainable use of natural resources. One important component of an enabling structure is to include constant monitoring, without which there is no way of assessing the results and for effective governance to improve necessary cooperation or remove bottlenecks.

The Government of **Djibouti** has created 'Djibouti Vision 2035', capturing the pressing needs of improved waste and agricultural practices in the country. It recommends alternative, more sustainable, and circular agricultural and food practices, such as diversification of food harvesting, recycling of agricultural wastewater, foraging crops, and planting gardens to combat desertification. To be able to develop in the described way, the country is recognizing the need for further integration in the WANA region to enable trade and to attract international investment.

In 2019, **Iraq's** Ministry of Planning created'The Future We Want: Iraq's Vision for Sustainable Development 2030'. The environmental goal is to find an environmental balance, for the current and future generations' possibility to live in a clean, safe, and sustainable environment. There are several circular processes and solutions mentioned to reach the goal, including developing sustainable cities, raising awareness about circular consumption, and building out the public transport network. This vision is described as a way forward, making an aspirational vision in which all ministries and local governments will work together to implement and to ensure the realization of the vision goals.

3.3 PPPs, CPP & EPR

Some policies that can help an economy transition to CE include the Polluter Pays Principle (PPP), sustainable public procurement policies like Circular Procurement Practices (CPP), and other fiscal policies like Extended Producer Responsibility (EPR).

Tunisia is the first state in the WANA region that, in 1996, delegated the responsibility of waste management and pollution control to the producer in the law. Unfortunately, these laws are not well implemented in Tunisia, and it is still mainly municipalities that are paying for the solid waste management (Fersi et al., 2021). **Palestine** has introduced PPP schemes into their environmental law to make companies degrading the environment pay for the damage created (UNEP, 2016). In **Lebanon's** Action Plan for the Industrial Sector, envisaged in 2015, several policies aimed towards improvements in the industry have been endorsed by the Ministry of Environment and the Ministry of Industry. According to Voluntary National Review (2021) of **Qatar**, the Planning and Statistics Authority planned to host the most environmentally friendly possible FIFA World 2022, stressing on the efficient use of natural resources, through the policies of sustainable procurement and using materials with recycled content or that can be reused or recycled after primary use (Taqi, 2017). Qatar is the third-largest exporter of natural gas globally and the leading producer of liquefied natural gas (LNG). It has been ranked as the richest country in the world, possessing one of the highest per capita gross domestic products (GDP) according to the IMF in 2019 (Ibid, 2017).

In the WANA region, some states have started to draft plans for sustainable or Circular Procurement Practices (CPPs). For instance, **Egypt's** Ministry of Environment, together with SwitchMed, has recommended such procurement practices in their 'National Action Plan for Sustainable Consumption and Production' (2016). Further development in this direction is limited to the GCC states, where **Qatar** and **United Arab Emirates** plan to introduce CPPs.

The Extended Producer Responsibility (EPR) schemes in the WANA region are almost non-existent, and in the few states where it is adopted, it only covers a couple of selected waste categories. Increasing the tax or decreasing subsidies on environmentally harmful resources, and decreasing the tax or introducing subsidies for circular processes or solutions are also being carried out in the WANA region. In **Egypt,** projects have been carried out to decrease the subsidies on energy, electricity, and oil. In addition, Morocco has been gradually eliminating subsidies for fossil fuels from the year 2012, particularly for liquid petroleum products like diesel. This is being done while closely monitoring the potential social repercussions and emphasizing the significance of solidarity with the most vulnerable parts of the population. **Palestine** and **Yemen** have other policies in place. Palestine will introduce Value Added Tax and increased income tax to target sustainable consumption, and Yemen will introduce tax exemptions to increase reuse and recycling and lower the use of virgin materials.

To reach this, the Yemeni government is planning to establish a certification framework that will allow for the verification, registration, certification, and recognition of environmental excellence in terms of energy efficiency, materials usage efficiency, and water efficiency. Another example is the 'Eco tax', which is a partnership agreement in Morocco, signed by the State Secretariat for Water and the Environment (SEEE), involving many ministries to solve the problem of plastic bags and other negative effects of resource use by implementing taxes on waste.

3.4 Regional Cooperation

Newly formed cooperation for a CE, circular processes, or solutions, can be found in several instances in WANA states. The regional cooperation was initiated in 2016 for the implementation of a CE in Africa, with the creation of 'The African Circular Economy Network'. The Arab states of **Morocco, Sudan, Tunisia, and Somalia** have joined the network with the goal of building a restorative African economy through knowledge sharing, research, networking, trainings, and awareness programs. Some examples of national documents focused on the importance and with the mindset of incorporating newfound cooperation for reaching a CE are created by the United Arab Emirates, Bahrain, Saudi Arabia, Lebanon, and Egypt.

International engagement and regional cooperation are key drivers in the 'Circular Carbon Economy National Program' developed by the Ministry of Energy in **Saudi**

Arabia (Almulhim & Al-Saidi, 2023). The focus is to take global leadership in this domain and help other states to follow. The **United Arab Emirates** 'Circular Economy Council' is focused on the national cooperation and coordination between different ministries, national and regional strategies, sectors, and the public and private domains. **Bahrain's** development of the 'Strategic and Sustainable Development Cooperation Framework 2021–2022' is a great example of an effort towards improving the structures needed to implement a CE. The government also emphasizes community partnership as a social responsibility for the industrial sector. Through collaborations with associations and parents of children with special needs, significant amounts of plastic waste have been collected and recycled since 2019. It's a partnership between the government and the UN, highlighting cooperation opportunities throughout the document. Similar efforts between national sectors or states could be a key to accelerating the progress through cooperation across the region or across sectors.

4 Challenges and Barriers

The adoption of the circular economy is not a very easy task. Several challenges have been found in the application of CE. The most obvious challenge is the lack of strong regional cooperation within WANA for enabling circular solutions. A CE cannot be achieved by sector sector-specific focus. It needs to be a cross-sectoral mainstreaming of circularity, giving CE the core space in development efforts in a multitude of sectors to be able to go from a limited transition of flows to a societal transformation. Insufficient knowledge and skills to promote a circular economy across businesses and industries, inadequate enforcement of environmental regulations due to the unclear legal framework, improper monitoring system, and unsatisfactory financial capacity are some of the challenges faced in the setup of the circular economy (Kirchherr et al., 2018).

4.1 Structural and Economic Bottlenecks

The WANA is known for its vast reserves of oil and natural resources, which have been the backbone of the region's economies for decades. However, this heavy reliance on non-renewable resources presents a significant obstacle to transitioning to a circular economy. Shifting away from fossil fuel extraction and towards renewable energy sources, such as solar and wind, requires substantial investments in infrastructure, technology, and capacity building. Transitioning to a circular economy involves significant upfront investments and may require businesses to restructure their operations around the green economy. (Bashir Khodaparasti & Bagheri Garabollagh, 2023) Some companies may be reluctant to adopt circular economy practices due to concerns about increased costs, reduced competitiveness, and potential job losses. This creates the problem of limited funding and financial support for CE projects and initiatives, which is reflected in inadequate infrastructure, especially for waste management and recycling. Governments can play a crucial role by providing financial incentives, tax breaks, and supportive policies to encourage the private sector's participation in the circular economy. Highlighting the potential for new job creation, especially green jobs and long-term cost savings, can help overcome these economic barriers (Gower & Schrofer, 2016). Additionally, diversifying the region's economies through the planned development of sustainable industries can help reduce the dependence on finite resources.

A combination of life cycles and industries may lead to the formation of intricate systems that hinder competitiveness and render the economy susceptible. In other words, complex systems may impede economic progress and result in a decline in national income. An economy frequently struggles to adopt a circular model due to the existence of a singular industry that cannot implement closed-loop processes. This is due to the existence of several collapses and complications. Furthermore, a sequence of occurrences may result in the collapse of the economic system.

4.2 Cultural and Social Barriers

In an EU study about the main barriers to a CE, it was found that technology or finance is not the binding barrier, but the strongest barrier was within the cultural realm, i.e., lack of consumer interest and awareness. Another study highlighted a limited awareness and understanding of the concept of CE in Saudi Arabia and linked it to a negative correlation with age and a positive correlation with educational level. Sustainable resource use does not mean the same thing for everyone; for many within the Arab region, it means consuming differently, and not necessarily less (Kirchherr et al., 2018).

There exists a strong resistance to change in consumer habits and acceptance of recycled products. Cultural and social norms can significantly influence the adoption of sustainable practices. In some Middle Eastern societies, the emphasis on convenience and rapid development can discourage individuals and businesses from prioritizing recycling and waste reduction. Promoting a culture of sustainability requires engaging communities, religious leaders, and influencers to create awareness and change mindsets. Public campaigns that emphasize the importance of preserving the environment and the long-term benefits of a circular economy can help overcome these cultural barriers.

Insufficient awareness and deep understanding of the principles of circular economy among businesses and the public in the WANA is a primary challenge hindering the transition to a circular economy. Many businesses, policymakers, and individuals in the region have yet to grasp the concept and its potential benefits. Effective education and awareness campaigns are needed to highlight the economic, social, and environmental advantages of adopting a circular economy approach. By showcasing successful case studies and promoting knowledge sharing, stakeholders can be encouraged to embrace sustainable practices.

4.3 Policy Loopholes and Regulatory Gaps

Absence of comprehensive policies and regulations supporting CE is noticeable in the WANA region. A study, mapping the main barriers to a circular food supply chain, highlighted the lack of a long-term shared vision among stakeholders as one of the main barriers (Preston et al., 2019). A recent study about the solid waste management in the WANA region highlighted inefficient governance as the main issue, including political factors, decentralization, and multi-level management and responsibilities. The lack of reliable data is considered the main hindrance to assessing a picture of the circularity in states or the region (Hemidat, 2022).

Effective waste management and recycling are crucial for a circular economy. However, the WANA region faces challenges in this regard. In many countries, the infrastructure for waste collection and disposal systems is inadequate, leading to inefficient waste management practices. Establishing comprehensive recycling facilities, improving waste segregation systems, and promoting waste-to-energy initiatives are essential steps to address these challenges. Collaboration between governments, private sector entities, and civil society organizations is vital to develop and implement sustainable waste management strategies (Keskes et al., 2022). The lack of legislation and weak implementation are considered two of the main challenges facing waste management in the WANA region countries (Achouri et al., 2022).

5 Opportunities and Benefits of Transition to CE

The use of a circular approach may enhance job creation, augment value extraction, boost productivity, decrease government spending, and improve the accessibility of public services in the region (Alhola et al. 2019). The expected outcomes for Saudi Arabia, for example, by the year 2035 are 73 million tons of reduced CO_2 emissions, 346 million BTU produced, 120 billion SAR added to the economy, and 77 thousand employment opportunities created. Egypt is one of several African countries that have benefited from the Circular Economy Investment Fund, which was established by the African Development Bank. The Small and Medium Enterprises Development Agency (MSMEDA) was founded in 2018 by the Ministry of Trade and Industry, among other domestic financing sources. Financial assistance and business development services are available through MSMEDA to small and medium-sized firms (SMEs), with a focus on those engaged in the circular economy. Most of the funding has been distributed between Morocco and Egypt for the project of renewable energy.

Composting has long been promoted as a method of treating biodegradable waste, creating new job opportunities and additional income generation for communities. There exists an immense potential for new jobs and industries emerging from circular economy initiatives. A shift to a greener economy could create 24 million new jobs globally by 2030 if the right policies are put in place. A green economy is generally focused on resource efficiency and sustainable consumption and production, and could reach great progress (Ghanem & Alamri, 2023). A transition to circularity will help fulfil the Environmental Social and Governance (ESG) goals (Al-Taai, 2021). The circular economy will result in a reduction of CO_2 emissions. The circular economic model offers opportunities for resource and energy conservation, thereby mitigating climate change and its related disasters.

Improvements in public health and quality of life through reduced pollution and better resource management are key to socio-economic upliftment of the WANA region. Adopting circular economy strategies can help the region build resilience to environmental challenges like climate change, water scarcity, and ecosystem degradation, ensuring long-term sustainability and economic growth. This will require enhanced community involvement and awareness regarding sustainable practices. Civil Society Organisations (CSOs) can play a pivotal role in advancing the circular economy within local communities and, in certain instances, may surpass the efficacy of public sector initiatives.

The construction sector in the WANA region is rapidly developing and refurbishing its infrastructure, and there seems to be a lot of opportunities for CE. This phase of vast construction presents an opportunity to future-proof the designs of the buildings as the planning is still underway (Geldermans, 2016). Contractors and builders can optimize the future use of buildings by planning and developing partnerships between owners and operators, enabling them to enhance the performance and value of the property with new, sustainable materials and equipping them with technology that can support sustainable operations for years to come (Morseletto, 2020).

In the field of energy sector, GCC countries have significant potential to use Carbon Capture and Storage (CCS) in energy-intensive industries. For example, Qatar has taken substantial steps toward the establishment of a national CCS framework through the launch of the largest carbon recovery and sequestration facility in the region (Tayeb, 2024). There is significant potential to utilize CCS in energy intensive industries and collaborate with other regions such as European Union (EU) through transferable best practices and lessons learned in the circular economy is desirable. This necessitates the assistance by national governments to bolster both green and circular enterprises through fiscal incentives and capacity-building initiatives, among other measures.

6 Future Prospects and Recommendations

Adopting a shared vision and developing cooperation between international companies, local private companies, and municipalities can enable the resolution of governance irregularities. For the successful inclusion of the circular economic model in the WANA region, new skills need to be developed, especially in the recycling and remanufacturing sectors. Cooperation with universities and research institutions is required in order to support municipalities facing challenges, provide sustainable solutions, and develop plans to improve the waste management system and provide better services (Okorie et al., 2018).

Climate finance, to support both mitigation and adaptation, can be used for the scaling up of the CE since circular solutions can both reduce emissions and increase resilience. The data on international public funding, from climate adaptation grants and funds, as well as from development banks, is easy to access. National private financial inputs can be developed by improving the bridging between the financial institutions and specific CE plans. Innovation in financial instruments for financing environmental sustainability, and especially budgetary support to scale up all pilot projects in the Arab region, is crucial.

Upgradation of both technical and human capacities is needed in the WANA region to transform to a CE model. Increased efficiency and technological advances are an important part for achieving relative decoupling, where the material intensity is reduced or the process emits less harmful emissions, but the products' purposes remain the same.

There is a need to expand the existing technical capacity. The outdated industrial sector in parts of the region is in urgent need of an update to encompass the latest energy efficiency technologies and to generate the capacity for circular processes. To build recycling facilities, extended sustainable public transport networks, renewable energy systems and other technically advanced infrastructure, some parts of the region need technical assistance and training programs to develop the necessary national capacity. New technical solutions can both increase the ease of circularity and create new circular solutions in a sector or overall system.

Human capital can be built at any stage of a human's life and needs to be expanded with greater inclusion of sustainability knowledge, skills, and practices created throughout their lifetimes. The knowledge on how to create eco-designs, build sustainable cities, and to maintain circular infrastructure can be transferred from leading national or international stakeholders. Appropriate policies include the inclusion of sustainability and circular solutions in curricula at all educational institutions and public trainings, and public campaigns targeting a shift in specific unsustainable practices to transform them into circular solutions. Workshops, forums, and conferences can also facilitate a space where the latest knowledge and best practices can be shared between stakeholders.

In the WANA region, the unutilized human capacity, both when it comes to women's and youth participation and empowerment, is considered a limiting factor for equity and economic growth. They are excluded from labour markets even though they are well educated. There is an enormous potential within the inclusion of this capacity in the job market, and new perspectives can fuel innovation. To harvest their unused potential, circular solutions such as extended public transportation can increase their mobility inclusion and fair chances to fully participate in the development story of the region.

7 Targeted Steps to Promote Circular Economy in WANA

Get everyone onboard: A shared vision of a sustainable WANA region is essential, where every stakeholder is on board since the CE is dependent on a system with strong cooperation, where resources are exchanged amidst high interdependence among actors. Linkages and synergies between new and old projects also need to be highlighted to make the transformation unified and efficient. A successful roadmap can create a shared mindset, engaging stakeholders with ownership on a clear path from strategy to action (Hemidat, 2022).

Create benefits with circular solutions: Clear-cut benefits should be created with circular solutions. Policies enabling circular enterprises can include different tax benefits and value-added tax reductions, certifications of circular products and services to increase their attractiveness. Fiscal regulations are required to mobilize financial instruments to absorb new enterprises into the market efficiently. There are three different models of the Circular Procurement Process (CPP) focusing on the system level (including product leasing and companies using EPR), supplier level (including product reparability and disassembly); and product level (including recyclable and recycled materials). A first step for the implementation of CPP is to establish clear guidelines of responsibilities, strategies for appropriate integration, and measuring indicators for validating chosen procurements. Circular certifications can help both encourage producers to limit their use of natural resources and to consumers to make more environmentally sustainable choices. They should be designed considering attractiveness and environmental benefits (Chizaryfard et al., 2021).

Make circularity desirable: A clear understanding of the link between how the linear mode of production and consumption of products and services contributes to environmental degradation and social unjust systems should be promoted in informational campaigns, educational institutions, and marketing (ESCWA, 2020). Encouraging sustainable consumption patterns and lifestyle choices with high utility and low resource intensity should be promoted (Barrie et al., 2022).

Build the needed capacity: New technical solutions in the industrial sector can both increase the ease of circularity and create new circular solutions in systems. Workshops, forums, and conferences can create an environment for the exchange of the latest knowledge and best practices among stakeholders. There is an enormous potential within the existing demographic pool. Inclusion of unutilised human capacity, such as empowerment of youth and women in the job market, should be taken up as the first step in this direction (Almulhim & Abubakar, 2021).

Enhance the finance: International public finance data, including climate adaptation grants and development bank resources, is readily accessible, whereas national budgets favouring climate change adaptation are more challenging to obtain. International funding is essential for the region, but it must be publicly owned and should be increasingly focused on capacity building. National private financial contributions can be enhanced by strengthening the connection between financial institutions and specific circular economy initiatives (ESCWA, 2022).

Measure the progress: A transition from one way of arranging an economy to another, as in the case of a CE, will benefit if it is systematically measured and tracked. A monitoring system should facilitate regional comparison, consistency, and comparability. Further, it is important that a CE index measures more than just efficiency. An ideal CE index should capture the effort of maintaining the value of resources while simultaneously monitoring social, economic and environmental impacts within a system perspective.

Address the information gaps: Since it is hard to maintain a reliable data set on circularity, proper processing of emerging data is as important as making existing data available for stakeholders to make informed decisions. In the WANA region, there are gaps in data reporting and data availability (PACE, 2021). For example, in SDG 12 'Responsible Consumption and Production', only two out of 13 indicators had enough data to be assessed in 2020.

Need more research to address and develop best practices: There must be a clear demonstration of the benefits, with a CE visible for everyone. More research is needed, but the available research will also have to be transferred from a small segment of the population to the wider public and transitioned from science or assessments to real pilot projects.

8 Conclusion

Considering the above information, a conclusion can be made that the WANA region cannot only reduces its environmental impact but also improves its economic and social well-being by implementing circular practices. The transition to circularity is an imperative dictated by the unsustainability of the currently prevailing linear economic models. Overall, there is a recognition in policy development documents, stressing the necessity of the transition. However, the disparities in levels of economic development make benchmarking at the regional level difficult across countries. In parallel, there are disparities in progress toward circularity between sectors. The waste management sector is commonly referred to as a priority for the transition. Other sectors are still at a pilot stage with the ambition to scale up.

A new generation of capacities is yet to be developed. Engaging with the youth and young entrepreneurs equipped with technological tools can realise the potential for expansion in green jobs. Moreover, the pool of competencies needed for the transition to CE is not sufficient in most of countries in the WANA region. The regional approach to the transition to CE will surely enhance the regional capacities, data availability, and its application, and innovation, particularly in emerging sectors of CE. Transition to CE offers an extraordinary opportunity for regional cooperation and economic integration with a clear potential for economic prosperity, social inclusion, and resilience, a sustainable use of resources, and a readiness to respond to shocks while mitigating risks at the same time. The paper also represents a significant advancement for future research to address the uncovered gap concerning the neglect of social and economic dimensions and their interrelationship.

Recognized policy deficiencies and suggested modifications will ideally foster sustainable corporate practices, encourage circular economy initiatives, and establish a conducive atmosphere for people-centered and lucrative circular models. Emphasizing a balance between social equality and economic sustainability is a comprehensive approach. It can result in more sustainable, profitable, ethical, and successful enterprises over time, thereby enhancing both local and global economies.

Circular enterprises are dependent on a framework of enabling policies that create the right environment for them to form, grow, and survive. Without appropriate regulations and laws, there is little capacity to encourage new circular enterprises and to regulate emissions in current businesses. The possibility for governance to follow up and act when needed is based on framework conditions where circularity or sustainability are measured and enforced.

References

Achouri, O., El Fels, L., Elagroudy, S., Hafidi, M., Chaouki, B., & Guo, J. (2022). Solid waste management in the context of a circular economy in the MENA region. *Sustainability, 14*(1), 480.

Acted Lebanon. (2020). *Towards a circular economy in Lebanon.* Acted Lebanon, June 2020. https//:www.acted.org/wp-content/uploads/2018/01/acted-lebanon-circular-economy-final.pdf

Alhola, K., et al. (2019). Exploiting the potential of public procurement: Opportunities for circular economy. *Journal of Industrial Ecology, 23*(1), 96–109.

Aljaradin, M., Persson, K. M., & Sood, E. (2015). The role of informal sector in waste management, a case study; Tafila-Jordan. *Resources and Environment, 5*(1), 9–14.

Almadhi, A., Abdelhadi, A., & Alyamani, R. (2023). Moving from linear to circular economy in Saudi Arabia: Life-cycle assessment on plastic waste management. *Sustainability, 15*(13), 10450.

Al-Masri, A., Al-Khasawneh, A., Al-Khateeb, N. (2020). Extended producer responsibility: A case study on the implementation of Jordan's EPR program for packaging Waste. *Sustainability Journal.*

Almulhim, A. I. (2024). Toward a greener future: Applying circular economy principles to Saudi Arabia's food sector for environmental sustainability. *Sustainability, 16*(2), 786.

Almulhim, A., & Abubakar, I. (2021). Understanding public environmental awareness and attitudes toward circular economy transition in Saudi Arabia. *Sustainability, 13*(18), 101–157.

Almulhim, I. A., & Al-Saidi, M. (2023). Circular economy and the resource nexus: Realignment and progress towards sustainable development in Saudi Arabia. *Environmental Development, 46*.

Alsaud, K., Assad, F., Patsavellas, J., & Salonitis, K. (2025). A comparative analysis of circular economy practices in Saudi Arabia. *Sustainability, 17*(8), 3281.

Al-Shihabi, F. (2025). *How are Gulf nations developing a circular economy?* Technical Review Middle East, 14 Mar 2025. https://technicalreviewmiddleeast.com/manufacturing/how-are-gulf-nations-developing-a-circular-economy

Al-Taai, S. H. H. (2021). Green economy and sustainable development. In *IOP Conference Series: Earth and Environmental Science, 779, Fifth International Scientific Conference on Environmental and Sustainable Development*, 1–2 June, Baghdad, Iraq, and Turkey.

Aoun, M. C. (2013). Oil and gas resources of the middle east and north Africa: A curse or a blessing? In J. M. Chevalier & P. Geoffron (Eds.), *The new energy crisis: Climate* (pp. 133–160). Springer.

Barrie, J., Schröder, P., & Schneider-Petsinger, M. (2022). *The role of international trade in realizing an inclusive circular economy.* Royal Institute of International Affairs.

Bashir Khodaparasti, R., & Bagheri Garabollagh, H. (2023). Examining circular economy practices and sustainability performance in knowledge-based companies in Iran. *Amfiteatru Economic, 25*(62), 196–212.

Chizaryfard, A., Trucco, P., & Nuur, C. (2021). The transformation to a circular economy: Framing an evolutionary view. *Journal of Evolutionary Economics, 31*(2), 475–504.

Corona, B., Shen, L., Reike, D., Carreón, J. R., & Worrell, E. (2019). Towards sustainable development through the circular economy—A review and critical assessment on current circularity metrics, resources. *Conservation and Recycling, 151.*

Dincer, I. (2000). Renewable energy and sustainable development: A crucial review. *Renewable and Sustainable Energy Reviews, 4*(2), 157–175.

El Hafdaoui, H., Khallaayoun, A., Ouazzani, K. (2024). Long-term low carbon strategy of Morocco: A review of future scenarios and energy measures. *Results in Engineering, 21.*

Ellen MacArthur Foundation. (2014). *Towards the circular economy: Accelerating the scale-up across global supply chains.* World Economic Forum 2014, https://www3.weforum.org/docs/WEF_ENV_TowardsCircularEconomy_Report_2014.pdf

ESCWA. (2020). Arab Sustainable Development Report. https://asdr.unescwa.org/sdgs/pdf/en/ASDR2020-Final-Online.pdf.

ESCWA. (2022). *Climate finance needs and flows in the Arab region.* Policy Brief. https://www.unescwa.org/sites/default/files/event/materials/Policy%20Brief-Climate-finance-needs-flows-2022.pdf

European Parliament News. (2022). *Circular Economy: Definition, Importance and Benefits.* European Parliament News, 26 Apr 2022. https//:www.europarl.europa.eu/news/en/headlines/economy/20151201STO05603/circular-economy-definition-importance-and-benefits

Ferdous, W., Manalo, A., Siddique, R., Mendis, P., Zhuge, Y., Wong, H. S., Lokuge, W., Aravinthan, T., Schubel, P. (2021). Recycling of landfill wastes (tyres, plastics and glass) in construction—A review on global waste generation, performance, application and future opportunities. *Resources, Conservation and Recycling, 173*, 105745, ISSN 0921-3449.

Fersi, C., et al. (2021). Circular economy in Tunisia. In S. K. Ghosh & S. K. Ghosh (eds.), *Circular economy: Recent trends in global perspective.* SpringerLink.

Geissdoerfer, M., Savaget, P., Bocken, N., & Hultink, E. J. (2016). The circular economy: a new sustainability paradigm? *Journal of Cleaner Production, 143*, 757–768.

Geldermans, R. J. (2016). Design for change and circularity-accommodating circular material & product flows in construction. In *SBE16 Tallinn and Helsinki Conference; Build Green and Renovate Deep, Tallinn and Helsinki Delft University of Technology*, pp. 301–311.

Ghanem, A. M., & Alamri, Y. A. (2023). The impact of the green middle east initiative on sustainable development in the Kingdom of Saudi Arabia. *Journal of the Saudi Society of Agricultural Sciences, 22*, 135–146.

Gower, R., & Schröder, P. (2016). *Virtuous Circle: How the circular economy can create jobs and save lives in low and middle-income countries.* Institute of Development Studies and Tearfund, UK.

Hemidat, S. (2019). *Feasibility assessment of waste management and treatment in Jordan.* Ph.D. Thesis, The Academic Board of Rostock University, Rostock, Germany.

Hemidat, S., Achouri, O., El Fels, L., Elagroudy, S., Hafidi, M., Chaouki, B., Ahmed, M., Hodgkinson, I., Guo, J. (2022). Solid waste management in the context of a circular economy in the MENA region. *Sustainability, 14*(1), 480.

Hosni, I. S. A., Amoudi, O., & Callaghan, N. (2020). An exploratory study on challenges of circular economy in the built environment in Oman. *Proceedings of the Institution of Civil Engineers - Management, Procurement and Law, 173*(3), 104–113.

Keskes, M. A., Zouari, A., Lehyani, F., & Houssin, R. (2022). Circular economy implementation within manufacturing companies at Sfax-Tunisia: Barriers and opportunities. *Environmental Engineering & Management Journal (EEMJ), 21*(3).

Kirchherr, J., Piscicelli, L., Bour, R., Kostense-Smit, E., Muller, J., Huibrechtse-Truijens, A., & Hekkert, M. (2018). Barriers to the circular economy: Evidence from the European Union (EU). *Ecological Economics, 150*, 264–272.

Mana, A. A., Allouhi, A., Ouazzani, K., & Jamil, A. (2021). Feasibility of agriculture biomass power generation in Morocco: Techno-economic analysis. *Journal of Cleaner Production, 295.*

Morseletto, P. (2020). Targets for a circular economy, resources. *Conservation and Recycling, 153.*

OECD. (2021). Towards a more resource-efficient and circular economy. https://www.oecd.org/environment/waste/OECD-G20-Towards-a-more-Resource-Efficient-and-Circular-Economy.pdf.

Okorie, O., Salonitis, K., Charnley, F., Moreno, M., Turner, C., & Tiwari, A. (2018). Digitisation and the circular economy: A review of current research and future trends. *Energies, 11*(11), 3009.

Olawuyi, D. (2018). Advancing innovations in renewable energy technologies as alternatives to fossil fuel use in the middle east: trends, limitations, and ways forward. In D. Zillman, M. Roggenkamp, L. Paddock, & L. Godden (Eds.), *Innovation in energy law and technology: Dynamic solutions for energy transitions* (pp. 354–370). Oxford University Press.

PACE: Platform for Accelerating the Circular Economy. (2021). *Circularity Gap Report 2021.* Circle Economy. https://www.circularity-gap.world/2021

Preston, F., Lehne, J., Wellesley, L., & Sugathan, M. (2019). *An inclusive circular economy: Priorities for developing countries.* Energy, Environment and Resources Department, The Royal Institute of International Affairs, Chatham House.

Reichel, A., De Schoenmakere, M., Gillabel, J., Martin, J., & Hoogeveen, Y. (2016). *Circular economy in Europe: Developing the knowledge base.* European Environment Agency Report, Copenhagen K Denmark.

Shehri, T. A., Braun, J. F., Howarth, N., Lanza, A., & Luomi, M. (2023). Saudi Arabia's climate change policy and the circular carbon economy approach. *Climate Policy, 23*(2), 151–167.

Suiko, Y., Elhadary, H. M. M. S., Doumani, F. M., Arif, S. K. F., Sakr, D. A. M., Ahijado Aparicio, P., & Abdelaal, T. A. T. E. (2024), *Egypt—Country environmental analysis: Promoting circular economy and blue economy for environmental sustainability.* Washington, D.C.: World Bank Group.

Suwaidi, H. (2023). 2031 and Beyond: UAE's journey towards prosperity and green dominance. *Gulf News.* https://gulfnews.com/opinion/op-eds/2031-and-beyond-uaes-journey-towards-prosperity-and-green-dominance-1.99455565

Taqi, J. E. (2017). *Sustainable Cities: PPPs and the emergence of the circular economy in the Middle East Norton Rose Fulbright.* https://www.nortonrosefulbright.com/en/knowledge/publications/ab20952f/sustainable-citiesppps-and-the-emergence-of-the-circular-economy-in-the-middle-east

Tayeb, S. (2024). *From scarcity to sustainability: The GCC's journey towards water security.* World Bank Group. https://www.worldbank.org/en/news/opinion/2024/03/26/from-scarcity-to-sustainability-the-gcc-s-journey-towards-water-security

UNEP. (2016). Sustainable Consumption and Production National Action Plan (SCP-NAP) in Palestine. Green Policy Platform. https://www.greengrowthknowledge.org/project/sustainable-consumption-and-production-national-action-plan-scp-nap-palestine

Beyond a Single Narrative: An Investigation of Non-western Indigenous Knowledge Systems and Their Application to Deliver an Effective Circular Economy

Augustus Kweku Sobeng, John Morrissey, and Julian Bloomer

Abstract The United Nation's (UN) Sustainable Development Goal (SDG) 12 advocates for responsible consumption and production, an idea that is encapsulated by the concept of Circular Economy (CE). Instead of using and disposing off resources, CE ensures that resources are used and re-used to prevent the generation of waste. While CE has gained prominence in the literature, CE research has been largely dominated by Western perspectives that have often overlooked non-Western perspectives. Ironically, while many Western CE approaches tend to pursue economic profitability ventures at the expense of the overall well-being of the environment, most non-Western indigenous approaches and perspectives have been found to prioritise the social and ecological well-being of the environment. To meet SDG 12 by 2030, it is imperative that CE principles include not only Western ideologies, but also non-Western discourses and perspectives. This chapter presents an overview of CE, and it proceeds to offer a critical review and discussion of sustainable non-Western indigenous knowledge systems and their contributions the CE discourse. The Hawai'ian ancestral traditional knowledge system is highlighted and explored as a case study of how non-Western indigenous knowledge systems can be applied to enhance CE. By critically examining non-Western indigenous perspectives, this chapter seeks to lay the foundation for a more inclusive CE discourse, with an ultimate goal of facilitating the inclusion of non-Western knowledge systems into CE conceptualisation.

Keywords Circular economy · SDGs · Non-western perspectives · Indigenous knowledge systems · Hawai'ian ancestral traditional knowledge system

---

A. K. Sobeng (✉) · J. Morrissey · J. Bloomer
Department of Geography, Mary Immaculate College, University of Limerick, Limerick, Ireland
e-mail: augustus.k.sobeng@mic.ul.ie

J. Morrissey
e-mail: john.morrissey@mic.ul.ie

J. Bloomer
e-mail: julian.bloomer@mic.ul.ie

P. Singh et al. (eds.), *Circular Economy and Sustainable Innovation*,
https://doi.org/10.1007/978-3-032-00437-6_3

1 Introduction

Circular Economy (CE) is a tool that advocates for a shift from the traditional way of using and disposing off resources, towards a more sustainable model where products and resources are used over and over to reduce waste generation (Geissdoerfer et al., 2017). CE has gained prominence over the last decade with studies strongly associating it with the attainment of the Sustainable Development Goal (SDG) 12 on responsible consumption and production (Friant et al., 2020; United Nations, 2024).

Despite its enormous potential, the CE discourse has been largely dominated by Global North perspectives and ideas that often overlook the context of the Global South (Friant et al., 2020; Marotti de Mello et al., 2022; Mawere, 2014). These biases have usually undermined non-Western sustainable development efforts and contributions. To realise its full potential, the CE discourse must move beyond the single narrative of the Global North to include unique non-Western contexts, knowledge systems, and perspectives. A context-sensitive approach to CE will go a long way to make it inclusive and a more potent tool towards achieving SDG 12 and other related SDGs.

This chapter presents a review of CE by highlighting both Global North and South narratives of the concept. The study then puts a spotlight on knowledge systems from the Global South and how they can be applied to strengthen the CE concept. This study makes significant contributions towards SDG 12, and it helps to address the dearth in CE research, particularly as regards the social, cultural and political dimensions of CE that have received little to no attention in the literature. Additionally, the study has broader implications for advancing SDG 7 (affordable and clean energy), SDG 10 (reduced inequalities), SDG 11 (sustainable cities and communities), and SDG 13 (climate action). All-in-all, the practical outcomes from this study can help to deliver real sustainability 'wins' not only in a Global South context, but in a global context.

1.1 An Overview of the Circular Economy Concept

CE has been strongly backed as a potent tool to help meet UN SDG 12 by the year 2030 (Dewick et al., 2022), as various studies consider it to be a self-sustaining model compared to the Linear Economy (LE) model. The LE is an unsustainable production and consumption model that encourages excessive generation and disposal of waste, resulting in significant losses along the value chain of products and materials (Ellen MacArthur Foundation, 2013, 2021; Friant et al., 2020). CE has attracted enormous attention ranging from the natural sciences, engineering, business and social science disciples. The concept has also garnered growing attention from governmental and non-governmental organisations globally (Korhonen et al., 2018a, 2018b; Leipold et al., 2023; World Economic Forum & Ellen MacArthur Foundation, 2016).

The growing attention received by CE has led to different interpretations and understanding of what the concept means. Some studies define CE to be the "realisation of a closed loop flow" (Geng & Doberstein, 2008, p. 231) that comes about through the repeated use of resources with the aim of maintaining their value so that they can be used for a longer period of time before they reach the end of their life cycle (Ghisellini et al., 2016; Webster, 2015; Yuan et al., 2006). One of the widely recognised definitions of CE was given by the Ellen MacArthur Foundation, which describes it as "an industrial economy that is restorative by intention and design" (Ellen MacArthur Foundation, 2013, p. 14). Based on the foregoing, Geissdoerfer et al., (2017, p. 759) also define CE as "a regenerative system in which resource input and waste, emission, and energy leakage are minimised by slowing, closing, and narrowing material and energy loops."

In a CE, manufacturers of products are encouraged to design their products with the intention that it would be easy for them to be used over and over again (Ellen MacArthur Foundation, 2013; Geissdoerfer et al., 2017; Gutberlet et al., 2017). CE, therefore, is often conceptualised to encapsulate a variety of the 3Rs (i.e., reduce, reuse, recycle) and 6R's (i.e., reuse, recycle, reduce, redesign, remanufacture, recover) seeking to ensure circularity of both biological (biodegradable) and technical (non-biodegradable) products and materials. This is best captured in the CE systems diagram (or the butterfly diagram) proposed by the Ellen MacArthur Foundation (2019) (Fig. 1).

However, the idea of fully circulating materials and products via processes such as recycling, as advocated by CE proponents, has been critiqued in the literature (Cullen, 2017). Studies show that providing for the needs of the world's population solely through recycled and recovered products is nearly impossible (Friant et al., 2020). Critics argue that restricting the usage of resources to only recycled materials could negatively impact the well-being of vulnerable populations whose survival needs might be overlooked, as they may be hindered from utilising resources around them that their livelihoods depend on (Bengtsson et al., 2018; Schröder et al., 2019).

The Ellen MacArthur Foundation stresses that CE can be achieved by focusing primarily on how products are designed (Ellen MacArthur Foundation, 2024). The Foundation spells out three key principles that undergird CE, namely: eliminate waste, circulate products and materials, and regenerate nature. They propose that through CE, waste can be eradicated when the designers and manufacturers of products make their products with the intent that the products, or the materials the products are made of, will continue to be used over and over, rather than being disposed of as waste when they near the end of their life cycle. Achieving the first principle of eliminating waste eventually leads to the achievement of the second principle which aims at retaining the inherent value within products and/or materials that are near the end of their life cycle, by using them repeatedly. Finally, the third principle indicates that as products and materials are used and reused, and waste is eliminated, it will eventually lead to a shift from the continuous extraction of resources to a healthier state of the environment where resources are used in a sustainable manner, and nature is regenerated (or restored) (Ellen MacArthur Foundation, 2024).

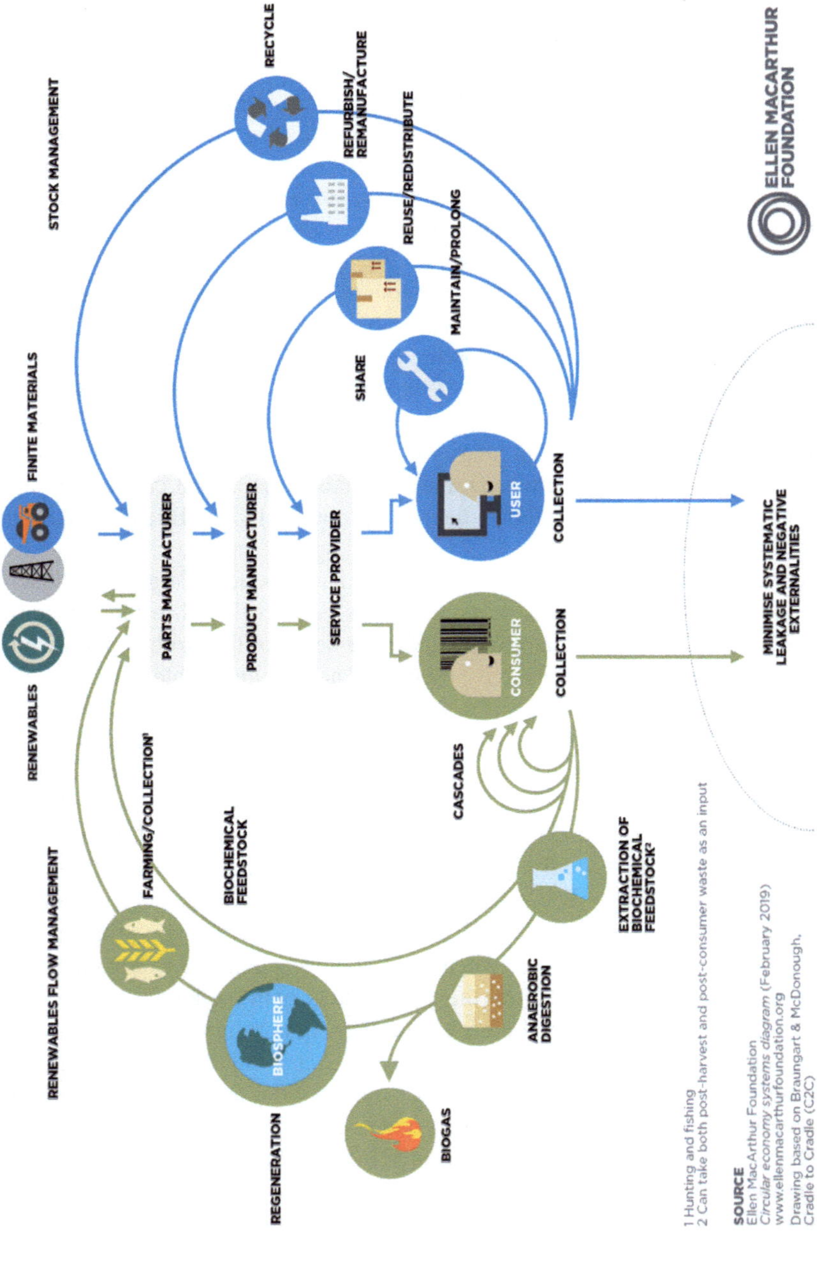

Fig. 1 The circular economy butterfly diagram (Copyright © Ellen MacArthur Foundation, Circular economy system diagram, 2019)

CE's attempt to deliver growth while ridding the environment and economy of any negative impacts has been questioned (Friant et al., 2020). Studies suggest that this approach is inherently contradictory (Vezzoni, 2023). While it advocates for the reuse of products and materials to keep the environment healthy, it promotes capitalism and profit-making, leading to a rebound effect that ultimately has a negative impact on the environment. The rebound effect underscores a significant weakness in the CE's ability to achieve true sustainability (Beamer et al., 2021, 2023; Friant et al., 2020). The CE approach appears to lean more towards making economic gains rather than true ecological well-being (Ellen MacArthur Foundation, 2015; Vezzoni, 2023). Many governments and renowned organisations have bought into the profit-making appeal of CE and thus still maintain economic growth as a core policy priority. This approach has perpetuated a concerning scenario where ongoing economic growth (and its associated ecological decline) are pursued under the guise of implementing a CE agenda (Vezzoni, 2023). This concern is corroborated in various studies that found that many production processes that are described to be circular can lead to much worse impacts on the environment than the traditional linear economy production processes (Hobson & Lynch, 2016; Velis, 2018). The concept of degrowth has been proposed by some studies to have enormous potential of helping shape the CE concept to help it depart from economic growth tendencies and make it more sustainable, however, it has not received much attention in the CE literature (Audier, 2019; Friant et al., 2020; Kallis et al., 2018).

1.2 Origins and Evolution of the Circular Economy Concept

CE has gained significant attention over the years. Consequently, various schools of thought have emerged that have played a pivotal role in shaping the concept as we have it (Ghisellini et al., 2016). In this section, the origin and evolution of CE is discussed to help appreciate how the concept has developed and evolved over the years.

The CE concept has been evolving since the late 1960's and 1970's. Some studies indicate that the initial idea about circularity emerged from an American Ecological Economist named Kenneth E. Boulding (Boulding, 1966) who introduced the notion of open and closed economy in his essay about the "spaceman economy" (Ghisellini & Ulgiati, 2020, p. 493), where he conveys the idea of circularity as a crucial necessity for ensuring that humans can live sustainably on Earth (Ghisellini et al., 2016). Industrial Ecology (IE) principles were later introduced in the late 1970's by Stahel and Reday (1976). The IE principles view circularity as constituting responsible industrial approaches that ensure that resources meant for industrial production are managed and utilised efficiently in a circularised system to avoid wastage (Erkman, 1997; Stahel & Reday, 1976). The IE idea opposes the view that the impact of industrial systems should be treated and studied separately from the environment. This stance is in line with the 'circular' or 'cyclical' idea, which indicates that environment and industrial systems are all characterised by flows of materials.

IE, therefore, encourages the shift from an open system, which encourages more waste production in industrial processes, to a closed system that aims at lessening the production of industrial waste (Erkman, 1997; Ghisellini et al., 2016). Stahel, who was among the early proponents of the IE theory, further introduced the idea of shifting from outright ownership of goods and resources to an IE model where one can only gain the right (by purchase) to utilise a product instead of owning it (Stahel, 1982).This idea was seen as a highly sustainable way to ensure the circularity of products and materials within the environment (Geissdoerfer et al., 2017).

Although significant progress was made in conceptualising what circularity is during the late 1960s and 1970s, many prominent studies agree that the foundational work of Pearce and Turner (1989) formalised the concept of the CE, with their efforts being notably inspired by Boulding's prior essay in 1966 (Andersen, 2007; Friant et al., 2020; Ghisellini et al., 2016; Su et al., 2013). In their study, Pearce and Turner (1989) examined the linear model of "contemporary economic systems," describing how natural resources influence the economy by providing the resources required for production and consumption while also acting as a recipient of waste materials (Geissdoerfer et al., 2017, p. 759).

The CE concept has gained significant momentum since its formalisation in the late 1980's (Ellen MacArthur Foundation, 2013). Ideas such as natural capitalism (Hawken et al., 1999), the natural step (Robèrt, 2002), and cradle to cradle were introduced into CE conceptualisation between the late 1990's and 2002. From 2010 onwards, the CE concept ushered in a period Friant et al. call the "validity challenge period" (2020, p. 8). This period marked a season of testing as contentions surrounding CE's conceptualisation grew, with its validity being questioned. During this period, concepts such as the performance economy were introduced, which significantly influenced the conceptualisation of CE and led to its widespread recognition and applicability, thus boosting its validity status (Stahel, 2010). Recent studies have also highlighted the impact of concepts such as biomimicry, blue economy, and regenerative design, which have together contributed to CE's prominence in the Twenty-first Century (Beamer et al., 2021; Ellen MacArthur Foundation, 2015, 2023; Friant et al., 2020; Geissdoerfer et al., 2017; Gutberlet et al., 2017).

1.3 Research and Policy Development in the Circular Economy

CE has received growing attention in the literature, with some studies suggesting that this growing research effort has likely been catalysed by the Ellen MacArthur Foundation's detailed report on CE in 2013 (Geissdoerfer et al., 2017; Kirchherr & van Santen, 2019). Despite the significant attention CE has received, the literature indicates a significant dearth in empirical studies. A study by Kirchherr and van Santen (2019), for instance, show that increased empirical studies on the CE would go a long way to provide more real-life experiences and perspectives on CE which could help improve how CE is understood and operates in different contexts.

Studies have revealed that discussions on CE within the literature are predominantly shaped by Western perspectives and the contexts of developed economies in the Global North. This one-dimensional focus often renders CE irrelevant or beyond reach by the less developed or developing regions of the world, as the issues the Western-aligned CE addresses may fail to align with their unique contexts and realities (Beamer et al., 2021, 2023; Friant et al., 2020; Gutberlet et al., 2017; Kirchherr & van Santen, 2019). These issues have raised concerns about how CE has been conceptualised and framed (Marotti de Mello et al., 2022). The dominance of Western perspectives in the CE discourse, in the guise of a universally applicable solution, has resulted in the widespread introduction and adoption of Western CE ideas and philosophies in many less-developed regions of the world, particularly Global South countries (Marotti de Mello et al., 2022). Marotti de Mello et al., (2022, p. 318) describe this mainstream influx of Western ideas of CE as an "epistemic injustice." Critics of this injustice have highlighted that CE would greatly benefit from incorporating perspectives rooted in non-Western indigenous knowledge systems and practices (Beamer et al., 2023; Friant et al., 2020; Kirchherr & van Santen, 2019). Indigenous knowledge systems from the non-Western part of the world are worth investigating as they possess the potential for providing solutions that could go beyond what the Westernised form of CE offers (Friant et al., 2020; Marotti de Mello et al., 2022).

The CE literature has also been critiqued for being largely dominated by technological and economic discussions at the expense of social, cultural, and political dimensions, which includes issues relating to governance and justice (Friant et al., 2020). It is important that while attempts are being made to advance technology and science for the furtherance of CE, social, cultural, and political components, which contribute significantly to the overall well-being of the environment, are not relegated to the background (Deléage, 2019; Friant et al., 2020; Kirchherr & van Santen, 2019; Korhonen et al., 2018a, 2018b).

In the policy-making arena, CE has received tremendous attention among governments, international organisations, and policy developers. Country's like Germany (1996 Closed Substance Cycle and Waste Management Act), Japan (2002 Basic Law for Establishing a Recycling-Based Society), and China (2008 Circular Economy Promotion Law of the People's Republic of China), are among countries that took early steps to include CE initiatives and thinking in their policy agendas (Geissdoerfer et al., 2017; Lieder & Rashid, 2016; METI, 2004; Su et al., 2013). The European Union (EU) has also, in the past few years, incorporated significant CE initiatives into its sustainable growth and development agenda, aiming to ensure responsible production and consumption within Europe (Deselnicu et al., 2018; European Commission, 2015). Despite attempts by government and private organisations to integrate CE into their development strategies, studies have revealed that some of these efforts are politically backed, with some governments criticised for using CE as a greenwashing tool to enhance their public perception while still pursuing business-as-usual ventures behind the scenes (Ampe et al., 2020; Kirchherr & van Santen, 2019; Korhonen et al., 2018a, 2018b; Nylén & Salminen, 2019). The EU, for instance, has publicly committed to a green growth agenda, however, there is limited empirical evidence of success in achieving these green targets, as many EU member states continue to invest heavily in fossil fuel and other raw material extraction that is detrimental to the sustainability of the environment (Vezzoni, 2023). China's CE approach has also been criticised as following a top-down strategy with the government controlling CE decision-making and implementation (Bleischwitz et al., 2022). This top-down approach often leaves little room for consensus-building and participation, and has been identified as the major factor behind the failure of some 49 sugar refinery CE projects in China (Huang, 2020).

2 Circular Economy Discourses in the Global North and South

The evolving conceptualisation of CE has given rise to diverse narratives and framings. This section explores these narratives, which have been grouped into two discourses: circularity discourses that align with the Global North and circularity discourses that align with the Global South.

2.1 Circularity Discourses Aligned with Global North Perspectives

The expression "Global North" describes developed countries in the world that are usually characterised by indicators such as high levels of industrialisation, economic stability, and high literacy levels, among others (Ukhami et al., 2022). CE is dominated by ideas from the Global North (Dewick et al., 2022), and these ideas and their related strategies are mostly characterised by the quest for economic profitability (Adamowicz et al., 1998; Beamer et al., 2021). CE discourses advanced in the literature that resonate with the Global North or Western ideologies can be grouped into three discourse types: Reformist Circular Society (Friant et al., 2020, p. 11), Technocentric Circular Economy (Friant et al., 2020, p. 12), and Fortress Circular Economy (Friant et al., 2020, p. 12). This grouping considers perspectives on social, technological, political, and ecological issues (Friant et al., 2020; Leipold et al., 2023).

Reformist Circular Society discourses support calls for the inclusion of social, environmental, and political dimensions in CE thinking. However, such discourses are characterised by more emphasis on the usage of technology for the pursuit of CE (Friant et al., 2020). To this end, technological innovation is pursued even to the detriment of the social and ecological health of the environment. Reformists do not see much difference between capitalism and sustainability, as they believe both capitalism and sustainability have a similar end-goal of bringing about improvement in the well-being of people (Friant et al., 2020). CE concepts such as regenerative design, natural capitalism, and the performance economy are considered to align with the Reformist Circular Society Discourses (Friant et al., 2020; von Weizsäcker & Wijkman, 2017). This discourse type has faced strong criticisms for its unwillingness to accept and address economic growth concerns raised about the CE (Friant et al., 2020).

Technocentric CE discourses, on the other hand, posit that advancing green growth through a heavy reliance on technology can achieve both economic growth and, at the same time, ecological health in a win-win scenario, without the need to sacrifice one for the other (Friant et al., 2020). This discourse type, like the Reformists, believes that capitalism and sustainability share common goals, as they both seek to enhance human wellbeing. Biomimicry and industrial ecology CE concepts are typical examples of Technocentric CE discourses. Studies indicate that the CE policies of many EU nations, along with the strategies of multinational corporations such as Apple Inc., and international organisations like the Organisation for Economic Cooperation and Development (OECD) and the World Economic Forum (WEF) are heavily influenced by Technocentric CE perspectives (Colombo et al., 2019; Kaźmierczyk, 2018; OECD, 2018; Valenzuela & Böhm, 2017; Vonk, 2018; World Economic Forum & Ellen MacArthur Foundation, 2016). Institutions and organisations that uphold the Technocentric CE vision have been criticised for prioritising advancements in technology as a key driver for attaining a CE, while overlooking the importance of a participatory governance system in the CE discourse (Friant et al., 2020, p. 12).

The Fortress CE discourse argues that neither the deployment of technological innovation nor the incorporation of socio-cultural and political dimensions to CE can effectively address the challenge of global environmental scarcity (Friant et al., 2020). The Fortress CE discourse posits that the scarcity crisis facing the globe can only be resolved by placing limitations on society and people's usage of resources. Institutions that argue in favour of this CE discourse type have been heavily criticised for appearing to promote frugal resource consumption measures while ironically engaging in the very actions that they criticise and frown on (Cavanagh & Benjaminsen, 2017; Dryzek et al., 2019). Studies suggest that this discourse type encourages scientists from the Global North to leverage the Fortress CE approach, emphasising resource scarcity and the need for conservation of limited resources, to justify imposing restrictions on populations—particularly in the Global South—who often have minimal or no involvement in contributing to the global resource scarcity issue in the first place (Dryzek et al., 2019). Historically, the principles of Fortress CE were applied in the Nazi era in Germany to enforce restrictions on people's usage of resources and protect resource overexploitation (Corvellec et al., 2022). This type of CE discourse presents significant risks to sustainability, as studies indicate it could worsen the development gap and intensify poverty in less developed or developing countries. At the same time, it has the potential to enable the wealthy to exploit sustainability agendas for profit, often at the expense of marginalised populations (Malleson, 2015; Valenzuela & Böhm, 2017).

2.2 *Circularity Discourses Aligned with Global South Perspectives*

The term "Global South" is often used to describe the less developed regions of the world, including countries in Africa, Latin America, Asia, and the Caribbean (Ukhami et al., 2022). Studies have identified a strong alignment between various non-Western indigenous knowledge systems from the Global South and the principles of the Transformational Circular Society discourse (Friant et al., 2020; Kothari et al., 2019). The Transformational Circular Society discourse prioritises the socio-cultural and political dimensions of sustainability, advocating for an inclusive and holistic society where there exists a healthy relationship between humans and the environment that surrounds them (Friant et al., 2020). Indigenous knowledge systems from the Global South, like Ubuntu (Shumba, 2011) and Buen Vivir (Kothari et al., 2014) have been found to be inherent with significant elements that align with the principles of the Transformational Circular Society discourse. Non-Western indigenous knowledge systems have, unfortunately, been critically underrepresented in the CE literature, and therefore, more empirical research is needed to explore the diverse range of these knowledge systems, particularly from the Global South, and to understand how their circularity principles can be harnessed and integrated into the CE.

3 Integrating Non-western Indigenous Knowledge Systems into the Circular Economy Concept

Indigenous knowledge refers to the practices, ways of thinking, and forms of knowledge unique to a particular group of people, typically passed down through generations by their ancestors (Adom, 2016; Noyoo, 2007). The terms traditional knowledge or local knowledge are usually used to describe indigenous knowledge, emphasising its distinction from foreign knowledge systems (Warren, 1992; Zewdu et al., 2024).

Non-Western indigenous knowledge systems are believed to have been in existence for hundreds of years (Beamer et al., 2023). Historically, many non-Western indigenous people and communities have developed unique ways of living and managing their resources that have helped them to overcome the numerous challenges they faced while keeping the environment they occupied in a good state of health (Adom & Kquofi, 2016; Akpabio & Subramanian, 2012). Studies have shown that incorporating non-Western indigenous knowledge ideologies into sustainability efforts can help ensure a safer and sustainable society (Anokye & Mohammed, 2024), where the inherent cultural practices and principles can help preserve and restore nature (Bannister & Hardison, 2006; Mawere, 2014).

In Africa, some studies have highlighted that indigenous societies are defined by strong community bonds, principles of sharing, and reciprocity—key elements that are often missing in mainstream CE discourses (Amuzu, 2022). Biodiversity conservation and waste management efforts in Ghana, for instance, have historically benefited from indigenous practices such as "sacred groves," "sacred trees and animals," "indigenous farming methods," and "cultural restrictions on fishing" (Anokye & Mohammed, 2024, p. 5) which efficiently helped to conserve Ghana's wildlife and forests in the past (Adjei & Sika-Bright, 2019; Anokye & Mohammed, 2024; Imoro et al., 2022).

Integrating more non-Western indigenous knowledge principles into the CE discourse could be highly transformative, however, there exists a significant dearth of empirical research on CE in non-Western contexts and how the circularity principles inherent in theses non-Western knowledge systems could be applied to the CE concept (Friant et al., 2020; Marotti de Mello et al., 2022). Findings from studies about how Hawai'i's Ancestral traditional knowledge principles were utilised to preserve nature and ensure circularity could help make a strong case for the need for CE research agenda to shift towards more empirical studies in non-Western regions of the World, particularly in the Global South, where there exists significant gaps in CE research (Friant et al., 2020; Gutberlet et al., 2017). The next section will explore these findings in detail.

3.1 Case Study: Hawai'i Ancestral Circular Economy

In Hawai'i, an island state in the USA, there exists a knowledge system called "aloha 'āina" which is a non-Western ancestral traditional knowledge (Beamer et al., 2021, p. 2) that promotes a relationship of respect between humans and nature (Beamer, 2020; Beamer et al., 2021, 2023; Low, 2020). The aloha 'āina traditional knowledge system encourages principles of environmental stewardship and reciprocity with the aim of ensuring responsible usage and protection of environmental resources (Beamer et al., 2021, 2023). This knowledge system applies both ancestral and community knowledge to achieve a healthy co-existence between humans and the environment. By so doing, the aloha 'āina knowledge system ensures circularity and promotes CE (Beamer et al., 2021, 2023; Low, 2020). Exploring the benefits of the Hawai'i Ancestral CE offers valuable insights and provides a blueprint for creating a more sustainable and transformative society that responsibly utilises the resources at its disposal. The aloha 'āina Ancestral CE of Hawai'i is founded on important circularity principles, including a balanced system of governance, reallocation of resources, and kinship of the environment (Beamer et al., 2023). These principles are discussed in the next subsections.

3.1.1 Balanced System of Governance

Within Hawai'i's Ancestral CE, there was the presence of a place allocation system, together with a robust governance structure where people were given shared responsibilities and access to manage a defined area of natural resources (Beamer et al., 2023;

Winter et al., 2018). This participatory structure of managing Hawai'i's resources helped to ensure that despite the island being geographically limited resources, in terms of resource availability, there was responsible management and stewardship, which ensured that the limited resources available were sustainably used (Beamer et al., 2023, p. 4). Because people shared collective responsibilities in the governance and management of Hawai'i's resources, they developed a strong sense of discipline and accountability, which led to the effective management of their scarce natural resources. Winter et al. (2018) highlight that the principle of place allocation (inherent in the aloha 'āina Ancestral CE of Hawai'i) enabled collective managers of coastal and marine areas, for instance, to implement measures that restricted people from exploiting fish and other aquatic organisms from the water bodies during their breeding seasons.

3.1.2 Reallocation of Resources

The Hawai'ian Ancestral CE system had a unique system that ensured equitable distribution of resources to members of its society. This system, referred to as "Kālai'āina", was conceived to replicate the cyclical nature of the water cycle that ensures that water is kept in circulation globally via an open system where water is released out of the earth through outflow processes such as evaporation, and returned back to the earth via inflow process like precipitation (Beamer et al., 2023, p. 4). Studies reveal that this reallocation system of Hawai'i was such that whenever a new "supreme sovereign" (Beamer et al., 2023, p. 5) assumed power, resources, and wealth accumulated were redistributed to people within various Hawai'ian communities (Beamer, 2014; Beamer et al., 2023; Kame'eleihiwa, 1992). The system of redistribution ensured that individuals and families within Hawai'i had a fair share of the island state's available resources, hence no one was left behind. This effectively discouraged the system of capitalism and material accumulation.

3.1.3 Kinship of the Environment

Finally, the Hawai'ian Ancestral CE was strongly undergirded by the principle of "environmental kinship" (Beamer et al., 2023, p. 5), where the environment was seen as a member of the family lineage system. This principle ensured that people subconsciously treated their surrounding environment with utmost respect and dignity, which helped to avoid over-exploitation and destruction of resources within the environment (MacKenzie et al., 2007). This kinship of the environment principle helped to ensure circularity within the Hawai'ian economy by discouraging and eradicating wasteful practices (Beamer et al., 2023).

3.2 *Integrating the Hawai'ian Ancestral Circular Economy Principles into CE*

Based on the lessons learnt from the Hawai'ian Ancestral CE model, the researcher proposes a modified CE (see Fig. 2) which incorporates social, cultural, and political dimensions into CE. The figure below is a graphical representation of how the socio-cultural and political elements of the Hawai'ian Ancestral CE can be incorporated into the CE model to make it an all-inclusive tool.

Figure 2 introduces socio-cultural and political dimensions as perspectives that non-Western indigenous knowledge systems bring to the CE concept. Studies indicate that incorporating socio-cultural dimensions into CE thinking and conceptualisation fosters greater consideration of societal needs while emphasising cultural and intrinsic values that can be leveraged to build a circular society. By integrating these elements, CE practitioners, researchers, and policymakers will move beyond a

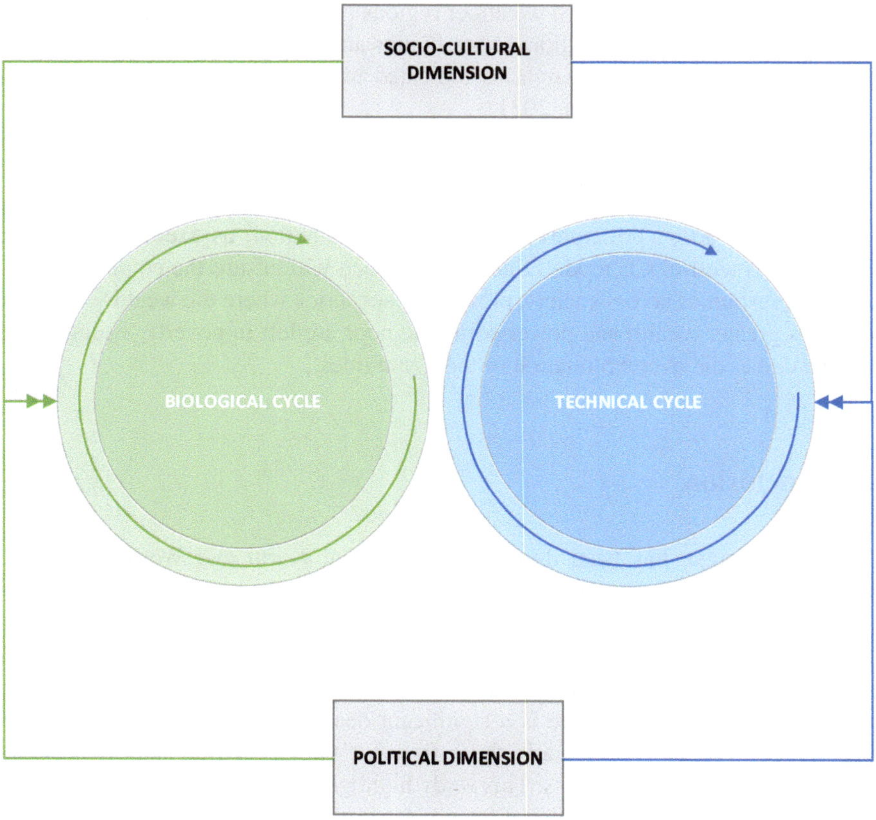

Fig. 2 A reconceptualisation of the CE model to incorporate socio-cultural and political dimensions (Source: Authors' construct based on Ellen MacArthur Foundation, 2019)

narrow focus on economic gains and technological advancements, ensuring a more holistic approach to CE development. This shift can significantly enhance CE practices and promote sustainable behaviours. For instance, the Hawai'ian Ancestral CE model demonstrates how cultural values that regard nature as kin foster a deep sense of care and respect for the environment. This perspective naturally translates into circular practices within the community, showcasing the transformative potential of integrating socio-cultural dimensions into the CE framework.

With respect to the political dimension of CE, as illustrated in Fig. 2, adopting a more participatory governance system, as exemplified by the Hawai'ian Ancestral CE model, could help foster a culture of stewardship and responsibility. This would encourage individuals to become accountable users and managers of environmental resources. A participatory approach also ensures that diverse societal perspectives are considered in CE decision-making. Achieving this requires a balanced combination of bottom-up and top-down governance strategies, enabling decisions to be made not just by wealthy institutions and governments but also by broader society and relevant stakeholders. Such inclusivity fosters a sense of involvement and ensures that CE discourse is not skewed toward wealthier regions (i.e., the Global North). Instead, it would reflect national and regional specificities and socio-cultural contexts in both policy prescriptions and research, as indicated in the Universal CE Policy Goals (Ellen MacArthur Foundation, 2021).

Finally, CE would greatly benefit from moving away from a capitalist discourse that promotes the accumulation of wealth and power through rampant extractivism disguised as green and circular practices. Instead, governments and institutions are encouraged to adopt principles of resource redistribution, as exemplified by the Hawai'ian Ancestral CE model. Such an approach will ensure the equitable distribution of resources across society, preventing scenarios where the wealthy continue to amass greater wealth and power while the poor are left in poverty, struggling to survive due to the overexploitation of their resources.

4 Conclusion

The CE discourse has been dominated by Global North and Western perspectives, often prioritising economic prosperity (for certain communities) over the social and ecological well-being of the environment. However, studies have shown that non-Western CE principles can foster a 'Transformational Circular Society' which sustains the environment more effectively than many Global North-centric approaches. Despite this, there is a significant dearth in research on non-Western Knowledge Systems—particularly those from the Global South—and their potential to enhance CE effectiveness. This study highlights the Hawai'ian Ancestral CE, rooted in the aloha 'āina traditional knowledge system, as an example of how non-Western indigenous knowledge systems can enrich the CE discourse. Hawai'i's aloha 'āina knowledge emphasises respect for the environment and embodies principles that can help make CE more inclusive and adaptable to diverse global contexts.

While the CE concept has achieved notable successes since its inception, the core idea must evolve to incorporate a broader range of worldviews, particularly those from underrepresented regions of the world. The researchers, therefore, recommend that future research focuses on conducting more empirical work on CE in less-developed regions, especially Global South regions, in order to uncover the lesser-known non-Western ideas and perspectives that can help inform CE practices. As studies empirically explore non-Western CE approaches, many unknown non-Western CE knowledge systems will begin to gain the needed recognition, ultimately shaping the CE discourse in a positive way. As Marotti de Mello et al., (2022, p. 319) argue, CE can greatly benefit from integrating "an understanding of how it is conceived from the point of view of indigenous epistemologies." The discussions in this chapter have revealed that non-Western CE principles extend beyond a narrow concern for economic profitability by emphasising environmental stewardship, respect, and participatory frameworks for managing scarce environmental resources. By embracing non-Western CE principles and ideologies, the CE concept has the potential to become more inclusive, transformational, and globally relevant.

Funding The corresponding author is a recipient of the Departmental Assistantship and the Department of Geography Postgraduate Awards, both provided by Mary Immaculate College, University of Limerick, Limerick, Ireland.

References

Adamowicz, W., Beckley, T., Macdonald, D. H., Just, L., Luckert, M., Murray, E., & Phillips, W. (1998). In search of forest resource values of indigenous peoples: Are nonmarket valuation techniques applicable? *Society & Natural Resources, 11*(1), 51–66. https://doi.org/10.1080/08941929809381061

Adjei, J. K., & Sika-Bright, S. (2019). Traditional beliefs and sea fishing in selected coastal communities in the Western Region of Ghana. *Ghana Journal of Geography, 11*(1), 1–19.

Adom, D. (2016). Asante indigenous knowledge systems: Repositories of conservation ethics for Ghana's biodiversity, 7(2).

Adom, D., & Kquofi, S. (2016). The high impacts of Asante indigenous knowledge in biodiversity conservation issues in Ghana: The case of the Abono and Essumeja Townships in Ashanti Region. *British Journal of Environmental Sciences, 4*(3), 63–78.

Akpabio, E. M., & Subramanian, S. V. (2012). *Water supply and sanitation practices in Nigeria: Applying local ecological knowledge to understand complexity* (ZEF Working Paper Series).

Ampe, K., Paredis, E., Asveld, L., Osseweijer, P., & Block, T. (2020). A transition in the Dutch wastewater system? The struggle between discourses and with lock-ins. *Journal of Environmental Policy & Planning, 22*(2), 155–169. https://doi.org/10.1080/1523908X.2019.1680275

Amuzu, D. (2022). Decolonizing African development education through indigenous knowledge. In O. A. T. Frimpong Kwapong, D. Addae, & J. K. Boateng (Eds.), *Reimagining development education in Africa* (pp. 39–54). Springer International Publishing. https://doi.org/10.1007/978-3-030-96001-8_3

Andersen, M. S. (2007). An introductory note on the environmental economics of the circular economy. *Sustainability Science, 2*(1), 133–140.

Anokye, K., & Mohammed, A. S. (2024). Indigenous traditional knowledge for cleaner waste systems and sustainable waste management system in Ghana. *Cleaner Waste Systems, 9*, Article 100165. https://doi.org/10.1016/j.clwas.2024.100165

Audier, S. (2019). *L'âge productiviste: Hégémonie prométhéenne, brèches et alternatives écologiques*. La Découverte.

Bannister, K., & Hardison, P. (2006). Mobilizing traditional knowledge and expertise for decision-making on biodiversity. *IMoSEB Case Study.*

Beamer, K. (2014). *No makou ka mana: Liberating the nation.* Kamehameha Schools Press.

Beamer, K. (2020). *An Aloha 'Āina economy—Give, take, regenerate.*

Beamer, K., Elkington, K., Souza, P., Tuma, A., Thorenz, A., Köhler, S., Kukea-Shultz, K., Kotubetey, K., & Winter, K. B. (2023). Island and indigenous systems of circularity: How Hawai'i can inform the development of universal circular economy policy goals. *Ecology and Society, 28*(1). https://doi.org/10.5751/ES-13656-280109

Beamer, K., Tuma, A., Thorenz, A., Boldoczki, S., Kotubetey, K., Kukea-Shultz, K., & Elkington, K. (2021). Reflections on sustainability concepts: Aloha 'āina and the circular economy. *Sustainability, 13*(5), 2984.

Bengtsson, M., Alfredsson, E., Cohen, M., Lorek, S., & Schroeder, P. (2018). Transforming systems of consumption and production for achieving the sustainable development goals: Moving beyond efficiency. *Sustainability Science, 13*(6), 1533–1547. https://doi.org/10.1007/s11625-018-0582-1

Bleischwitz, R., Yang, M., Huang, B., Xu, X., Zhou, J., McDowall, W., Andrews-Speed, P., Liu, Z., & Yong, G. (2022). The circular economy in China: Achievements, challenges and potential implications for decarbonisation. *Resources, Conservation and Recycling, 183*, 106350. https://doi.org/10.1016/j.resconrec.2022.106350

Boulding, K. (1966). The economy of the coming spaceship earth. In H. Daly & W. H. Freeman (Eds.) (1980), *Economics, ecology, ethics: Essay towards a steady state economy, San Francisco.*

Cavanagh, C. J., & Benjaminsen, T. A. (2017). Political ecology, variegated green economies, and the foreclosure of alternative sustainabilities. *Journal of Political Ecology, 24*(1), 200–216.

Colombo, L. A., Pansera, M., & Owen, R. (2019). The discourse of eco-innovation in the European Union: An analysis of the eco-innovation action plan and horizon 2020. *Journal of Cleaner Production, 214*, 653–665. https://doi.org/10.1016/j.jclepro.2018.12.150

Corvellec, H., Stowell, A. F., & Johansson, N. (2022). Critiques of the circular economy. *Journal of Industrial Ecology, 26*(2), 421–432. https://doi.org/10.1111/jiec.13187

Cullen, J. M. (2017). Circular economy: Theoretical benchmark or perpetual motion machine? *Journal of Industrial Ecology, 21*(3), 483–486. https://doi.org/10.1111/jiec.12599

Deléage, J.-P. (2019). L'insurrection des «gilets jaunes», et après? *Écologie & Politique, 1*, 5–8.

Deselnicu, D. C., Militaru, G., Deselnicu, V., Zăinescu, G., & Albu, L. (2018). *Towards a circular economy—A zero waste programme for Europe* (pp. 563–568).

Dewick, P., de Mello, A. M., Sarkis, J., & Donkor, F. K. (2022). The puzzle of the informal economy and the circular economy. *Resources, Conservation and Recycling, 187*, Article 106602.

Dryzek, J. S., Bächtiger, A., Chambers, S., Cohen, J., Druckman, J. N., Felicetti, A., Fishkin, J. S., Farrell, D. M., Fung, A., & Gutmann, A. (2019). The crisis of democracy and the science of deliberation. *Science, 363*(6432), 1144–1146.

Ellen MacArthur Foundation. (2013). *Towards the circular economy. Vol. 1: An economic and business rationale for an accelerated transition.*

Ellen MacArthur Foundation. (2015). Delivering the circular economy: A toolkit for policymakers. *Delivering the Circular Economy: A Toolkit for Policymakers, 177.*

Ellen MacArthur Foundation. (2019). *The butterfly diagram: Visualising the circular economy.* Available at: https://www.ellenmacarthurfoundation.org/circular-economy-diagram (Accessed October 18, 2024).

Ellen MacArthur Foundation. (2021). *Universal circular economy policy goals.* Ellen MacArthur Foundation.

Ellen MacArthur Foundation. (2023). *Schools of thought that inspired the circular economy*. https://www.ellenmacarthurfoundation.org/schools-of-thought-that-inspired-the-circular-economy

Ellen MacArthur Foundation. (2024). *Circular economy principles*. https://www.ellenmacarthurfoundation.org/circular-economy-principles

Erkman, S. (1997). Industrial ecology: An historical view. *Industrial Ecology, 5*(1), 1–10. https://doi.org/10.1016/S0959-6526(97)00003-6

European Commission. (2015). *Closing the loop—An EU action plan for the circular economy, Com(2015) 614 communication from the commission to the European Parliament, the council, the European economic and social committee and the committee of the regions. European Commission, Brussels.* https://arquivo.pt/wayback/20170822151328mp_/.umic.pt/images/stories/publicacoes6/20110413-communication_en.pdf

Friant, M. C., Vermeulen, W. J. V., & Salomone, R. (2020). A typology of circular economy discourses: Navigating the diverse visions of a contested paradigm. *Resources, Conservation and Recycling, 161*, Article 104917.

Geissdoerfer, M., Savaget, P., Bocken, N. M., & Hultink, E. J. (2017). The circular economy—A new sustainability paradigm? *Journal of Cleaner Production, 143*, 757–768.

Geng, Y., & Doberstein, B. (2008). Developing the circular economy in China: Challenges and opportunities for achieving "leapfrog development." *International Journal of Sustainable Development & World Ecology, 15*(3), 231–239. https://doi.org/10.3843/SusDev.15.3:6

Ghisellini, P., Cialani, C., & Ulgiati, S. (2016). A review on circular economy: The expected transition to a balanced interplay of environmental and economic systems. *Journal of Cleaner Production, 114*, 11–32.

Ghisellini, P., & Ulgiati, S. (2020). Managing the transition to the circular economy. In *Handbook of the circular economy* (pp. 491–504). Edward Elgar Publishing.

Gutberlet, J., Carenzo, S., Kain, J.-H., Martiniano, M., & de Azevedo, A. (2017). Waste picker organizations and their contribution to the circular economy: Two case studies from a global south perspective. *Resources, 6*(4), 52. https://doi.org/10.3390/resources6040052

Hawken, P., Lovins, A. B., & Lovins, L. H. (1999). *Natural capitalism: Creating the next industrial revolution*. Little, Brown Boston.

Hobson, K., & Lynch, N. (2016). Diversifying and de-growing the circular economy: Radical social transformation in a resource-scarce world. *Futures, 82*, 15–25. https://doi.org/10.1016/j.futures.2016.05.012

Huang, H. (2020). *The emergence, evolution and frequent failure of the industrial symbiosis: The evidence from the sugar refinery industry, Chongzuo City*. University of Cambridge.

Imoro, A. Z., Boaka Hlordze, R. G., Duwiejuah, B. A., Abukari, A., Alidu, S. M., Acheampong, N. A., & Imoro, Z. A. (2022). Chapter 26—Indigenous technologies: Knowledge and practices for sustainable development. In U. Chatterjee, A. Kashyap, M. Everard, G. K. Panda, & D. Mahata (Eds.), *Indigenous people and nature* (pp. 593–612). Elsevier. https://doi.org/10.1016/B978-0-323-91603-5.00013-0

Kallis, G., Kostakis, V., Lange, S., Muraca, B., Paulson, S., & Schmelzer, M. (2018). Research on degrowth. In *Annual review of environment and resources* (Vol. 43, Issue Volume 43, pp. 291–316). Annual Reviews. https://doi.org/10.1146/annurev-environ-102017-025941

Kame'eleihiwa, L. (1992). *Native land and foreign desires*. Bishop Museum Press.

Kaźmierczyk, P. (2018). Implementing resource efficiency in Europe—Overview of policies, instruments and targets in 32 European countries. In H. Lehmann (Ed.), *Factor X: Challenges, implementation strategies and examples for a sustainable use of natural resources* (pp. 185–198). Springer International Publishing. https://doi.org/10.1007/978-3-319-50079-9_11

Kirchherr, J., & van Santen, R. (2019). Research on the circular economy: A critique of the field. *Resources, Conservation and Recycling, 151*.

Korhonen, J., Honkasalo, A., & Seppälä, J. (2018a). Circular economy: The concept and its limitations. *Ecological Economics, 143*, 37–46. https://doi.org/10.1016/j.ecolecon.2017.06.041

Korhonen, J., Nuur, C., Feldmann, A., & Birkie, S. E. (2018b). Circular economy as an essentially contested concept. *Journal of Cleaner Production, 175*, 544–552. https://doi.org/10.1016/j.jclepro.2017.12.111

Kothari, A., Demaria, F., & Acosta, A. (2014). Buen Vivir, degrowth and ecological Swaraj: Alternatives to sustainable development and the green economy. *Development, 57*(3), 362–375. https://doi.org/10.1057/dev.2015.24

Kothari, A., Salleh, A., Escobar, A., Demaria, F., & Acosta Espinosa, A. (2019). *Pluriverse: A post-development dictionary*. Tulika Books.

Leipold, S., Petit-Boix, A., Luo, A., Helander, H., Simoens, M., Ashton, W. S., Babbitt, C. W., Bala, A., Bening, C. R., & Birkved, M. (2023). Lessons, narratives, and research directions for a sustainable circular economy. *Journal of Industrial Ecology, 27*(1), 6–18.

Lieder, M., & Rashid, A. (2016). Towards circular economy implementation: A comprehensive review in context of manufacturing industry. *Journal of Cleaner Production, 115*, 36–51. https://doi.org/10.1016/j.jclepro.2015.12.042

Low, L. (2020). *Decolonizing natural resource management: Kānaka Maoli Struggles for Aloha 'Āina.*

MacKenzie, M. K., Serrano, S. K., & Kaulukukui, K. L. (2007). Environmental justice for indigenous Hawaiians: Reclaiming land and resources. *Natural Resources & Environment, 21*(3), 37–79. JSTOR. http://www.jstor.org/stable/40924828

Malleson, T. (2015). A community-based good life or eco-apartheid. *Radical Philosophy Review.*

Marotti de Mello, A., Mdiniso, J., Ndlovu, M., & Donkor, F. K. (2022). Call for papers–indigenous epistemologies and circular economy: Critical reflections from the periphery. *Revista de Gestão, 29*(4), 318–320.

Mawere, M. (2014). *Culture, indigenous knowledge and development in Africa: Reviving interconnections for sustainable development*. Langaa Rpcig.

METI. (2004). *Handbook on resource recycling legislation and 3R initiatives.*

Noyoo, N. (2007). Indigenous knowledge systems and their relevance for sustainable development: A case of Southern Africa. *Tribes and Tribals, 1*, 167–172.

Nylén, E.-J.A., & Salminen, J. M. (2019). How does the circular economy discourse affect policy-making? The case of streamlining waste utilisation in Finnish earthworks. *Resources, Conservation and Recycling, 149*, 532–540. https://doi.org/10.1016/j.resconrec.2019.06.029

OECD. (2018). *Realising the circular bioeconomy*. OECD. https://www.oecd-ilibrary.org/industry-and-services/realising-the-circular-bioeconomy_31bb2345-en?crawler=true&mimetype=application/pdf

Pearce, D. W., & Turner, R. K. (1989). *Economics of natural resources and the environment*. Johns Hopkins University Press.

Robèrt, K.-H. (2002). *The natural step story: seeding a quiet revolution*. New Society Publishers.

Schröder, P., Anantharaman, M., Anggraeni, K., & Foxon, T. J. (2019). The circular economy and the Global South. In *The circular economy and the Global South*. Routledge.

Shumba, O. (2011). Commons thinking, ecological intelligence and the ethical and moral framework of Ubuntu: An imperative for sustainable development. *Journal of Media and Communication Studies, 3*(3), 84–96.

Stahel, W. (1982). The product life factor. In *An inquiry into the nature of sustainable societies: The role of the private sector (Series: 1982 Mitchell Prize Papers), NARC* (pp. 74–96).

Stahel, W. (2010). *The performance economy*. Springer.

Stahel, W., & Reday, G. (1976). *The potential for substituting manpower for energy, report to the commission of the European communities.*

Su, B., Heshmati, A., Geng, Y., & Yu, X. (2013). A review of the circular economy in China: Moving from rhetoric to implementation. *Journal of Cleaner Production, 42*, 215–227.

Ukhami, E. I., Kumapayi, S. B., & Lassana, D. (2022). *The North and the South Dichotomy: Understanding the disparity in a "Global Village."*

United Nations. (2024). *The sustainable development goals report 2024*. UN.

Valenzuela, F., & Böhm, S. (2017). Against wasted politics: A critique of the circular economy. *Ephemera: Theory & Politics in Organization, 17*(1), 23–60.

Velis, C. (2018). No circular economy if current systemic failures are not addressed. *Waste Management & Research, 36*(9), 757–759. https://doi.org/10.1177/0734242X18799579

Vezzoni, R. (2023). Green growth for whom, how and why? The REPowerEU Plan and the inconsistencies of European Union energy policy. *Energy Research & Social Science, 101*, Article 103134. https://doi.org/10.1016/j.erss.2023.103134

von Weizsäcker, E. U., & Wijkman, A. (2017). *Come on!: Capitalism, short-termism. Population and the destruction of the planet*. Springer. https://doi.org/10.1007/978-1-4939-7419-1

Vonk, L. (2018). Paying attention to waste: Apple's circular economy. *Continuum, 32*(6), 745–757. https://doi.org/10.1080/10304312.2018.1525923

Warren, D. M. (1992). Indigenous knowledge, biodiversity conservation and development. In *Sustainable development in third world countries: Applied and theoretical perspectives* (pp. 81–88).

Webster, K. (2015). The circular economy: A wealth of flows.

Winter, K. B., Beamer, K., Vaughan, M. B., Friedlander, A. M., Kido, M. H., Whitehead, A. N., Akutagawa, M. K. H., Kurashima, N., Lucas, M. P., & Nyberg, B. (2018). The Moku system: Managing biocultural resources for abundance within social-ecological regions in Hawai'i. *Sustainability, 10*(10). https://doi.org/10.3390/su10103554

World Economic Forum & Ellen MacArthur Foundation. (2016). *The new plastics economy—Rethinking the future of plastics.*

Yuan, Z., Bi, J., & Moriguichi, Y. (2006). The circular economy: A new development strategy in China. *Journal of Industrial Ecology, 10.*

Zewdu, A., Girma, Z., & Belay, T. (2024). Traditional knowledge and attitude of the local communities towards wildlife conservation in and around fragmented Forest of Germeba Mountain in southern Ethiopia. *Journal of Forestry and Natural Resources, 3*(2), 61–76.

Sustainable Innovation and the Circular Economy: Regulatory and Economic Drivers

Daniel Balsalobre-Lorente, Niu Rong, Syed Ale Raza Shah, Rena Huseynova, and Naila Abbas

Abstract Recently, practitioners have paid momentous attention to climate change and have proposed interesting themes to address these growing challenges. However, a practical study is needed to provide a clear message to policy analysts about the real challenges and solutions. Thereupon, the present work attempts to break this suspense and portrays key problems and their best alternatives for 24 OECD economies covering the period of 2012–2023. As part of an empirical strategy, this study uses reliable estimators and obtains straightforward outcomes. For example, it finds total factor productivity, population aging, circular economy, and environmental tax revenue as leading problems to ecological sustainability. In lieu, digital transformation and sustainable innovations knowingly help to tackle those sustainability matters. Also, this work determines the green role of digital transformation and

D. Balsalobre-Lorente
Department of Applied Economics I, University of Castilla La Mancha, Ciudad Real, Spain

UNEC Research Methods Application Center, Azerbaijan State University of Economics (UNEC), Baku 1001, Azerbaijan

Economic Research Center (WCERC), Western Caspian University, Baku, Azerbaijan

D. Balsalobre-Lorente
e-mail: Daniel.balsalobre@uclm.es

N. Rong
College of Economics and Management, Northwest Agricultural and Forestry University, Yangling, P. R. China
e-mail: niurong911@hotmail.com

S. A. R. Shah (✉)
Department of Economics, Government College University, Faisalabad 38000, Pakistan
e-mail: syedaaleraza56@gmail.com

R. Huseynova
Department of Digital Technologies and Applied Informatics, Azerbaijan State University of Economics (UNEC), Baku 1001, Azerbaijan
e-mail: r.huseynova@unec.edu.az

N. Abbas
School of Economics and Finance, Xi'an Jiao Tong University, Xian 710061, China
e-mail: nailaabbaskml@gmail.com

P. Singh et al. (eds.), *Circular Economy and Sustainable Innovation*,
https://doi.org/10.1007/978-3-032-00437-6_4

artificial intelligence in environmental sustainability. The results illustrate a decay in emissions by 0.1538 and 0.0025, validating the robust role of digital transformation in environmental management. Finally, the green implications are suggested to minimize sustainability problems.

Keywords Sustainable innovations · Circular economy · Environmental regulations · Technology · Sustainability

1 Introduction

An unexpected escalation of carbon emissions has become a topmost global challenge. Although experts have presented several green strategies to cope with it, the problem remains stagnant. It is essential to draw an accurate picture and guide policymakers to understand the true causes and solutions for environmental betterment. Under the existing environmental theme, most researchers have blamed rising income levels as a crux of all negative externalities and emissions. In order to disclose this connection, Grossman and Krueger (1995) introduced the non-linear linkage of income with emissions and confirmed an inverted U-shaped Environmental Kuznets Curve (EKC-hypothesis). In late, this theme has been validated by numerous scholars [refers: (Jiang et al., 2022; Zhao & Li, 2024; Zheng et al., 2023)]. Considerable income variations also encourage boosting the social factor's role in environmental deterioration (ED). Rising population growth has remained another factor contributing to emissions, and practitioners have proposed several measures to overcome these environmental harms (Erdogan, 2021). Over time, anthropogenic activities have increased, reducing the environmental quality and disturbing nature as a whole. For example, excessive waste in public places may directly cause human health problems and greenhouse gas emissions. Therefore, waste and its serious consequences for nature remain the main threat to nature (Costa & Ribeiro, 2020).

Beyond the main problems, analysts have introduced core solutions and believe these can help manage environmental concerns. Here, technology has received colossal attention from them, which helps in efficient resource allocation and minimizing emission levels. Since the Industrial Revolution, technology has transitioned, and nations tried to update it from energy-intensive to energy-efficient (Biely & Chakori, 2024). It did not perform well in obtaining the desired output level without innovative methods. As a result, consultants have proposed a sustainable innovation theme that indirectly encourages all kinds of technology to maximize their efficiency and enhance the environmental quality without harming the growth level (Afeltra et al., 2021). Both solutions may be directly associated with government revenue allocated to boost the technical and research and development activities for sustainability. Different tax forms are imposed to hurt consumers using emission-intensive products, and environmental tax is one of these. In recent years, environmental tax revenue has become a green strategy, and the world has considered it to finance green projects (Li et al., 2022). These key solutions would help to manage the environmental quality and may introduce appropriate implications to become clean and green shortly.

This empirical research considers the 24 OECD economies covering 2012–2023 and contributes to the existing literature in the following ways. Firstly, economies have faced a common problem: trying to compete globally and setting their targets to magnify development targets. In this race, most economies have voluntarily ignored their sustainability matter and have operated accessible resources for development progress. In addition, the existing literature has used different development measures to portray its deep connection with emissions. This study utilizes the most updated development proxy, the total factor productivity (TFP) within the development subject, and evaluates its response to emissions. From this inclusion, policymakers can understand the true association of TFP with emissions and may suggest appropriate implications for specified economies. Secondly, under social factors, most studies have used the overall population, urbanization, rural population, and gender base population, but they have passed over the response of population aging for selected nations. Population aging (15–64) is vital at the domestic level because it may deal with all kinds of activities and be directly associated with accessible resources. Excessive exploitation of resources for the sake of rapid progress may cause environmental stress. In order to investigate the role of population aging in sustainability, the current study considers it an environmental factor and suggests green implications for becoming green shortly. Thirdly, technology is another dominant environmental factor because it can be used resourcefully and expand overall growth. Over time, there has been a technological transition, and economies have adopted various technologies to meet their desired output level, and have left out the eco-consciousness. But in recent years, rising environmental stress has urged nations to reconsider their technology setup and transform it for sustainability. Resultantly, the globe has adopted digital transformation (DT) and artificial intelligence (AI) as a decent measure of technology and believes these measures would significantly reduce emissions. Thus, to measure their response to environmental sustainability, this empirical research uses an explanatory variable and offers a comparative analysis to portray which factor is more efficient. Also, such a response may compel policy analysts to rethink the policy perspective under the sustainability theme.

Fourthly, most nations have approved environmental taxes as a green strategy, and these have a common belief that environmental issues can be resolved. This strategy has two advantages: it hurts the emission-intensive activities and collects revenue for environmental betterment. For that reason, this research considers the environmental tax revenue as a measure of regulation to sketch this strategy adoption that is beneficial for sustainability. This connection would accurately guide selected nations' interest in allocating revenue to green projects. Fifthly, sustainable innovation (SI) is being familiarized as another green measure by specified nations because it can deal with environmental matters through innovative methods. Under Goal 17: global cooperation to deal with environmental problems, the selected nations have tried to foster international cooperation to develop eco-friendly technologies. That cooperation in research and development would effectively help reduce ecological concerns. To answer this question, the present work includes sustainable innovation as an environmental factor and opens a new direction for forthcoming studies across different regions. The indispensable topic is that nations have always faced excessive waste at the domestic level. This wastage burdens nature, and its negative impact on humankind is debatable. To address similar problems, the affected economies have adopted the theme of the circular economy, which is responsible for using this waste and producing solid products or green energy. Commonly, the belief is that only green strategies in development plans are not enough solutions. It is essential to check whether this strategy helps implement environmental solutions. To answer this question clearly, this research chooses the circular economy as an environmental factor and would help to put forward the green implications under the specified theme. Finally, the current study uses the most consistent estimators, such as the Fully Modified Ordinary Least Squares (FMOLS), Dynamic Ordinary Least Squares (DOLS), and Instrumental Variable 2SLS, which can deal with panel data problems.

2 Literature Review

This section is an effort to overview the past studies related to sustainable innovation (SI) and circular economy (CE) under the shadow of sustainability. It is divided into two sub-segments, i.e., SI and sustainability and CE and sustainability.

The first segment aims to cover core empirical studies relevant to sustainable innovation and sustainability. For example, Petrović and Lobanov (2020) considered the 16 OECD economies to assess an inverse association of SI with emissions. They demonstrated that such an association was observed because specified economies have made significant efforts in research and development activities to introduce innovative methods. Due to innovation, concerned economies have maximized their output without environmental harm. Shao et al. (2021) inspected a deep connection between SI and emissions and found a significant decline in emissions due to novelties. Under the logical debate, they exhibited that the US economy had worked proficiently on innovations, indirectly promoting energy efficiency and technology involvement in domestic productivity and ensuring carbon neutrality. Hailemariam et al. (2022) analyzed a key role of SI in sustainability by focusing on advanced and emerging economies and built a rise in environmental quality due to innovations. They claimed that innovations were the leading solution that significantly encouraged green activities for rapid growth without harming environmental quality. It is recommended that quantified economies' legislators try to encourage government expenditures in R&D quests to introduce new methods. Gonenc and Poleska (2023) evaluated the situation of 44 economies by considering innovation as an environmental determinant and tried to compare MNCs' performance with domestic ones. Outcomes showed that international companies played a dominant role in environmental protection with inventions. As per the given arguments, they claimed that international companies have more competent methods of developing advanced technologies, successfully reducing ecological anxieties. Mehmood et al. (2024) focused on Pakistan's case study and tried to estimate the long-run SI role in sustainability and determine a significant upsurge in emissions due to a rise in modernizations. Logical reasonings explain an entire relationship behind the scenes well. For example, profitability has driven Pakistan's economy, and inventions have gradually increased growth alongside massive energy consumption. The shift from conventional energy sources to innovative methods has created environmental pressure. They also claimed that Pakistan's economy has not accepted innovative technology well and is still based on outdated technologies that eliminate new performance. A case study of BRICS economies Amin et al. (2025) attempted to explore the negative role of SI in the emissions ratio. They illustrated that the selected nations might focus on energy-conserving technologies and innovative methods in the manufacturing process, which indirectly promote sustainability. From the policy perspective, they guided the BRICS economies should try to promote revolutions on the domestic level to expand growth at lower environment cost. Here, this study offers a bibliometric graph that expresses 7548 studies from 2020–2025 have been probed under theme of SI and emissions (See Fig. 1).

The second segment focuses especially on studies relevant to circular economy (CE) and sustainability and summarizes interesting arguments. For example, Gallego-Schmid et al. (2020) strained to establish the key assembly of CE with emissions by considering 20 articles from 2018–2019. They grouped all the studies into a single frame and offered an innovative method to overview any specific problem. In concluding remarks, they explained that the circular economy was an effective method to reduce the garbage from society and utilize it in useable products, ultimately reducing environmental strains. Magazzino et al. (2021) focused on Denmark's economy and found a contrary linkage between circular economy and emissions. As per the given rational discussion, they exemplified that efficiently employing wastage in final products and energy supply may reduce pollution. Such enclosure was abridged on behalf of two advantages: firstly, the stated economy was designed with waste reuse policies to lessen the emissions and yield useful products for the entire populace. These strategies were beneficial for environmental affairs and reduced the over-exploitation of resources. Pao and Chen (2022) summarized the behavior of CE in sustainability for EU economies with different circular economy measures i.e., per capita wastage and reuse rate as environmental determinants, and revealed that changes in per capita waste significantly increased emissions. On the other hand, the reuse of waste is significantly declared a green solution to address environmental questions. In concluding remarks, as per the increase in population growth, most nations have faced severe challenges, and debris was one of them, which not only caused biodiversity loss but also disturbed human health. That's why waste management was highly promoted to protect the biosphere. Mawutor et al. (2023) inspected the harmful role of CE in sustainability in view of Ghana's economy and the reasoning they explained in different arguments. For instance, they manifested that the Ghanaian economy does not effectively manage waste in productive activities and put it in other places, contributing to emissions. The Ghanaian economy was

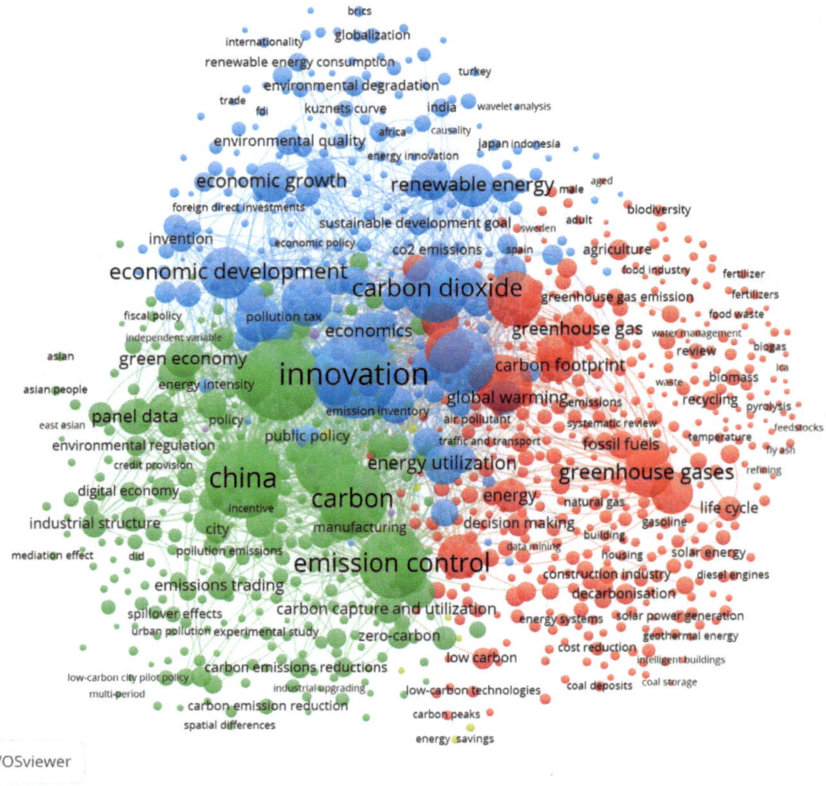

Fig. 1 Bibliometric graph for innovation and emissions

still in the transition phase and had not updated its waste recycling technology that eventually help to reduce emissions. Sahadevan and Mishra (2024) displayed the role of CE in mobile phone markets and its indirect role in sustainability. The extensive investment in carbon capture technology and waste reuse technology were leading factors that reduced the overall digitalization wastage and condensed environmental pressure. Finally, they designated that specified technologies support sustainability and increase green supply chain profit. Ren et al. (2024) concise a surprising alliance of CE with sustainability by taking into account 282 Chinese cities. They applied DID model to gauge a hidden response of treated vs non-treated cities and studied outcomes labelled that treated cities with CE's facilities have more performance to deal with environmental matters, and CE was taken as more preferable instrument to manage rising emissions effects on nature. This empirical research offers a bibliometric graph that expresses 5350 studies from 2020–2025 have been explored under theme of CE and emissions (See Fig. 2).

Some flaws must be well-thought-out to overcome existing ambiguities for future theorists. First, previous studies have used research and development expenditures to measure sustainable innovations, which has become outdated and may not provide

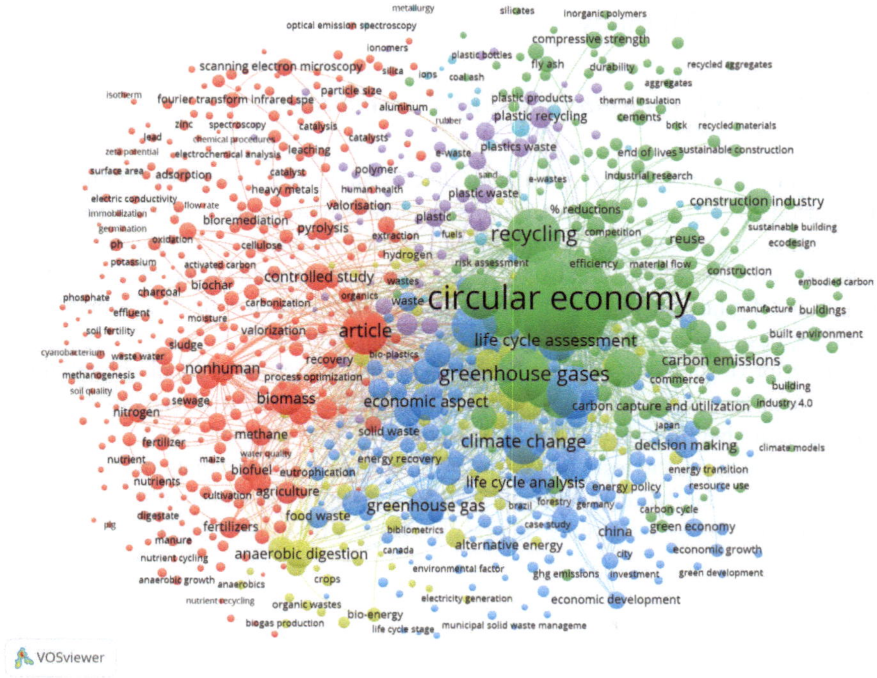

Fig. 2 Bibliometric graph for circular economy and emissions

comprehensive opinions to exponents. The present work performs an additional step to acquaint with more effective SI measures that help in sustainability and understand the current OECD interest in international collaboration for innovations. Secondly, some prevailing studies have taken OECD economies to calculate the SI's response to sustainability but have not reviewed enough countries to propose accurate suggestions; the present study uses 24 OECD economies to help propose implications for the whole OECD group. In addition, the outlined research has voluntarily given up the role of technology under the SI-CE-emissions nexus because advanced technology can help tackle ecological problems. To fill this gap, current research uses digital transformation and artificial intelligence as environmental measures to create a sound message. Including environmental regulation revenue is a substantial theme that has not been presented yet.

3 Data and Methods

3.1 Data Information and Initial Screening

This study presents the environmental determinants of 24 OECD economies over the period 2012–2023. These determinants include total factor productivity, population aging, sustainable innovation, revenue from environmental regulations, and the circular economy, along with two different measures of technology: artificial intelligence and digital transformation. The selected data has been collected from different sources; basic information about it is provided in Table 1.

Descriptive statistics, correlation, and VIF tests were run to check the reliability of the data for this research work. Table 2 provides descriptive analysis outcomes that validate that there is no chance of an outlier in the selected data.

Table 3 exhibits the results of correlation analysis and finds that not even a single variable correlates with more than 0.80%, indicating no multicollinearity in the selected data. Table 3 also gives the results of the VIF test to validate it further.

Table 1 Variable description

Symbol	Description	Unit	Source
CO_2	Carbon emissions	Kt	WDI
TFP	Total factor productivity	Index	OWD
DT	Digital transformation (information and communication technology)	Individuals using the internet (% of population)	WDI
AI	Artificial intelligence	Triadic patent families	OECD
SI	Sustainable innovation (green patents)	International collaboration in development of environment-related technologies (Index)	OECD
ER	Regulation tax revenue	Environmentally related tax revenue (US dollar)	OECD
CE	Circular economy	Municipal waste recycled (Tonnes)	UNEP

Note For AI (https://data.oecd.org/rd/triadic-patent-families.htm)

Table 2 Descriptive statistics

	$LNCO_2$	LNTFP	LNPA	LNDT	LNSI	LNER	LNCE
Mean	5.0447	0.000066	1.8149	1.9283	2.8006	0.2753	6.4411
Median	4.9341	0.000003	1.8152	1.9431	2.9540	0.3222	6.4501
Maximum	6.7081	0.0524	1.8502	2.0098	4.3463	0.6104	7.8144
Minimum	3.8406	− 0.0651	1.7688	1.6900	0.9242	− 0.9277	3.4129

Table 3 Correlation analysis

Correlation	$LNCO_2$	LNTFP	LNPA	LNDT	LNSI	LNER	LNCE	VIF	1/VIF
$LNCO_2$	1.000							–	–
LNTFP	–0.0312	1.000						1.12	0.8952
	0.590	–						–	–
LNPA	–0.1655	–0.1428	1.000					1.44	0.6956
	0.004	0.013	–					–	–
LNDT	–0.0711	0.1567	–0.2664	1.000				1.31	0.7636
	0.219	0.006	0.000	–				–	–
LNSI	0.7557	–0.0428	–0.3486	0.3011	1.000			4.57	0.2186
	0.000	0.460	0.000	0.000	–			–	–
LNER	–0.3676	–0.1261	–0.2907	–0.0043	–0.1304	1.000		1.23	0.8160
	0.000	0.028	0.000	0.940	0.023	–		–	–
LNCE	0.7679	0.0006	–0.3378	0.1075	0.8522	–0.1083	1.000	4.16	0.2406
	0.000	0.990	0.000	0.062	0.000	0.061	–	–	–
Mean VIF								2.30	

3.2 *Empirical Model Construction*

Over the decades, researchers have tried to sort out a common empirical function that can elaborate on the connection between socio-economic and technical factors in response to environmental sustainability. For this purpose, the IPAT model was introduced by Ehrlich and Holdren (1971, 1972) to investigate socioeconomic and technical factors' response to sustainability. Numerous studies utilized this environmental model, but at a later stage York et al. (2002) claimed that this method cannot capture all other variables that can influence environmental themes. York et al. (2003) presented the STIRPAT model as an extension of the IPAT model, which can add additional environmental determinants, and its general form can be expressed as,

$$I_i = aP_i^b A_i^c T_i^d e_i \tag{1}$$

Equation (1) displays the non-linear form of the STIRPAT model and can transform it into a log–log model by taking natural logs on both sides.

$$\ln I_{it} = a + b(\ln P_{it}) + c(\ln A_{it}) + d(\ln T_{it}) + e_i \tag{2}$$

Equation (2) delivers the linear form of the STIRPAT model, which contains the natural log of sustainability (lnI), population (lnP), affluence (lnA), and technology (lnT). After that, this study follows the theme of (York et al., 2003) and added sustainable innovation, environmental tax revenue, and circular economy. This extended form of the STIRPAT model can be exhibited as,

$$\begin{aligned}\ln CO_{2,it} &= \beta_0 + \beta_1 \ln TFP_{i,t} + \beta_2 \ln PA_{i,t} + \beta_3 \ln T_{i,t} \\ &\quad + \beta_4 \ln SI_{i,t} + \beta_5 \ln ER_{i,t} + \beta_6 \ln CE_{i,t} + \mu_{i,t}\end{aligned} \tag{3}$$

Equation (3) describes $\ln CO_2$, lnTFP, lnPA, lnT, lnSI, lnER, and lnCE, which states the natural log form of environment, total factor productivity, population aging, technology, sustainable innovations, environmental regulations, and circular economy. Furthermore, the present study has some core hypotheses that must be discussed here. First of all, TFP is being used as a development measure. Due to the rise in development activities, all economic actors, i.e., consumers, producers, and government, try to consume for the betterment of the economy. At the beginning level, no economy can guarantee to provide green products to their populace, which indirectly spurs energy consumption and increases emissions ($\beta_1 > 0$). In the same way, population aging (15–64) has become another core environmental indicator that contributes to overall growth exertions but can deteriorate the environment badly due to anthropogenic exercises. Then, its slope would be positive for specified regions ($\beta_2 > 0$). Technology is being measured here by digital transformation and artificial intelligence. This section presents technology in two different models, and the reader can observe at a later stage. Sustainable innovation has become an urgent need that brings innovative methods to boost growth and increase sustainability. Due to its immense contribution to sustainability, this study imagines its coefficient would be negative ($\beta_4 < 0$). Environmental regulations and revenue from such policies have become the most suitable strategies for supporting green energy projects. However, studied nations may use their earnings to improve their development plans, and it may spur environmental stress ($\beta_5 > 0$). Finally, the rising trend in garbage and its harmful role in biodiversity has become a hot debate. Most advanced nations have adopted the circular economy theme to turn all waste into usable products. But quantified nations have not yet reached the threshold level to apply it efficiently; thus, this study imagines its coefficient would be positive ($\beta_6 > 0$).

If we look at the technology theme, the current study initially focuses on digital transformation offered to abate anthropogenic actions by humans and productive doings. Due to its responsive behavior, it may significantly help to reduce the pollution level ($\beta_{3,DT} < 0$).

$$\begin{aligned}\ln CO_{2,it} &= \beta_0 + \beta_1 \ln TFP_{i,t} + \beta_2 \ln PA_{i,t} + \beta_3 \ln DT_{i,t} \\ &\quad + \beta_4 \ln SI_{i,t} + \beta_5 \ln ER_{i,t} + \beta_6 \ln CE_{i,t} + \mu_{i,t}\end{aligned} \tag{4}$$

Here, the role of artificial intelligence is imperative because it ably treats all productive activities and may determine which sector is not eco-friendly. This study reflects its slope negative due to its green behavior in sustainability via efficient resource allocation ($\beta_{3,AI} < 0$).

$$\ln CO_{2,it} = \beta_0 + \beta_1 \ln TFP_{i,t} + \beta_2 \ln PA_{i,t} + \beta_3 \ln AI_{i,t} + \beta_4 \ln SI_{i,t} + \beta_5 \ln ER_{i,t} + \beta_6 \ln CE_{i,t} + \mu_{i,t} \quad (5)$$

3.3 Estimation Strategy

It is necessary to offer some of the best estimators to obtain reliable outcomes, but before choosing long-term estimators, the empirical strategy utilized some advanced tests for data reliability. For instance, it uses the Augmented Dickey-Fuller (ADF), and I'm Pesaran and Shin (IMPs) data integration tests to validate data properties, and these tests were proposed by Dickey and Fuller (1981) and Im et al. (1997). With strong evidence related to data integration properties, it is essential to check out cointegration among the selected variables, and Padroni and Kao cointegration tests (Kao, 1999; Pedroni, 2001) apply for this purpose.

Finally, the long-run relationship of explanatory variables with sustainability is being measured with the most consistent estimators, the FMOLS and DOLS, as suggested by (Pedroni, 2000; Stock & Watson, 1993). The general form of both estimators can be articulated as,

$$\beta_{FMOLS} = \left[\frac{1}{N}\sum_{i=1}^{N}\sum_{t=1}^{T}(Ait - barAi))^2\right]^{-1} \times \sum_{t=1}^{T}(Ait - barAi)^2)barYit - Tbar\Delta e\mu \quad (6)$$

and,

$$\beta_{DOLS} = \left[\frac{1}{N}\sum_{i=1}^{N}\sum_{t-1}^{T}(AitA'it))^2\right]^{-1}\sum_{i=1}^{N}(\sum_{t-1}^{T}(AitYit)] \quad (7)$$

Also, this study develops the Instrumental Variable-2SLS to validate the prior results given by Theil (1953).

4 Results and Discussion

Table 4 designates the conclusions of the data integration test and displays all selected variables are integrated at first difference. These outcomes allow this study to move toward long-term cointegration.

Table 5 shows the results of the cointegration tests, and both tests validate the long-term cointegration among particular variables.

Table 4 Data integration results

Variable	IMPs		ADF	
	Level	1st difference	Level	1st difference
$lnCO_2$	–	– 5.9681***	–	– 6.9381***
lnTFP	–	– 7.1517***	–	– 7.9104***
lnPA	–	– 1.6634**	–	– 1.7217**
lnDT	–	– 5.6480***	–	– 5.6432***
lnAI	–	– 8.4279***	–	– 8.6531***
lnSI	–	– 9.3710***	–	– 7.2343***
lnR	–	– 5.7737***	–	– 6.5191***
lnCE	–	– 7.8355***	–	– 7.9276***

Note *** and ** show levels of significance at 1% and 5%, respectively

Table 5 Cointegration tests

Within-dimension				
	Stat	P-value	Stat	P-value
Panel v-statistic	– 6.5678	1.000	– 7.9290	1.000
Panel rho-statistic	8.1906	1.000	8.0226	1.000
Panel PP-statistic	– 16.794	0.000	– 17.257	0.000
Panel ADF-statistic	– 2.6291	0.004	– 3.2261	0.000
Between-dimension				
Group rho-statistic	10.012		1.000	
Group PP-statistic	– 21.044		0.000	
Group ADF-statistic	– 2.6551		0.004	
Kao cointegration				
ADF	– 1.8081		0.035	

4.1 Long-Run Outcomes by FMOLS and DOLS Estimators

Table 6 labels the findings of the long-term response of explanatory variables with carbon emissions and summarizes fascinating results. For example, the total factor productivity is employed as an environmental determinant. It shows a positive connection with emission and deteriorates the sustainability level by 0.310% and 0.297%, respectively, under FMOLS and DOLS. Three different aspects can explain this association because it is an index that proposes behalf of growth, capital accumulation, and compensation and wages to the labor force. From the growth perspective, three economic actors, i.e., the government, producer and consumer, directly affect the environmental quality due to human-caused happenings. It is a common behavior with any improvement in growth level; the government tries to invest in development plans such as institutions, roads, hospitals, etc., which may be highly associated with energy-intensive workouts and contribute to environmental corrosion. The producer mostly tries to maximize their output at any cost of negative externalities and utilizes emission-intensive inputs that worsen environmental quality. Consumer behavior regarding consumption is uncertain because, with a rise in incomes, consumers always go to consume luxury items, and particular items may have energy-intensive behavior, which further causes emissions. From a capital accumulation perspective, most government officials strive to invest in capital to accelerate growth, but these legislators do not believe its long-term benefits affect sustainability. Such an escalation of capital formation and growth returns is considerable; however, less attention to sustainability encourages energy-intensive activities that contribute to emissions. Finally, compensation to the labor force and wages significantly increase consumption. Most laborers try to invest in their living standards, which may increase by means of outdated products and pay for biodiversity loss. These are a few logical reasons that directly influence environmental quality via total factor productivity. This outcome can tally with the findings of Amri et al. (2019).

The given coefficient value of population aging also describes a positive connection with emissions and significantly reduces sustainability by 3.798% and 3.906%, respectively. This study ponders population aging, which ranges from 15 to 64 years old, and it is a common belief that this population may contribute to rapid growth.

Table 6 Outcomes by FMOLS and DOLS Estimators

Variable	FMOLS estimator		DOLS estimator	
LNTFP	0.3107***	0.0552	0.2979**	0.1524
LNPA	3.7981***	0.1476	3.9065***	0.3896
LNDT	− 0.1538***	0.0232	− 0.1426**	0.0628
LNSI	− 0.0406***	0.0081	− 0.0244*	0.0315
LNER	0.0153**	0.0060	0.0095*	0.0207
LNCE	0.0333***	0.0051	0.0373**	0.0153

Note *** and ** show levels of significance at 1% and 5%, respectively

Different channels through defined populations can influence environmental quality. Over time, work-related incomes have increased significantly, and the entire population is consuming to simplify their lives. In doing so, they seek to reduce consumption and maximize satisfaction. This behavior can hinder the consumption of environmentally friendly and energy-intensive products, contributing to emissions. Leisure activities are meaningful actions for the population, as with a rise in income, people seek to visit different places and utilize human and natural resources, which leads to biodiversity loss. Household construction activities have also become a significant challenge, but most countries still fail to control increasing pollution and natural resource depletion, which indirectly contribute to emissions. Finally, the preference for private cars could be another reason why most people try to earn their main income from their vehicles. Undeniably, OECD economies have attempted to optimize green transportation, but these nations have not reached a threshold level. The excess supply of conventional transport systems and their energy use could contribute to emissions. This outcome can be verified by the results of a recent study (Huang et al., 2024).

Given that the coefficient value of digital transformation (DT) is interesting, it efficiently reduces environmental stress and contributes to sustainability by 0.153% and 0.142%, respectively. These results encourage interpreters to debate the mechanism. Over the last two decades, the OECD economies have immensely focused on digital transformation and bound their telecommunication authorities to provide digital services at lower prices and ensure supply at an optimum level. These nations have provided enough facilities, indirectly changing consumption behavior, and most people purchase goods online. That variation has effectually condensed time, transaction, and traveling costs, indirectly dropping emission-intensive happenings. Its performance in the energy sector is not negatable; when production sectors apply digital services, it enhances resource efficiency and increases overall growth. The rebound effect remains dominant, and it improves energy productivity to control environmental deterioration. These arguments would be enough to understand the core details behind the scenes, and this outcome aligns with the findings of Ganda (2024).

Sustainable innovation has become a foremost dream, and this study uses it as an environmental determinant for OECD economies. Its response to sustainability is considerable because it reduces the emission by 0.040% and 0.024% under the concerned specifications. A few judgments can explain the perception behind the scene. The OECD nations have been considered leading economies that brought numerous innovative green methods to maximize output without environmental harm. These nations strive to take full advantage of technology consumption in economic activity and ensure green energy utilization under the Paris Agreement. The deputies have obligated all stakeholders to ensure the green technologies and their compatible energy to combat rising emissions, which may be impossible if these nations do not focus on green inventions. Besides the general logic, the OECD economies have adopted the theme of SDG 17 and prolonged the collaboration across borders to green technologies in daily productive activities. That kind of progress could be possible when all economies work together and enhance the share of green patents at the domestic level, which may spur growth exertions at lower environmental costs. That being so, this study highly promoted the adoption of sustainable innovations to fight for a green environment. This result can be tested from the recent conclusions of a study (Gong et al., 2024).

Environmental regulation (ER) is a smart strategy by proponents who guide higher authorities in adding it to government plans. To demonstrate its role in sustainability, the present study considers it an environmental factor, but its role is against expectations. There could be several reasons, but some are summarized here. In terms of environmental tax, the ER may collect massive amounts due to carbon emissions, which is a tremendous action. The studied economies may still face problems with the efficient allocation of environmental tax revenue because most nations consume for development plans, further deteriorating environmental quality. Maybe their legislators have other strategies, and their guidance to government officials is inappropriate. Suppose these economies add them into income revenues for eco-friendly projects such as green technology adoption, green energy investment, subsidies on green products, and campaigns to raise green knowledge. In that case, it may help to reduce emissions. Finally, institutional performance in environmental tax revenue collection cannot be ignored. There may be two reasons: domestic institutions' performance is low, or they cannot collect revenue up to the desired level. Therefore, collective efforts are needed to strengthen the role of environmental tax revenue in sustainability. This implication is in line with the findings of Kausar et al. (2024).

Finally, the circular economy (CE) is the last environmental determinant, and well-organized economies have implemented this strategy to manage environmental concerns. According to the relevant specifications, it does not work well in the current situation, degrading the sustainability level by 0.033% and 0.037%. This link can be explained by concrete logic. For example, the concerned nations have established a circular economy to recover all waste and produce usable products. Many loopholes may still be responsible for emissions, and consumer behavior is irrational regarding environmental quality and adverse effects of waste on stainability; therefore, they don't care about putting their waste in suitable places. Overall, garbage has two different effects on nature: it causes greenhouse gas emissions and brings health issues. The industrial sector in the urban part greatly contributes to wastage through garbage and industrial material flow to rivers, affecting life on land and creating problems for life below the water. Another leading reason could be that specified nations have traduced circular economy concepts but are still in the transition phase, and the technology adoption rate may not reach the desired level to reduce overall garbage and bestow efficient products. Hence, selected nations should increase awareness concerning garbage dumps at suitable places and ensure sustainable recycling. This result is consistent with the findings of Bhattacharjee et al. (2023) and Chen et al. (2020).

Table 7 Robust check by IV-2SLS

Variable	Coefficient	Std. error	t-statistic	Prob
LNTFP	0.2979***	0.0873	3.4094	0.000
LNPA	3.9065***	0.4298	9.0881	0.000
LNDT	− 0.1426***	0.0536	− 2.6608	0.008
LNSI	− 0.0244*	0.0330	− 0.7394	0.460
LNER	0.0095*	0.0134	0.7126	0.476
LNCE	0.0373***	0.0125	2.9821	0.003
C	− 1.9450**	0.7638	− 2.5462	0.011
R-squared			0.997932	
Adjusted R-squared			0.997702	
F-statistic			4327.487	
Prob (F-statistic)			0.000000	

Note ***, **, and *, show level of significance at 1%, 5%, and 1%, respectively

4.2 Robustness Analysis

Table 7 depicts the results of IV-2SLS to validate the prior upshots obtained by FMOLS and DOLS estimators. This test finds exciting sequels; for instance, total factor productivity and population aging contribute to emissions by 0.297 and 3.906%. Digital transformation and sustainable innovations are inverse to emissions levels over the selected period. Finally, the given coefficient value of environmental tax revenue and circular economy depreciate the sustainability level by 0.009% and 0.073%, respectively. The potted outcome clearly illustrates that the IV-2SLS estimator's fallouts strongly authenticate the prior outcomes.

4.3 Comparative Analysis

The current study employs the FMOLS estimator to conduct the comparative analysis. It considers AI, an indicator recently used by Rasheed et al. (2024). With constant data, digital transformation significantly reduces pollution by 0.153% (See Table 8). This study has already addressed its logical debate. In the case of artificial intelligence, it shows an inverse correlation with emissions and significantly raises the sustainability level by 0.002%. The main objective of OED economies proposing AI at the domestic level is to stimulate innovation, green industries, and sustainability knowledge. The ecological role of AI in this study indicates that the countries concerned have adopted AI strategies at the national level to increase overall economic growth without environmental pollution. These tools contribute to productive activities and improve workers' knowledge of how to achieve their production activities sustainably. They can also contribute to introducing innovative methods

Table 8 AI and digital transformation comparative analysis

Variables	LNDT		LNAI	
LNTFP	0.3107***	0.0552	0.2592***	0.0113
LNPA	3.7981***	0.1476	3.8753***	0.0293
LNSI	− 0.0406***	0.0081	− 0.0463***	0.0016
LNDT	− 0.1538***	0.0232	–	–
LNAI	–	–	− 0.0025**	0.0011
LNER	0.0153**	0.0060	0.0268***	0.0011
LNCE	0.0333***	0.0051	0.0133***	0.0008

Note ***, and **, show level of significance at 1%, 5%, and 1%, respectively

through green R&D activities and significantly change emission-intensive firms' behavior. The sustainable role of AI in labor force training may reduce the information asymmetry in the labor market and effectively increase labor skills to expand output levels. Finally, its green role in the energy sector could ideally increase energy efficiency across the different sectors and promote environmental quality. These upshots can match the effects of Cao et al. (2025) and Zhong et al. (2023).

5 Conclusion and Policy Recommendations

This empirical research introduces the innovative environmental determinants for 24 OECD economies covering the period of 2012–2023. In order to explore the long-run connection among variables (total factor productivity, population aging, digital transformation, artificial intelligence, sustainable innovation, environmental regulation, and circular economy), this study chooses the FMOLS, DOLS, and IV-2SLS estimators. After applying all the tests, it obtained fascinating outcomes, such as total factor productivity, population aging, environmental regulations, and circular economy, all of which are positively connected with emissions. Whereas digital transformation, AI, and sustainable innovations significantly reduce environmental pollution. Based on the outcomes obtained, this study concluded that in the present situation, digital transformation has more impact on environmental correction than digital conversion.

There are a few implications that OECD economies must consider to manage environmental concerns.

- In response to total factor productivity, the OECD economies should try to focus on the given suggestions. It is necessary to reallocate growth share in green activities such as green technology adoption, green energy installation, and subsidies for green imports. Capital formation is necessary to boost the overall economy, and legislators should compel domestic investors and government officials to allocate their revenue resources to green infrastructure development to fight for sustainability. Finally, the labor force should try to consume their major income share on green projects and cooperate in green consumption.

- Population aging (15–64) positively connects with emissions and deteriorates environmental quality. That being so, specified economies should try to manage their construction activities, which are causative to emissions and depleting natural resources. The government should take action to set environmental standards and impose heavy plenty if the populace does not follow environmental rules in leisure activities. The government should add enough public transport facilities to reduce consumers' interest in their vehicles and provide green energy to the transport sector. Finally, the supply of green energy should be increased to overcome environmental stress caused by conventional energy use.
- Digital transformation and artificial intelligence perform well in managing environmental quality, but there is still a need to work on the following points. For instance, higher authorities should coerce telecommunication companies to offer innovative services that eventually help to maximize human satisfaction and diminish environmental stress. The selected economies should push all digitalization companies to use green energy services to strengthen the rebound effect. It would be a rational choice if these nations shifted their conventional digital economy and promoted the inclusion of artificial intelligence. AI is highly recommended for managing the market situation, improving skills, and developing green methods to combat emissions.
- Sustainable innovations illustrate an interesting response to sustainability, and it would be a rational choice for higher authorities to stretch their overall development plans as an environmental solution. Legislators should take into account the strict rules for R&D teams to propose innovative methods for production processes at lower environmental costs. In the framework of this theme, higher authorities should attempt to augment energy-efficient technologies that secure current generations from environmental harm and minimize energy thirst.
- Environmental regulation remains an attractive approach for higher authorities to manage sustainability subjects, but it doesn't behave well in the present situation. The government should collect environmental taxes and use them eco-friendly to deal with the concerned hitches. Additionally, all institutions should be independent if government officials want to deal with environmental fears. Finally, collected revenue should be employed in green projects, i.e., energy installation, import technology, and providing green products at a subsidized rate.
- A circular economy could be the best solution to all kinds of garbage in reuse products. The government should start a campaign to guide the long-term spillover effect if consumers dump their waste in suitable places and it is convenient for labor to collect it. Also, it would be rational for representatives to focus on advanced technologies that directly support CE's theme. Finally, the government should offer a handsome amount concerning these job activities, which not only help in labor performance but also manage well waste from collection to dumping point.

Besides all the conclusions and policy recommendations, the current study is not free from future shortcomings. Firstly, the present study chooses only 24 OECD economies due to data constraints, while upcoming studies can make an effort to add other nations to expand this study's upshots, as well as to compare these outcomes with new ones. Next, this study uses only one environmental measure, but forthcoming research can apply other measures, such as ecological footprint, load capacity factor, sustainability index, and $PM_{2.5}$, to revalidate this study's outcomes. Finally, imminent empirical works should try advanced data and utilize non-linear estimators to obtain reliable results.

References

Afeltra, G., Alerasoul, S. A., & Strozzi, F. (2021). The evolution of sustainable innovation: From the past to the future. *European Journal of Innovation Management, 26*(2), 386–421. https://doi.org/10.1108/ejim-02-2021-0113

Amin, N., Sharif, A., Shabbir, M. S., & Pan, Y. C. (2025). Evaluating the impact of natural resource rents, R&D expenditures, green finance and energy efficiency on carbon emissions in BRICS economies: Greening the path to carbon neutrality in the post-COP 27 era. *Technology in Society, 81*, Article 102826.

Amri, F., Ben Zaied, Y., & Ben Lahouel, B. (2019). ICT, total factor productivity, and carbon dioxide emissions in Tunisia. *Technological Forecasting and Social Change, 146*, 212–217. https://doi.org/10.1016/j.techfore.2019.05.028

Bhattacharjee, S., Panja, A., Majumder, D., Hindorya, P. S., & Kumar, R. (2023). From garbage to global warming: A bibliometric analysis of greenhouse gas emissions from municipal solid waste and its relationship with climate change. *Environment, Development and Sustainability, 26*(5), 10973–10998. https://doi.org/10.1007/s10668-023-04145-5

Biely, K., & Chakori, S. (2024). Sustainability transition theories: Perpetuating or breaking with the status quo. *Sustainable Development, 33*(1), 52–68. https://doi.org/10.1002/sd.3101

Cao, Q., Chi, C., & Shan, J. (2025). Can artificial intelligence technology reduce carbon emissions? A global perspective. *Energy Economics, 143*, Article 108285. https://doi.org/10.1016/j.eneco.2025.108285

Chen, S., Huang, J., Xiao, T., Gao, J., Bai, J., Luo, W., & Dong, B. (2020). Carbon emissions under different domestic waste treatment modes induced by garbage classification: Case study in pilot communities in Shanghai, China. *Science Total Environment, 717*, Article 137193. https://doi.org/10.1016/j.scitotenv.2020.137193

Costa, F. N., & Ribeiro, D. V. (2020). Reduction in CO_2 emissions during production of cement, with partial replacement of traditional raw materials by civil construction waste (CCW). *Journal of Cleaner Production, 276*, Article 123302. https://doi.org/10.1016/j.jclepro.2020.123302

Dickey, D. A., & Fuller, W. A. (1981). Likelihood ratio statistics for autoregressive time series with a unit root. *Econometrica, 49*(4), 1057–1072. https://doi.org/10.2307/1912517

Ehrlich, P. R., & Holdren, J. P. (1971). Impact of population growth. *Science, 171*(3977), 1212–1217. https://doi.org/10.1126/science.171.3977.1212

Ehrlich, P. R., & Holdren, J. P. (1972). One-dimensional economy. *Bulletin at Science, 28*(5), 16–27.

Erdogan, S. (2021). Dynamic nexus between technological innovation and building sector carbon emissions in the BRICS countries. *Journal of Environmental Management, 293*, Article 112780. https://doi.org/10.1016/j.jenvman.2021.112780

Gallego-Schmid, A., Chen, H. M., Sharmina, M., & Mendoza, J. M. F. (2020). Links between circular economy and climate change mitigation in the built environment. *Journal of Cleaner Production, 260*, Article 121115. https://doi.org/10.1016/j.jclepro.2020.121115

Ganda, F. (2024). The influence of democracy, corruption, economic growth and ICT on carbon emissions in Sub-Saharan African countries: Does FDI matter? *Journal of Open Innovation: Technology, Market, and Complexity, 10*(3), Article 100324. https://doi.org/10.1016/j.joitmc.2024.100324

Gonenc, H., & Poleska, A. (2023). Multinationals, research and development, and carbon emissions: International evidence. *Climate Policy, 23*(8), 959–974. https://doi.org/10.1080/14693062.2022.2135484

Gong, Y., Cao, H., & Yuan, L. (2024). Does patent pledge reduce pollution and carbon emissions? Evidence from China. *Environment Research, 247*, Article 118274. https://doi.org/10.1016/j.envres.2024.118274

Grossman, G. M., & Krueger, A. B. (1995). Economic growth and the environment. *The Quarterly Journal of Economics, 110*(2), 353–377. https://doi.org/10.2307/2118443

Hailemariam, A., Ivanovski, K., & Dzhumashev, R. (2022). Does R&D investment in renewable energy technologies reduce greenhouse gas emissions? *Applied Energy, 327*, Article 120056. https://doi.org/10.1016/j.apenergy.2022.120056

Huang, Y., Elahi, E., You, J., Sheng, Y., Li, J., & Meng, A. (2024). Land use policy implications of demographic shifts: Analyzing the impact of aging rural populations on agricultural carbon emissions in China. *Land Use Policy, 147*, Article 107340. https://doi.org/10.1016/j.landusepol.2024.107340

Im, K. S., Pesaran, M. H., & Shin, Y. (1997). *Testing for unit roots in heterogeneous panels*. Department of Applied Economics, University of Cambridge.

Jiang, S., Yang, J., Yu, M., Lin, H., Li, C., & Doty, H. (2022). Strategic conformity, organizational learning ambidexterity, and corporate innovation performance: An inverted U-shaped curve? *Journal of Business Research, 149*, 424–433. https://doi.org/10.1016/j.jbusres.2022.05.023

Kao, C. (1999). Spurious regression and residual-based tests for cointegration in panel data. *Journal of Econometrics, 90*(1), 1–44. https://doi.org/10.1016/S0304-4076(98)00023-2

Kausar, A., Rasul, F., & Asghar, N. (2024). How to get green with agricultural footprint: A global analysis of carbon emissions, environmental taxes, and agrochemical use. *Journal of Environmental Management, 370*, Article 122665. https://doi.org/10.1016/j.jenvman.2024.122665

Li, Y., Zhou, M., Sun, H., & Liu, J. (2022). Assessment of environmental tax and green bonds impacts on energy efficiency in the European Union. *Economic Change and Restructuring, 56*(2), 1063–1081. https://doi.org/10.1007/s10644-022-09465-6

Magazzino, C., Mele, M., Schneider, N., & Sarkodie, S. A. (2021). Waste generation, wealth and GHG emissions from the waste sector: Is Denmark on the path towards circular economy? *Science of the Total Environment, 755*(Pt 1), Article 142510. https://doi.org/10.1016/j.scitotenv.2020.142510

Mawutor, J. K. M., Sogah, E., & Gborse, F. C. (2023). Circular economy and carbon emissions: Threshold effect of quality of governance. *Management of Environmental Quality: An International Journal, 36*(4), 948–970. https://doi.org/10.1108/meq-04-2023-0110

Mehmood, S., Zaman, K., Khan, S., Ali, Z., & Khan, H. U. R. (2024). The role of green industrial transformation in mitigating carbon emissions: Exploring the channels of technological innovation and environmental regulation. *Energy and Built Environment, 5*(3), 464–479. https://doi.org/10.1016/j.enbenv.2023.03.001

Pao, H. T., & Chen, C. C. (2022). The dynamic interaction between circular economy and the environment: Evidence on EU countries. *Waste Manag Res, 40*(7), 969–979. https://doi.org/10.1177/0734242X211057015

Pedroni, P. (2000). Fully modified OLS for heterogeneous cointegrated panels. In B. H. Baltagi, T. B. Fomby, & R. Carter Hill (Eds.), *Nonstationary panels, panel cointegration, and dynamic panels* (Vol. 15, pp. 93–130). Emerald Group Publishing Limited. https://doi.org/10.1016/s0731-9053(00)15004-2

Pedroni, P. (2001). Purchasing power parity tests in cointegrated panels. *The Review of Economics and Statistics, 83*(4), 727–731. https://doi.org/10.1162/003465301753237803

Petrović, P., & Lobanov, M. M. (2020). The impact of R&D expenditures on CO_2 emissions: Evidence from sixteen OECD countries. *Journal of Cleaner Production, 248*, Article 119187. https://doi.org/10.1016/j.jclepro.2019.119187

Rasheed, M. Q., Yuhuan, Z., Haseeb, A., Ahmed, Z., & Saud, S. (2024). Asymmetric relationship between competitive industrial performance, renewable energy, industrialization, and carbon footprint: Does artificial intelligence matter for environmental sustainability? *Applied Energy, 367*, Article 123346. https://doi.org/10.1016/j.apenergy.2024.123346

Ren, Q., Li, J., & Albrecht, J. (2024). Toward circular economy: Implementing circular economy strategies to reduce carbon emissions in Chinese cities. *Business Strategy and the Environment, 34*(1), 914–931. https://doi.org/10.1002/bse.4019

Sahadevan, V. K., & Mishra, U. (2024). A green manufacturer–retailer circular economic production model with carbon emissions and waste control. *Environment, Development and Sustainability, 26*(5), 12367–12406. https://doi.org/10.1007/s10668-023-03805-w

Shao, X., Zhong, Y., Li, Y., & Altuntas, M. (2021). Does environmental and renewable energy R&D help to achieve carbon neutrality target? A case of the US economy. *Journal of Environmental Management, 296*, Article 113229. https://doi.org/10.1016/j.jenvman.2021.113229

Stock, J. H., & Watson, M. W. (1993). A simple estimator of cointegrating vectors in higher order integrated systems. *Econometrica, 61*(4), 783–820. https://doi.org/10.2307/2951763

Theil, H. (1953). Repeated least squares applied to complete equation systems. *Central Planning Bureau.*

York, R., Rosa, E. A., & Dietz, T. (2002). Bridging environmental science with environmental policy: Plasticity of population, affluence, and technology. *Social Science Quarterly, 83*(1), 18–34. https://doi.org/10.1111/1540-6237.00068

York, R., Rosa, E. A., & Dietz, T. (2003). STIRPAT, IPAT and ImPACT: Analytic tools for unpacking the driving forces of environmental impacts. *Ecological Economics, 46*(3), 351–365. https://doi.org/10.1016/s0921-8009(03)00188-5

Zhao, R., & Li, L. (2024). Does digitalization always benefit cultural, sports, and tourism enterprises quality? Unveiling the inverted U-shaped relationship from a resource and capability perspective. *Humanities and Social Sciences Communications, 11*(1), 1066. https://doi.org/10.1057/s41599-024-03545-w

Zheng, R., Wu, G., Cheng, Y., Liu, H., Wang, Y., & Wang, X. (2023). How does digitalization drive carbon emissions? The inverted U-shaped effect in China. *Environmental Impact Assessment Review, 102*, Article 107203. https://doi.org/10.1016/j.eiar.2023.107203

Zhong, J., Zhong, Y., Han, M., Yang, T., & Zhang, Q. (2023). The impact of AI on carbon emissions: Evidence from 66 countries. *Applied Economics, 56*(25), 2975–2989. https://doi.org/10.1080/00036846.2023.2203461

From Fossil Fuels to Green Finance: Economic Indicators and Circular Economy Transitions in India's Renewable Energy Landscape

Shalini Singh, Shubhanker Yadav, Kanika, and Sumedha Sharma

Abstract Previous studies have explored the economic and environmental factors that affect the growth of renewable energy, suggesting that crude oil prices, coal prices, and foreign investment are crucial for promoting green energy within the circular economy. This model promotes sustainability, resource efficiency, waste reduction, and the continuous use of energy, which fits hand in hand with the goals of renewable energy development. The relationship between carbon emissions and investments in renewable energy has been both positive and negative results according to studies based on their research in i.e. a positive relationship due to government conduct of low carbon footprints and circular economies in their research. Moreover, the effects of per capita income on the growth in the field of green energy is often oriented towards higher economic resources to promote sustainable development on this path strengthening the principles/circular economy. Broader application of these insights, this study investigates certain determinants of growth in the Green Energy Index of India and analyzes various economic and environmental factors over the past five years. By means of stationarity test, correlation analysis and multiple regression techniques, the relationship between these variables is studied. It helps them focus on specific obstacles and drive forward the transition to renewable energy, an essential element of the circular economy, which the study finds is positively impacted by per capita income and crude oil prices, clear evidence of the interplay between economic resources and investments in renewables. Nonetheless, the index demonstrates a negative co-relationship with coal prices, indicating the negative effect of being reliant on coal on the advance of green energy. The Green Energy Index was also found to have no meaningful relationship with foreign capital and carbon emission. The paper concludes by emphasizing that Green energy index of India may significantly improve through increased investments in energy generation through renewable resources in circular economy context. It recommends that further study should consider a longer time series analysis and more companies in this field

S. Singh (✉) · Kanika · S. Sharma
School of Commerce, Finance and Accountancy, CHRIST University, Bengaluru, India
e-mail: shalini_dec26@yahoo.co.in

S. Yadav
Jaipuria Institute of Management, Lucknow, India

P. Singh et al. (eds.), *Circular Economy and Sustainable Innovation*,
https://doi.org/10.1007/978-3-032-00437-6_5

to provide a clearer picture of the relationship between these variables and green energy expansion.

Keywords Green energy index · Renewable energy · Circular economy · Sustainable development · Crude oil and coal prices · Foreign investment · Carbon emissions · Per capita income

JEL Classification O13 · Q01 · Q42 · Q56 · Q58

1 Introduction

The transition to a low-carbon economy which is resource efficient, sustainable is indeed moving at a global level. The circular economy, a restorative process aimed at trimming waste, recycling items and extending the life of products, is among the top transformative strategies helping lead the change. Green finance is the catalyst to facilitate this process—to channel money into projects that encourage progress and ensure our (shared) future. It comprises diversified financial products such as green bonds, investments in sustainable infrastructure, funding for renewable power, and contributions with E, S and G called ESG lines (Zhang et al.) Each of these fiscal weapons is firmly rooted in circular economy thinking. As highlighted by Geissdoerfer et al. (2020), the circular economy is consistent with the UN's Sustainable Development Goals (SDGS), such as SDG 13 (Climate Movement), SDG 8 (Greater than Average Work and Financial Development), and SDG 7 (Sensible and Clean Imperativeness). In a way, green financing promotes resource capacity, carbon reduction, environmentally responsible innovation, and the development of clean vitality—all of which are critical elements of a circular economy.

Generally speaking, the necessity to address climate change has prompted a move toward renewable energy and realistic financing strategies. Because it suggests assistance for typically valuable wanders that contribute to a circular budgetary system, this approach has expanded charm in the field of green finance (Kirchherr et al., 2020). Green bonds have been developed as a key tool among the budgetary disobedient. These specialist budgetary resources are specifically designed to support regular and climate-focused activities (Ellen MacArthur Foundation, 2021). The Green Support has been crucial in the establishment of advancements like solar wind, hydro, and geothermal systems within the renewable vitality segment. By reducing reliance on fossil fuels, these advancements reduce typical harm and advance imperativeness faithful quality (Murray et al., 2021). Businesses and governments are constantly embracing flexible, low-emission, and rational green vitality solutions as they work to fulfil their climate pledges, preparing for the circular economy's long-term objectives (Oliveria et al., 2023; Su et al., 2022).

The growing market for green energy technology is also creating new chances to make money. Better access to renewable options and help from policies and money for new ideas have opened up ways for circular business plans. These include energy-as-a-service, turning waste into energy, and small local power grids (McKinsey and Company, 2022; Pauliuk et al., 2023).

India is at a turning point in how it handles energy and sustainability. The government wants to grow clean energy and fight climate change as part of its bigger plan for a circular economy. India aims to have 175 GW of renewable energy by 2022. It grows through green bond markets, tax breaks, and team-ups between public and private groups (Ministry of New & Renewable Energy, 2023; Reddy & Ghosh, 2022). Groups like the Reserve Bank of India have made green lending easier by putting green assets in priority sector lending. This puts green finance more into India's economic scene. This policy has led to more green energy companies in India's stock markets. Investors who care about ESG issues are interested (Bhattacharya & Arora, 2022; Sachs et al., 2023). The NSE Green Index helps measure how these companies do. It shows market trends, how people invest, and if green businesses can make money.

More than INR 5.9 to 6.1 lakh crores were invested in renewable energy in India between 2014 and 2023, mostly in wind and solar power (Mohanty, 2024). While programs like the National Action Plan on Climate Change (NAPCC) and Renewable Purchase Obligations have reinforced regulatory support, tools like carbon pricing and green bonds have assisted in luring private investment (Khemnar & Pandey, 2024). High capital requirements, short loan terms, and investment risks in decentralized energy systems are ongoing difficulties, though (Khule et al., 2024).

Going forward, the need for creative financial strategies, cutting-edge technologies, and cooperative governance is highlighted by the alignment of green finance with SDGs and circular economy objectives (Bilal & Shaheen, 2023; Sahu & Khatri, 2024). Financial and regulatory obstacles must be removed to realize the potential of India's green transition fully.

Using the NSE Green Index as a case study, this chapter investigates the economic and policy factors that influence the development of green energy in India. It analyzes and discusses the impact of macroeconomic variables such as coal and oil prices, carbon dioxide emissions, foreign direct investment, per capita income on the green finance dynamics in a circular economy. The chapter analyzes such dynamics and contributes to the discussion around climate-resilient development, financial inclusion and sustainable innovation.

1.1 Green Finance and Circular Economy Transition

As noted in the outset, the principles of the circular economy are closely tied to India's transition to a low-carbon, resource efficient economy. That the green financing is in the middle of this transition is because ecological funds can be secured for ecological projects, and the ecological integrity is balanced with economic development. By concentrating investments on clean technologies, sustainable infrastructure, renewable energy, and responsible consumption, green finance plays a crucial role in operationalizing the circular economy in both policy and practice (Geissdoerfer et al., 2020). Green finance takes into account long-term environmental and social factors, in contrast to traditional funding vehicles that prioritize short-term advantages. Green bonds, sustainability-linked loans, and green credit lines mobilize funds to fund projects that lower resource extraction, waste reduction, and closed-loop systems—pillars of the circular economy (Ellen MacArthur Foundation, 2021). In India, these instruments are being used increasingly to fund solar parks, electric transport infrastructure, energy-efficient houses, and distributed clean energy systems. The proactive approach of the Indian government towards renewable energy has been complemented by green finance models that actively promote circular outcomes. Steps taken by the Indian government, for example, including the National Action Plan on Climate Change, Perform, Achieve and Trade scheme, and bringing green investments as a priority sector lending by the Reserve Bank of India, have established effective institutional support in environmentally aligned lending (Khemnar & Pandey, 2024; Reddy & Ghosh, 2022). These schemes reflect a solid understanding of how finance can stimulate sustainable innovation and circularity of matter.

Among the strongest financing vehicles for this transition is the green bond. India's green bond market has grown significantly in recent years, with more than USD 10 billion raised to finance projects tracking circular economy goals—such as renewable energy growth, low-carbon transportation for cities, and sustainable water management (Climate Bonds Initiative, 2023). They bridge the green capital gap for capital-heavy green technologies while bringing on board environmentally aware investors. In addition to large-scale infrastructure, green finance facilitates the development of decentralized and community-based circular systems. For instance, funding rooftop solar systems, community biogas plants, and local waste-to-energy facilities promotes grassroots resource efficiency and increases energy equity (Su et al., 2022). Decentralized systems are important for constructing localized circular economies that are resilient and inclusive. Additionally, including environmental, social, and governance (ESG) factors in conventional investment choice-making has further spurred the circular innovation wave. Financial intermediaries and institutional investors increasingly place a premium on sustainability performance, causing businesses to adopt green technologies, enhance material efficiency, and adopt life-cycle thinking across their business models (Bilal & Shaheen, 2023; Pauliuk et al., 2023). At its core, green finance does more than supply capital—it incentivizes systems change by infusing sustainability into the financial architecture. It de-risks circular economy investments, draws blended finance from international stakeholders, and encourages public-private partnerships. As India continues toward climate neutrality and sustainable development, green finance will remain a pillar of strength, driving innovation, lowering environmental footprints, and fast-tracking the use of circular practices across sectors.

2 Literature Review

2.1 Transition from Fossil Fuels to Renewable Energy: Global and Indian Perspectives

The global transition from fossil fuels to renewable energy sources may be a key step in achieving change and developing circular economy principles. It addresses a financial and natural need for nations seeking to reduce their reliance on natural resources and prepare for emergencies. Greiner et al. (2013) highlight the macroeconomic factors involved in switching from non-renewable to renewable energy sources and underscore the need for economies to restructure their energy use and generation systems to achieve long-term sustainability. By connecting the development of capital markets to shared effects, Paramati et al. (2017) examine how financial mechanisms such as stock market index and further arrange hypotheses (FDI) influence CO_2 emissions. Their disclosures suggest that green hypotheses, including those that are balanced with circular economy objectives, can be developed by well-regulated budgetary systems. In addition, Punzi (2019) depicts a small open market in which imperativeness is a fundamental input for both time and use. The thought process illustrates how global essentialness shifts, which are largely influenced by global show elements, impact family essentialness actions, and outline the route of action. Methodologically, Dutta (2019) uses the GARCH-jump method to analyze the influence of commodity market volatility, that is, silver prices, on global solar energy stock indexes. This highlights how financial market volatility can indirectly affect investor confidence in clean sectors, thereby influencing the rate of clean energy uptake. Supporting evidence by Shafiullah et al. (2021), using data from 1986 through 2019, provides preliminary evidence on the impact of economic policy uncertainty on renewable energy use in the United States and suggests that there is a necessity for regular and prolonged policy schemes to favor circular energy models. Cedic et al. (2021) study different forms of market uncertainty and how they affect stock returns in the oil, gas, and renewable energy markets. Their evidence is that, although new, renewable energy markets are more sensitive to world financial and geopolitical changes. Liu et al. (2021) also examine the effect of economic uncertainty brought about by the pandemic through time-frequency domain methods and conclude that clean energy stock indexes in the US, Europe, and the world are highly volatile during times of crisis, calling for strong financial strategies to protect green transitions. In South Asia, where the dependence on fossil fuels is still significant, Sharma et al. (2021) evaluate the dynamics between stock market development, per capita income, trade openness, technological progress, and the use of renewable energy. Their analysis highlights how indicators of socio-economic growth can either support or clash with environmental goals, depending on policy orientation. Likewise, Lei et al. (2021) use nonlinear ARDL models to investigate long-run connections between energy consumption and economic variables, highlighting the potential of renewable energy to decouple growth from emissions—a central tenet of the circular economy.

2.2 *Green Finance: Definitions, Mechanisms, and Global Trends*

Based on the developing relationship between energy transitions and financial systems, green finance has become an important tool to pursue low carbon and circular economy agendas. It thus enables the allocation of financial resources towards sustainable energy projects that contribute to the reduction of carbon emissions and achieving long-term climate- resilient development (Ali et al., 2023; Sule et al., 2024). Green finance is also a key enabler for sustainable agriculture that is critical for rural economic resilience through financing of organic farming and water conservation practices and sustainable resource management, each of which helps to protect the environment and support circular use. The performance of clean energy companies is highly interdependent with macroeconomic factors such as oil prices, influencing the behavior of investors and capital movements into renewable markets. Pham (2019) highlights that the nonstop relationship between oil fetched changes and the execution of clean essentialness stocks carries basic comes about for policymaking, portfolio broadening, and the progression of green budgetary defiant. Developing on this, Kocaarslan et al. (2019) address the doubt of symmetry inside the joins between captivated rates, oil costs, and clean imperativeness stock returns, fighting that theorist wants alter over particular budgetary cycles and imperativeness divisions.

Ma et al. (2019) and Bildirici et al. (2019) explore how changes in oil costs influence the returns of U.S.-listed essentialness companies, particularly considering customer conclusion and macroeconomic vulnerabilities, utilizing advanced econometric approaches like Markov Trading models. So too, Zhao (2020) and Zhang et al. (2021) apply wavelet examination and causality-in quantiles methodologies to assess the effect of course of action defencelessness and exterior oil dazes on clean vitality stock markets, underlining the fundamental portion of strong money related educate in keeping up unfaltering wanders in renewable imperativeness. Huang et al. (2020) empower observe that since 2011, supply-side staggers, especially from tight oil markets, have progressively impacted stock return instability in records a bit like the S&P Energy List. Other than, Li et al. (2020) see at carbon trading systems as a market-based course of action device for diminishing outpourings, progressing the course of action of temperate finance with green establishment headway. Such system, tallying viable transport and imperativeness frameworks, is noteworthy for moving forward efficiency and bringing down defilement levels (Ali et al., 2023; Liu, 2024). In any case, green back still faces noteworthy obstacles, such as clashing authoritative frameworks, moo money related pro mindfulness, and missing disclosure rules. Tending to these challenges is essential to totally opening its potential to drive circular economy exercises and low-carbon changes (Sule et al., 2024; Чала & Demidov, 2024).

2.3 Part of Green Fund in Supporting Circular Economy in India

Inside the Indian circumstance, green finance plays a crucial portion in advancing the move to a circular economy by planning capital toward renewable essentialness wanders and temperate establishment. It acts as a key money related enabler for finishing long-term common targets such as reducing carbon radiations, optimizing resource utilize, and bracing climate flexibility. By supporting wanders in sun fueled, wind, and other clean imperativeness propels, green support propels money related advancement that adjusts with biological sustainability—both vital measures of the circular economy (Khemnar & Pandey, 2024).

Government policies such as the National Action Plan on Climate Change (NAPCC) have greatly improved India's renewable energy sector investment inflows. Such policies promote green technologies and provide a stable regulatory environment that provokes local and foreign investors to invest in clean practices (Khemnar & Pandey, 2024). In addition, banks are more often factoring in environmental sustainability when making lending and investment choices, encouraging industries to invest in energy-efficient technologies and low-impact manufacturing systems (Mukhopadhyay, 2024). These changes are indications of the growing realization of the economic and strategic benefits of adopting the circular economy. Financial products such as green bonds have become increasingly popular in India to mobilize funds for circular economy-focused projects—such as waste energy plants, water recycling facilities, and energy-efficient transport infrastructure (Devi, 2023). These investments reduce environmental damage and yield socio-economic returns such as job opportunities, local development, and the development of clean technologies. The development of a healthy green finance system thus directly assists with enhancing material circularity, containing environmental risk, and enabling a more inclusive and durable economic structure (Devi, 2023).

2.4 Implications of Fiscal and Financial Policies for Unlocking Green Finance and Green Investment

As per Nassary (2019), three primary fields exist in which fintech can be employed to facilitate green finance: blockchain solutions for sustainable development, blockchain applications for renewable energy, decentralized electricity markets, carbon credits, and climate finance, and financial instrument innovation, i.e., green bonds. Schumacher et al. (2020) examine the potential of sustainable investment and finance in Japan and how the financial sector in Japan could mitigate worsening climate risks and support Japan in its shift towards a sustainable, low-carbon economy. The authors argue for better analysis and understanding of finance needs as a basis to guide international funding and offer ways to improve the capacity needed to attract and secure funding. It identifies potential causes for the gap, including donor priorities and a lack of clear signals from national policymakers. They note that broader policies, in addition to financial and fiscal ones, are necessary; energy, industrial, and transportation policy are shown to be particularly crucial for mitigation, and nearly every government agency would be involved in responses to adaptation (Winkler et al., 2021). Purkayastha et al. (2021) go over how to set up a blended finance option so it works. Find numerous conclusions from the quarterly data of Chinese listed companies from 2019Q1 to 2021Q2 (Dongyang, 2021). The government can use (Gu et al., 2021) as a guide to intensify policy regulation and create a system of varied environmental governance. Zahoor et al. (2021) analyze, using manufacturing value-added and urbanization as moderator variables, the effects of renewable energy investment and financial development on environmental sustainability and China's economic growth from 1970 to 2016. Ali et al. (2022) want to perform a bibliometric analysis of 363 publications from the Scopus database that were written between 1976 and 2021. Among other significant works are (Sachs et al., 2019).

2.4.1 Time and Frequency Dynamics of Connectedness Between Renewable Energy Stocks and Crude Oil Prices

Song et al. (2019) using the connectedness network approach, investigate the dynamic information spillover of return and volatility between the fossil energy market, investor perceptions of renewable energy, and the renewable energy stock market (Yahya et al., 2019). Look into the relationship between the price of crude oil and 10 important agricultural commodities in terms of time and frequency. The dynamics of interconnectedness between the realized volatility indices of the 16 clean energy stocks included in the SPGCE and the implied volatility indices of the S&P 500 and the STOXX50 and the crude oil and gold markets are examined (Fuentes et al., 2020). Naeem et al. (2020) examine the frequency and time dynamics of spillover between energy, power, carbon, and renewable energy markets and oil price shocks (supply and demand) using the approach formulated by Diebold and Yilmaz (2012) and Barunik and Krehlik (2014, 2018). To ascertain if green bonds represent a new asset class (Ferrer et al., 2021), investigate the time-frequency relation between the international green bond market and various conventional financial and energy markets. Maghyereh et al. (2021) analyze the time-frequency dependence and quantile dynamics between the stock prices of clean energy firms, oil and gas firms' stock prices, and crude oil price shocks. The spillover frequency between crude oil futures contracts and the Middle East and North African (MENA) stock markets is investigated by Mensi et al. (2021). The Bombay Stock Exchange (BSE) energy index is employed by Agarwalla et al. (2021) to identify the dynamic relationship between Indian stock prices and global crude oil prices. To find out the short- and long-term dynamic relationship between the oil price shock and stock price movement, the Granger causality test is accompanied by the vector error correction (VEC) model, impulse response function, and variance decomposition test (Agarwalla et al., 2021). To measure the correlation between the variables and try to come up with a valid conclusion, vector autoregression estimation (Agarwalla et al., 2021) is considered. Zhang et al. (2021) use two empirical methods for connectedness—the time-domain approach created by Diebold and Yilmaz (2012) and the method based on frequency dynamics created by Barunik and Krehlik—to analyze the return and volatility spillover between the COVID-19 pandemic in 2020, the crude oil market, and the stock market (2018). Li et al. (2020) explain the dynamics and portfolio diversification from a novel cryptocurrency-renewable energy perspective using weekly data from 07 August 2015 to 26 March 2021.

2.5 Role of Green Finance in Improving Energy Efficiency and Renewable Energy Development

Zhang et al. (2019) offer a thorough analysis of the methods for assessing renewable energy used in green building/neighborhood rating systems, which would be highly beneficial to comprehend and subsequently enhance the rating systems. Investigate the drivers necessary to accomplish the Sustainable Development Goals with respect to lowering environmental pollution in EU member nations in order to provide policy guidance in achieving the emission objectives (Alola et al., 2019). In order to do this (Ulucak, 2020), employ sophisticated panel data estimate approaches that yield results that are resistant to concerns with endogeneity, heteroskedasticity, and cross-sectional dependence. Khattak et al. (2021) investigate the asymmetrical links between environmental sustainability and research on green and sustainable technologies among the BRICS states, using exports, FDI, and the utilization of renewable energy sources as control variables. Investigating the spatial implications of energy consumption and green production in Shanghai Cooperation Organization member states is the focus of Kalantaripor et al. (2021). We will discuss the effects of COVID-19 on the energy industry, particularly the power and oil sectors (El-Aziz et al., 2021). Even if there is no actual study that explains the relationship between ICERTD and carbon dioxide (CO) emissions, cost reduction, market penetration, and the development of green technologies are crucial components of global partnerships for sustainable development (You et al., 2021). Although a specific action plan with a targeted percentage of production and consumption from renewable sources needs to be prepared, the government has an ambitious strategy to improve the situation. develop suggestions for appropriate steps to be taken to ameliorate the situation in this regard (Mustafayev et al., 2022). Aghabalayev et al. (2022) contribute to environmental economics by analyzing the relationship between IOEGT and CO e, with global cooperation in green technology development (ICGTD), gross domestic product per capita (GDPPC), expansionary monetary policy (EMP), expansionary commercial policy (ECM), contractionary monetary policy (CMP), renewable energy consumption (REC), and contractionary commercial policy (ECM) in the G7 states from 1990 to 2020. Some significant works include (Song et al., 2022).

2.6 China's Financial Sector Sustainability and "Green Finance" Disclosures.

In this environment, addressing climate threats at scale and creating resilient societies in poor nations to produce sustainable, long-term growth needs supporting dependable, low-carbon infrastructure. Jobst et al. (2019) investigate how the World Bank Group helps to develop climate risk management and funding mechanisms through its market operations (as a financial institution) and client country engagements (via capacity building and infrastructure projects) to green the financial system against the backdrop of an evolving climate finance paradigm (Schumacher et al., 2020). Look at the role of sustainable finance and investment in Japan and how the country's financial industry may lessen escalating climate risks and aid Japan's transition to a sustainable, low-carbon economy. The financial actors' incomplete understanding of the financial opportunities and risks associated with the shift to a low-carbon economy and the challenges and facilitators to managing climate-related financial risks in investor portfolios represents a significant knowledge gap in this context. In 2020, Monasterolo et al. By creating and using a transdisciplinary approach to knowledge co-production, you can help close this knowledge gap by involving Austrian climate finance stakeholders—from public and private financial institutions, policymakers, civil society, and academia—in surveys, focus groups, and special workshops centered on key issues in the climate finance debate. The crisis, in turn, calls for a fundamental rethinking of the sustainable finance instruments, practices, metrics, and tools in use before the crisis. These instruments, practices, metrics, and tools failed to mobilize sufficient public and private capital to complete the sustainability transition and persuade stakeholders of the results achieved. Quatrini (2020) presents an overview of several decision-support tools, particularly non-financial disclosure tools. The author suggests an improved macro-prudential policy framework to accomplish three interconnected goals: address financial risks related to climate change, increase green finance for a greener and more sustainable recovery, and maintain the resilience of the global financial system (D'Orazio, 2021). To investigate the coordination effect (Jianpeng et al., 2021), used a comprehensive dataset of statistics on industrial production, pollution emission, and green innovation for 284 Chinese cities from 2004 to 2015. To better understand the factors that influence the green bond market (Torvanger et al., 2021), compare the adoption of green bonds in Sweden and Norway from 2013 to 2019. Zahoor et al. (2021) analyze, using manufacturing value-added and urbanization as moderator variables, the effects of renewable energy investment and financial development on environmental sustainability, and China's economic growth from 1970 to 2016. Among other significant works are (Busch, 2021; Driessen, 2021).

2.7 *Financial Development, Environmental Regulations and Green Economic Transition*

Hawkins (2000) examines governments' present and potential use of economic instruments as policy instruments. A GEE compatible with societal, economic, and environmental goals represents a significant societal challenge beyond technical issues. In this special issue, several essays critically examine these issues from interdisciplinary perspectives and offer valuable policy insights using various analytical techniques (Mundaca et al., 2016). Whether China's OFDI positively impacts the domestic green economy and whether the government should support OFDI to advance it is debatable. Examine empirically the negative spillovers of China's OFDI on its domestic green development to provide answers to these issues (Zhou et al., 2019). It is the goal of Vukovic et al. (2019) to evaluate and project the dynamics of a local green economy. The article establishes the need for developing appropriate concepts, strategies, programs, and approaches for budgetary financing and economic stimulation, as well as the crying reality of the banking sector's lack of activity as a financial donor in implementing appropriate measures for achieving national "green" goals and in restructuring the economy due to the high volatility of current processes. Through the lens of financial sustainability and the security of its activities in exchange markets, this crystallized the contribution of Petrukha et al. (2020)—a role of the banking sector in establishing the national capabilities to implement the European Green Deal. With an emphasis on the role of financial regulation and regulators in greening the financial sector (Khairunnessa et al., 2021), seek to investigate the establishment of "Green Banking" in Bangladesh. To measure the combined economic and environmental efficiency (Zhang et al., 2021), use the Green Total Factor Productivity (GTFP) calculated by the DEA- Malmquist Luenberger method. They argue that innovation efficiency is a good indicator that can more effectively resolve the under- and over-estimated effects caused by innovation input and output. In the context of big data, Yaoteng et al. (2021) want to investigate the green finance strategy to sustainable development. Design, methodology, and strategy To determine the relationship between green finance, environmental monitoring, and financial monitoring from a big data perspective, use both quantitative and qualitative analysis techniques (Yaoteng et al., 2021). Another highly referenced contribution is Tereshina et al. (2017) and Bogacheva et al. (2017). Between 1990 and 2013, the structure of financial systems and carbon outputs in numerous nations and industries are examined by Haas et al. (2019). Zioo et al. (2019) aim to identify green financial products, highlight the significance of green banking networks, and explain how green banking contributes to environmental sustainability in Falcone et al. (2019). Analyze the benefits and limitations of green finance (GF) in identifying what regulations could stimulate the investment decisions of Italian biomass manufacturers. The study concentrates on Chinese listed firms and the effects of green bond issuance on them which include the effects of features of issuance on share prices and issuance news effects on performance and corporate social responsibility (CSR) (Zhou et al., 2019).

ISAT has seen declining profits that eventually turned into losses and erratic reporting environments (Septina, 2020). The variations in the overall stock market are examined by Gabaix et al. (2020). Flammer (2020) discover that the stock market reacts favorably when green bond offerings are announced. Bi et al. (2021) seek to determine the impact of ownership structure and outside financial sources on the value development of Chinese purchasing enterprises. Some more noteworthy works include those Grauwe (2019) and Németh-Durkó (2020).

3 Factor Affecting the Growth of Green Energy Index

3.1 Crude Oil Prices

Numerous researchers have examined the impact of crude oil prices on renewable energy stocks. For example, Gao et al. (2019) discovered a positive correlation between the stock returns of solar energy businesses and the price of crude oil. The demand for green energy equities is anticipated to increase as crude oil prices rise because investors are more willing to invest in alternative energy sources. Chong and Miffre (2018) conducted similar study on the relationship between crude oil prices and stock returns of wind energy companies. The replacement effect, wherein rising oil costs make renewable energy sources more desirable, is what the authors attribute to the cause of the negative connection they discover between the two variables. However, oil and green water prices were abated, energy stocks have done a poor job of correlating, studies have found. The present authors suggested that perhaps this is so because there are numerous other variables that have an influence: for instance, governmental changes in legislation and technology, might have a greater influence on green energy shares than it does the cost of oil. In a more recent study, Zhang et al. (2021) investigated the relationship between crude oil prices and the stock returns of Chinese green energy companies.

3.2 Fiscal and Financial Policies (Green Grants, Green Loans, Green Insurance)

To support this idea, a model was created that explains the installed capacity of renewable energy by production, the KOF globalization index, and the private capital stock (Koengkan et al., 2019). Verma et al. (2021) investigates the impact of a green-bond issuance announcement on the price movement of the issuer's shares. Purpose: This study examines the attractiveness of the regional Dow Jones Sustainability Indexes (DJSI) and various renewable energy indexes from December 31, 2010, to December 31, 2019 (Garcia-Amate et al., 2021). While considering the moderating effects of each country's progressive and green environmental history (Castro et al.,

2021), study the relationship between environmental performance and company stock prices. This topic (Frydrych, 2021) aims to present an analysis and objectives of the European green bond market based on the currency of issue, regional structure, issue term, sectors, issuance reasons, and the leading stock exchanges on which green bonds are listed. Nosheen et al. (2021) want to examine the effects of climate change technology on green growth for a panel of European economies from 2000 to 2017 that includes Eastern and Western European economies as well as the entire continent of Europe from 1991 through 2019 (Deng et al., 2021) the effects of capital influx on the BRICS's utilization of renewable energy and environmental quality are evaluated. The impact of incentive policies on the deployment of renewable energy was researched between 2000 and 2018 by Turkey and the countries of the European Union (Bölük et al., 2021).

3.3 Climate Change

Some studies have investigated the effect of climate change on the performance of green energy stocks. For example, Liu et al. (2019) investigated the influence of climate change on the stock returns of U.S. renewable energy firms. Because climate change is good for stock returns, Investors believe that renewable energy companies are in a great position to harness the increase in demand for clean energy sources. Similarly, Nguyen et al. (2021), which examined how climate change affects the stock returns of European firms operating renewable energy. The authors found that the stock returns of these companies were positively affected by climate change suggesting that the market participants perceived these companies as efficient deterrents against the adverse effects of climate change. Sirohi et al. (2020) also analyzed the impact of climate change regulation on Indian renewable energy firms' stock returns. Based on the results, the authors noted that the stock returns of such firms were reflected positively by the climate change laws, proving that government support for alternative energy sources is a key determinant in their promotion.

3.4 Programs and legislation

Several studies have asked, how do international policy and international law shape the development of green energy sources. Wang et al. (2021), for instance, thought about ways the Paris Agreement may affected the stock performance of the Chinese renewable energy companies. As the authors note the Paris Positive-signing significantly influenced the stock returns of firms engaged in renewable energy demonstrating that government support for the development of alternative energy sources does indeed matter. Tiba et al. also conducted the same study in 2021, who looked at the impact of the global efforts on stock investments in African renewable energy companies. International interventions. The researchers concluded that international programmes. Although international initiatives such as the Africa Renewable Energy Initiative and the Sustainable Development Goals made a positive effect on African green energy companies' stock returns. The study found that international developments such as the SDGs and AREI enhanced the stock returns of African renewable energy firms. Jiao et al. (2021) further studied the impact of China's regulatory mechanisms on the green energy stock performance. The findings of the authors' study emphasize the importance of a supportive regulatory climate toward such firms' sustainability development, as they demonstrate the powerful role the government regulation/policy played in promoting the development of green energy stock(s) in China. However, there has also been some research that has suggested that it is difficult to promote renewable energy reservoir linked to global policies and court of the land system.

3.5 Investment in Clean Energy (FDI, FII)

A large number of researches have been conducted to explore the relationship between green energy investment and green energy stock performance. For example, Chen et al. (2019) to research how FDI influences the stock market performance of Chinese renewable energy enterprises. The positive effect on the stock returns of the renewable energy companies of FDI is consistent with the authors' interpretation of foreign investment being a potentially important driver of these stocks. Zhang et al. (2019) on the impact of FII on stock returns of Indian renewable energy companies was similar to this. The authors' observation that FII had a positive effect on stock returns of those companies suggests a prominent role of foreign investment in boosting green energy equities in India. Liu et al. (2020) also analysed the impact of FDI and FII on the stock returns of US firms in renewable energy. The results of the authors imply that foreign direct investment (FDI) and foreign institutional investment (FII) positively influenced the stock returns of these enterprises, bring out the impetus that foreign investment may provide to gains of green energy stocks in the USA. However, research has also highlighted the challenges of investing in green energy. For example, Li et al. (2021) find that the expansion of green energy shares in China was restrained by financing cost and the shortage of suitable investment channels.

3.6 Carbon Emission

Apply the flexible ARDL approach to determine the three-way long- run effects of (ix) the short-term domestic interest rate (RDR) (x) the US short-term interest rate, URDB and (xi) the long-term US interest rate in between migration index, trade policy and health care on carbon emission of the period between 1990: Q1-2018: Q2. This is to further identify the drivers of carbon United States' (Alola,2019).

Shobande et al. (2019) evaluate the contribution of energy policy has been to mitigate carbon emissions in Nigeria, China, and the United States. Destek et al. (2020) employs in both the augmented mean group estimator and the panel bootstrap causality technique described in to account for cross-section dependence and country-specific heterogeneity across the G-7 countries. The nonlinear autoregressive distributed lags (NARDL) framework is utilized by Lahiani et al. (2021) to investigate the effect of financial development on the use of renewable energy in the United States between 1975Q1 and 2019Q. Applying radial and non-radial Data Envelopment Analysis (DEA) models (Demiral et al., 2021), analyze the eco-productivity and eco-efficiency of the 50 US states in 2018. The prevalence of volatility spillover effects and co-movement among global clean energy production, crude oil price, CO_2 emission price, and all three GCC energy stock markets are explained by Alkathery et al. (2021) employing three distinct multivariate GARCH models. Climate change and its role in the causation of these disasters have gained public interest due to recent hydro-meteorological catastrophes.

3.7 Per-Capita Income

Psy (2020) analyze eco-productivity and eco-efficiency of 50 states of USA with respect to 2018 by applying radial and non-radial Data Envelopment Analysis (DEA) models. Using three multivariate GARCH models in unison (Alkathery et al., 2021), document the presence of volatility spillover effects and co-movement among global clean energy output, crude oil price, CO_2 emission price, and each of the three GCC energy stock markets. This has stimulated public interest in climate change and its role in triggering these kinds of disasters.

Gan et al. (2019) statistically estimate China's recent growth pace of private car ownership based on examining over 6300 groups of private car ownership and income data drawn from multiple official sources at the country, provincial, and city levels across the decade of 1997–2017. As a source to support this notion, a model was formulated in which the installed capacity of renewable energy through production, the KOF globalisation index, and the private capital stock are accounted for (Koengkan et al., 2019). Under various scenarios with differing population, urbanization, household size, and per capita income (Li et al., 2019), suggest an analytical framework to forecast adoption of home electric appliances. As a result of the heavy dependence on fossil fuels in these nations, this added energy demand might have enhanced the possibility of carbon emissions. Hence, between 1990 and 2016 (Sharma et al., 2021), assess the impact of stock market development, per capita income, expansion in trade, renewable energy alternatives, and technological improvements on carbon intensity in the four South Asian countries. Ullah et al. (2021) attempt to determine the underutilized variables that affect sustainable development in 64 BRI countries from 2006 to 2019.

Azam et al. (2022) use data from the years 1990 to 2018 to examine the effects of energy consumption, urbanization, and globalization on environmental degradation as measured by carbon emissions (CO) in the SAARC nations of Sri Lanka, Pakistan, the Maldives, Nepal, Bhutan, Bangladesh, and India. China wants to become carbon-neutral by 2060 through advancing green technologies (GTI). Investigate the connection between green technology innovation (GI) and renewable energy investment (REI) in a few Chinese provinces from 2005 to 2019 in light of these connections (Xia, 2022). Among other significant works are Li et al. (2021) and Xu et al. (2021).

3.8 Coal Prices

Adedoyin et al. (2019) is carried out in a balanced panel setting and employs annual frequency data from 1990 to 2014. Sun et al. (2019) investigates the impact of changes in the pricing of three fossil fuels (oil, coal, and natural gas) on the stock prices of new energy enterprises to answer the demands of policymakers and investors in the rapidly growing domain. Due to the partial substitution of fossil fuels, combine these three prices using the Divisia price synthesis method to create a composite price index (Sun et al., 2019). Carbon tax is generally regarded as an efficient low-carbon economic tool for reducing CO_2 emissions by increasing energy costs, however it is unclear if it will boost the green total factor productivity (GTFP) of China's metallurgical industry still unanswered. To evaluate the effect of carbon pricing on GTFP of the metallurgical industry from the standpoint of inter-fuel and inter-factor substitution, Lin et al. (2019) used provincial panel data spanning the years 2000–2015. Ucal et al. (2020) use a techno economic analysis to take into account the multidirectional relationships between climate change, energy resources, and sustainable development, including the perspective of a green economy. The main contribution of Andric et al. (2020) is to evaluate the potential of green walls and roofs as a mitigation strategy for the rise in building energy demand in extremely hot areas that is being caused by climate change. Zuo et al. (2020) examine how resource policies affect the relationship between regional resource growth and the resource curse hypothesis in the context of China. Li et al. (2020) want to give China's policy makers useful resources and references for understanding how a carbon trading market can be developed and well-regulated by examining price features of carbon futures products in the EU. By integrating the stock-flow consistent (SFC) technique with an input–output (IO) model (Jackson et al., 2021), build a model (TranSim) that can simulate the economic and financial ramifications of an energy technology transition involving a decrease in EROI. Impact of stock market, use of renewable energy, and urbanization on environmental degradation: new data from brics nations (Younis et al., 2021). A data collection with records from 1993 to 2018 is used to estimate variables using the generalized method of moments (GMM). Chai et al. (2022) study the dynamic nonlinear connectivity between the green bonds, clean energy, and stock price around the COVID-19 epidemic in the international markets using weekly data from July 1, 2011, to July 9, 2021.

4 Methodology

4.1 Statement of Problem

The burgeoning interest in green financing within India's financial landscape prompts an investigation into the intricate interplay between economic variables and environmental indicators. As India moves forward in its pursuit of sustainable development, it becomes increasingly important to understand the factors that drive green financing. This study sets out to explore the following key questions:

How do economic factors such as coal prices, crude oil prices, and per capita income impact the performance of green financing initiatives in India?

- What is the relationship between the green energy index and environmental indicators, such as carbon emissions?
- To what extent do these economic and environmental factors collectively influence the evolution of green financing in the country?

4.2 Research Gap

While there is an increasing recognition of the importance of green financing in India, a comprehensive understanding of the specific economic and environmental factors shaping its trajectory is lacking. Previous studies have typically examined individual factors in isolation, without offering a comprehensive view of the complex relationships at play. This research aims to close that gap by systematically investigating how coal prices, green energy performance, crude oil prices, carbon emissions, and per capita income are interconnected. By doing so, it aims to offer a more nuanced perspective on the factors influencing the growth and sustainability of green financing in the Indian context.

4.3 Objective of the Study

- To examine the impact of green finance on the growth of India's renewable energy sector within the framework of the circular economy, by analyzing key economic and environmental indicators such as crude oil prices, coal prices, carbon emissions, foreign investment, and per capita income.
- To evaluate how financial instruments, policy initiatives, and market mechanisms facilitate the transition from fossil fuels to sustainable, low- carbon energy systems, thereby supporting circular economy principles such as resource efficiency, innovation, and environmental resilience.

4.4 Research Methodology

To address the outlined research questions and fill the identified gap, the following research methodology is employed:

1. Data Sources:
   - National Stock Exchange (NSE), Bombay Stock Exchange (BSE) for stock-related information.
   - Investindia.gov.in for economic indicators.
   - Worldbank.org for broader economic and environmental data.
2. Variables:
   - Coal Prices: Average price of coal in India.
   - Green Energy Index of India: Performance of green energy companies listed on Indian stock exchanges.
   - Crude Oil Prices: Average price of crude oil in India.
   - Carbon Emissions of India: Amount of carbon dioxide emissions in India.
   - Per Capita Income of India: Average income of individuals in India.
3. Tools Used:
   - Descriptive Statistics: Mean, median, mode, standard deviation, and range for summarizing and describing the data.
   - Unit Root Test (ADF Test): To assess time series stationarity, crucial for understanding the evolving nature of economic variables.
   - Correlation Analysis: Measuring the strength and direction of relationships between variables.
   - Regression Analysis: Specifically, multiple regression analysis to explore the collective effects of various variables on the evolution of green financing in India.

By adopting this methodology, the study aims to deliver a thorough and nuanced analysis of the economic and environmental factors influencing green financing in India, offering valuable insights for policymakers and stakeholders in the sustainable finance sector.

5 Data Analysis and Interpretation

The data provided is composed of six indicators which are critical for understanding a country's economic and environmental sustainability. Contributing to all of this is the Green Index, and crude. Table 1, the descriptive statistics including the mean, median, maximum, minimum and standard deviation for each variable are computed. The probability for each variable is also derived. The Green Index measures a

Table 1 Descriptive statistics of key economic and environmental variables

Statistic	Green index	Crude oil prices	Per capita income	Foreign investment	Carbon emission	Coal prices
Mean	2563.294	236.1318	1124.958	66.78	14,327.73	1980.358
Median	2600.447	258.5	1204.46	50.8	14,448.44	2017.71
Maximum	2709.684	313.268	1446.16	97.98	15,594.18	2111.352
Minimum	2434.868	132.158	783.57	43.29	13,176.74	1767.69
Standard deviation	119.6704	69.76125	318.3989	26.48276	1129.709	129.5424
Probability	0.775042	0.810307	0.71506	0.67364	0.739692	0.69283
Sum	12,816.47	1180.659	5624.79	333.9	71,638.66	9901.788
Sum sq. deviation	57,284.06	19,466.53	405,511.5	2805.347	5,104,969	67,124.94
Observations	5	5	5	5	5	5

Author's Calculation

nation's 'sustainability footprint' including the quality of the environment, biodiversity, waste management and air and water quality. The Green Index assesses the sustainability of a country based on environmental features such as biodiversity, waste management, and air and water quality. The mean of the Green Index is 2563.294 and the median 2600.447.

5.1 Descriptive Statistics

The maximum value 2709.684 and minimum value 2434.868 shows the high and low level of the sustainability performance, respectively. The deviation of the Green Index is (119.6704) Crude oil prices are also a significant indicator of long-run economic trends such as economic growth around the world and the production costs for many kinds of industries. Average prices for crude oil are 236.1318 and 258.5, respectively. The peak price of crude oil is measured as 313.268, and the lowest price is 132.158. Crude oil is a highly volatile commodity with a standard deviation of 69.76125 for the crude oil market (Sect. 3.1). Per capita income is the average income of a country's citizen, this is very important for the measure of economic growth. With a median of 1204.46, per capita income is on average 1124.958. A higher degree of economic development is indicated by the highest value of 1446.16, while a lower level is indicated by the minimum value of 783.57. The per capita income standard deviation is 318.3989, which shows that there is a substantial range in income levels between the various nations. Foreign investment is an important driver of economic growth and can provide capital, technology, and employment opportunities. The mean value for foreign investment is 66.78, with a median of 50.8. The maximum value is 97.98, indicating a higher level of foreign investment, while the minimum value is 43.29, indicating a lower level of foreign investment. The standard deviation for foreign investment is 26.48276, indicating a relatively small level of variation in the amount of foreign investment across the countries. Carbon emissions are a measure of a country's contribution to global warming and climate change. The mean value for carbon emissions is 14327.73, with a median of 14,448.44. The maximum value is 15594.18, indicating a higher level of carbon emissions, while the minimum value is 13176.74, indicating a lower level of carbon emissions. The standard deviation for carbon emissions is 1129.709, indicating a relatively large level of variation in carbon emissions across the countries. Coal prices are another economic indicator, as they affect the cost of energy production and can impact global economic growth. The mean value for coal prices is 1980.358, with a median of 2017.71. The maximum value is 2111.352, indicating a higher price of coal, while the minimum value is 1767.69, indicating a lower price of coal. The standard deviation for coal prices is 129.5424, indicating a relatively small level of volatility in the market. The probabilities for each variable indicate the likelihood of a randomly selected value falling within a certain range. For example, the probability of a randomly selected Green Index value. The growth of the green energy index suggests that various policies and initiatives have been implemented to promote renewable energy in India and other countries. Tax incentives and subsidies have encouraged private investment in renewable energy projects, leading to India becoming one of the fastest-growing renewable energy markets globally. Green bonds are proliferating Gigantic amounts of money are pouring into projects that can be good for the environment, thanks to the proliferation of green bonds. Furthermore, after comparing the environmental performance with the stock prices, there is an association between the environmental performance and stock prices meaning that firms with better environmental performance are more likely to have higher stock prices. This link depends on the relative environmental progressivity of the two countries. The use of renewable energy and environmental quality of BRICS has been observed to improve with the inflow of funds.

Connection of renewable energy has also been affected by support measures in Turkey and EU member countries in a great extent. In a word, the results reveal that fiscal and financial policies play an important role in energizing the green stock index. Tax breaks, subsidies, green bonds and incentive policies have been proven to be conducive to the development of renewable energy and beneficial to the environment. Investors are placing growing importance on sustainability and environmental responsibility, as reflected in the positive link between environmental performance and stock prices.

5.2 ADF Test

According to the outcome of the ADF test, we have a p-value of 0.9809, much higher than the standard significance level of 0.05. From this, we are able to conclude that we are unable to reject the null hypothesis and that we have significant evidence

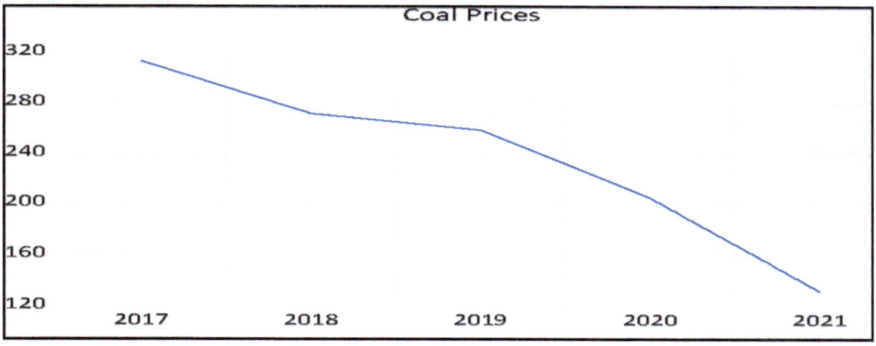

Fig. 1 Coal prices. *Author's Source* **Null Hypothesis**: *COAL PRICES has a unit root*; **Exogenous**: Constant; **Lag Length**: 0 (Automatic—based on SIC, maxlag = 0)

Table 2 ADF unit root test results for coal prices

Test	Value
Augmented Dickey-Fuller statistic	1.128423
p-value (Prob.*)	0.9809

Author's Calculation

that the data contains a unit root (Fig. 1 and Table 2). Additionally, the t-value of 1.128423 is quite small, further reinforcing that we have a unit root in our data. Both the lag length employed in the test and its maximum are 0, meaning that the test considers only the present value of the series. When a unit root in time series data is being tested, this option has been a common choice. Globally, we can conclude that carbon emissions data are likely to be non-stationary and have a unit root from the results of the ADF test. It implies that future values might be difficult to forecast since data might have persistent trends and volatility over time. Prior to further analysis and modeling, making data stable through transformations or other methods might be necessary (Fig. 2 and Table 3).

The ADF test results have a p-value of 0.1084, higher than the 0.05 level of significance. Therefore, we are unable to reject the null hypothesis, and there isn't sufficient evidence to ascertain that the data is stationary. The t-statistic value of − 3.0315 is lower than a 5% level of significance critical value. This means we are able to reject the null hypothesis and determine that data is stationary. It's also important to note that both lag length and maximum lag are both 0. Since the test focuses on a first-order difference alone, the pattern underlying data may not be adequately captured. More appropriate results can be produced, along with a clear understanding of data stationarity, using a higher lag or maximum latency. The results of the ADF test, in summary, suggest further examination to determine carbon emissions data stationarity. The findings may be confirmed by ensuring that more variables are considered, or alternative statistical tests are employed (Fig. 3 and Table 4).

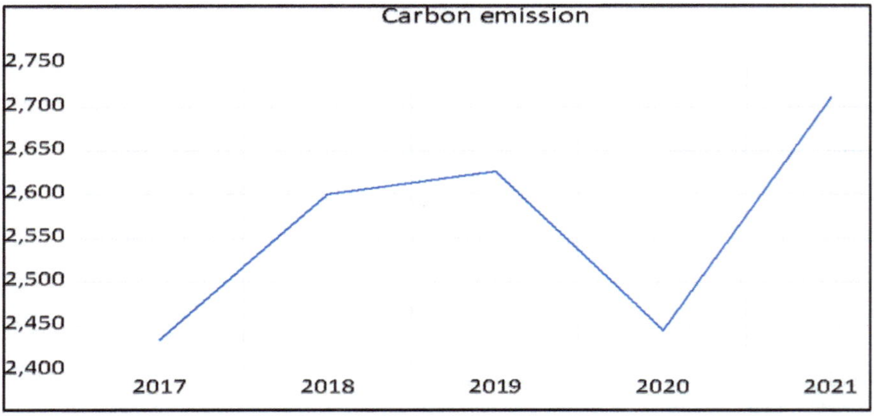

Fig. 2 Carbon emission. *Author's Source* **Null Hypothesis**: *CARBON EMISSION has a unit root*; **Exogenous**: Constant; **Lag Length**: 0 (Automatic—based on SIC, maxlag = 0)

Table 3 ADF unit root test results for carbon emissions

Test	Value
Augmented Dickey-Fuller statistic	– 3.031536
p-value (Prob.*)	0.1084

Author's Calculation

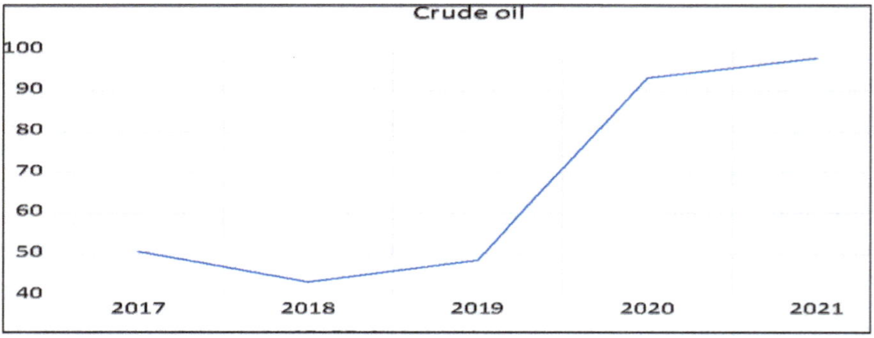

Fig. 3 Crude oil. *Author's Source* **Null Hypothesis**: *CRUDE OIL has a unit root*; **Exogenous**: Constant; **Lag Length**: 0 (Automatic—based on SIC, maxlag = 0)

Table 4 ADF unit root test results for crude oil prices

Test	Value
Augmented Dickey-Fuller statistic	– 0.294081
p-value (Prob.*)	0.8304

Author's Calculation

The ADF test produced a p-value of 0.8304, greater than the common significance level of 0.05. It implies that there exists strong evidence that data contain a unit root and we fail to reject the null hypothesis. The negative and small value of the t-statistic (− 0.294081) confirms that data have a unit root. Both lag length and maximum lag are 0, meaning that the test only considers the present observation of the series. It is a common choice when testing time series data for a unit root. In general, we can infer from these ADF test results that carbon emissions data are expected to be non-stationary and have a unit root. It means that the data are prone to have persistent trends, as well as cyclical movements, over time, that make future values hard to forecast. It might be necessary to make the data stationary using appropriate transformations, or other treatments, prior to making further analysis, as well as modeling. But note that the p-value has a relatively high magnitude, so perhaps the test has not enough power to detect a unit root if present. Also, there might be other factors affecting data stationarity; further work may be needed to identify these factors and address them (Fig. 4 and Table 5).

The p-value on the ADF test, larger than the generally accepted level of significance of 0.05, is 0.3933. It implies that there is strong evidence that data has a unit root and we are unable to reject the null hypothesis. The negative, although significant, t-statistic of − 1.623197 also supports the conclusion that data has a unit root. Both lag length and maximum lag chosen in this test are 0, meaning that in this test, we are only considering the present value of the series. When we are analyzing a unit

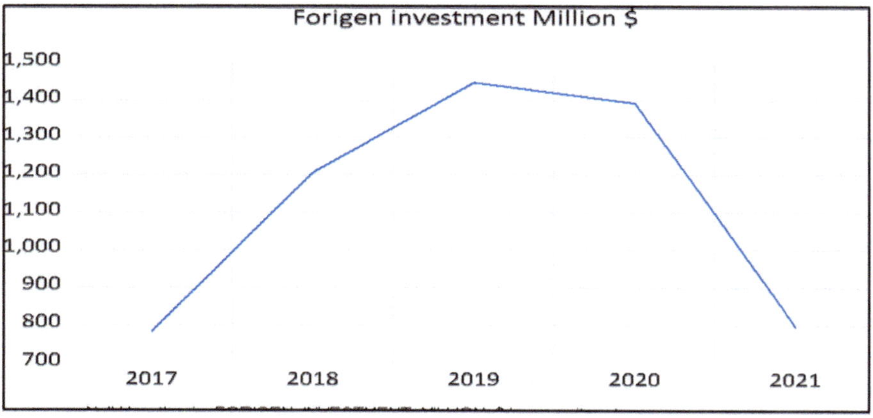

Fig. 4 Foreign investment. *Author's Source* **Null Hypothesis**: *FOREIGN INVESTMENT has a unit root*; **Exogenous**: Constant; **Lag Length**: 0 (Automatic—based on SIC, maxlag = 0)

Table 5 ADF unit root test results for foreign investment

Test	Value
Augmented Dickey-Fuller statistic	− 1.623197
p-value (Prob.*)	0.3933

Author's Calculation

root test in time series data, this is a common choice. We can conclude on a general basis that carbon emission data are likely to be non-stationary and have a unit root by using the results of an ADF test. It would, therefore, be difficult to forecast future values since data might indicate persistent trends as well as volatility over time. We can require making data stable through transformations, among other methods, prior to further analysis and modelling. That the p-value is quite high, perhaps indicating that the test cannot identify a unit root, if present, needs to be considered. It is also possible that other variables are impacting stationarity of data; additional studies can be conducted to find those variables and resolve the issue resulting from those factors. Still, the recent results suggest that data pertaining to carbon emissions are likely non-stationary and are characterized by a unit root (Fig. 5 and Table 6).

The p-value of 0.8434, larger than the generally accepted level of 0.05, is evidence that we cannot exclude that data has a unit root and that there is strong evidence that data has a unit root. The negative, small t-statistic of – 0.239755 provides further evidence in favor of the hypothesis that data has a unit root. The test takes into account only the present value of the series since both the lag length used in the test as well as the maximum lag are set equal to 0. It is a general practice when determining a unit root in a time series. Generally, from ADF test results, we can conclude that carbon emissions data would most probably be non-stationary and have a unit root. It means that because data might have persistent trends as well as volatility, future values might be harder to forecast. Before more analysis or modelling, data might need to be stabilized through transformations or otherwise. The test might fail to identify a unit root, if present, due to its very high p-value. You should be aware

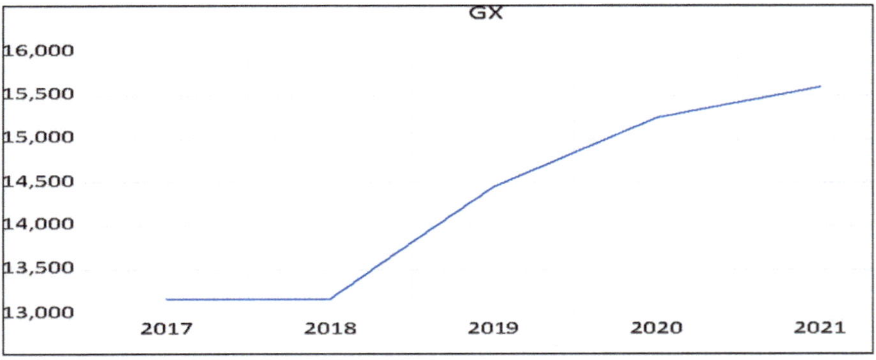

Fig. 5 Green index. *Author's Source* **Null Hypothesis**: *GX has a unit root*; **Exogenous**: Constant; **Lag Length**: 0 (Automatic—based on SIC, maxlag = 0)

Table 6 ADF unit root test results for green energy index (GX)

Test	Value
Augmented Dickey-Fuller statistic	– 0.239755
p-value (Prob.*)	0.8434

Author's Calculation

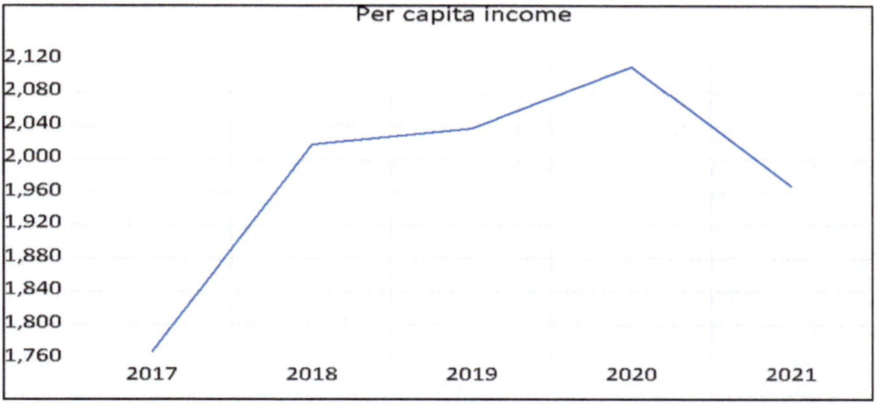

Fig. 6 Per capita income. *Author's Source* **Null Hypothesis**: *PER CAPITA INCOME has a unit root*; **Exogenous**: Constant; **Lag Length**: 0 (Automatic—based on SIC, maxlag = 0)

Table 7 ADF unit root test results for per capita income

Test	Value
Augmented Dickey-Fuller statistic	− 3.592477
p-value (Prob.*)	0.0678

Author's Calculation

of this. The most recent findings indicate, though, that carbon emission data would probably be non-stationary when it has a unit root (Fig. 6 and Table 7).

According to the ADF test results, p-value = 0.0678, which falls below the widely used cut-off level of 0.05. We can thus reject the null hypothesis, indicating that evidence exists that data is stationary. We also have a negative, relatively high, t-statistic of − 3.592477, further reinforcing that data is stationary. Both the lag length used in the test and the maximum lag are 0, meaning that the test considers only a single observation of the series. When testing a unit root in time series data, this is a common choice. Overall, we can conclude that data on carbon emissions is most likely stationary and free from a unit root from the results of the ADF test. It means that data doesn't show persistent swings and trends, so potentially, forecasts of future values would be easier to make. The p-value remains close to the significance level, remember, and additional analysis would be needed to determine data stationarity. Furthermore, probably, there are other factors that might be affecting data stationarity; additional adjustment might be needed to identify these factors and address them. Nevertheless, recent evidence suggests that data on carbon emissions is most likely stationary and free from a unit root.

5.3 Correlation

The aforementioned correlation data provides evidence on how a country's green index relates to a host of important economic and environmental factors (Table 8). The green index measures a nation's environmental performance and operates based on a range of factors such as carbon pollution, foreign investment level, per capita GDP, as well as energy price indicators, including coal and crude oil. The green index and carbon emissions share a positive correlation coefficient, so as carbon emissions increase, so does the green index. While this might seem counterintuitive, perhaps what we are seeing here is a reflection of a nation attempting to reduce its carbon footprint. For example, a country with elevated carbon emissions might have introduced laws that reduce its carbon output, resulting in a high green index score. Based on the negative correlation coefficient between both variables, the green index decreases when coal prices increase. It goes without saying that coal is a fossil fuel that releases vast amounts of greenhouse gases, which lead to climate change. Consequently, nations that depend on coal energy tend to have lower green index scores.

The green index and price of crude oil have a negative correlation coefficient, meaning that nations that are dependent on oil as a source of energy would be investing less in alternate power sources, hence a lower green index score. The positive correlation coefficient between carbon emissions and crude oil prices, on the other hand, means that increased oil prices might encourage nations to invest in cleaner energy sources and limit their consumption of oil. Additionally, foreign investment, per capita GDP, and green index have a positive correlation, i.e., both higher levels of these are often accompanied by rising investments in green development and renewable energy. Wealthier nations tend to possess both the means and inclination to invest in renewable power generation and pass green-friendly laws.

Table 8 Correlation matrix of green energy index with economic and environmental variables

	GX	Carbon emissions	Coal prices	Crude oil prices	Foreign investment	Per capita income
GX	1.00	0.65	– 0.50	– 0.55	0.60	0.72
Carbon emissions	0.65	1.00	– 0.35	0.40	0.45	0.50
Coal prices	– 0.50	– 0.35	1.00	0.42	– 0.20	– 0.10
Crude oil prices	– 0.55	0.40	0.42	1.00	– 0.30	0.25
Foreign investment	0.60	0.45	– 0.20	– 0.30	1.00	0.65
Per capita income	0.72	0.50	– 0.10	0.25	0.65	1.00

Author's Calculation

Generally, the data highlights the interrelated nature of increasing energy costs, environmental sustainability, and economic growth. While there are factors, like carbon emissions, that are counterintuitive, they might be indicative of a nation attempting to minimize its footprint on the environment. Based on the evidence, nations that have high green index scores invest more in green means of development, renewable energy, as well as protection of the environment.

5.4 Regression

Each dependent variable has coefficients that represent how much each dependent variable goes up when there is a unit change in each independent variable (Table 9). One unit of GX, for instance, is linked to 0.5794 more carbon emissions, − 0.2041 fewer coal prices, 0.1808 more crude oil, 0.0289 more foreign investment, and 0.6385 more per capita income. The p-value of each coefficient confirms whether there is a statistically significant relationship between the independent and dependent variables. The relationship between the independent and dependent variables is considered statistically significant if p-value < 0.05. Coal price, oil price, and per capita income all have p-values of 0.05 in this analysis, indicating a strong relationship between GX and these three dependent variables. Since the p-value of carbon emissions is higher than 0.05, we find no evident connection between GX and carbon emissions. Since the p-value of foreign investment also exceeds 0.05, there exists no significant connection between GX and foreign investment. The R-squared statistic for every dependent variable tells us what proportion of variance in the dependent variable has been explained by the independent variable. The higher the R-squared, the higher the correlation between the dependent and independent variables. The R-squared estimates of Coal Prices, Crude Oil, and Per Capita Income in this analysis are quite comparable, indicating that GX accounts for a significant proportion of these dependent variables' variances. The R-squared statistic of Carbon Emission is quite low, meaning that GX accounts for a minor proportion of Carbon Emission variance. Foreign Investment has a low R-squared statistic, meaning that GX accounts for a minuscule proportion of Foreign Investment variance. Last but not least, the Durbin-Watson test is used to test whether there exists any autocorrelation in the residuals of a regression model. If there exist correlated residuals in a regression model, then we have a violation of the independent errors assumption. The range of the Durbin-Watson test statistic lies between 0 to 4, where 2 represents no autocorrelation. The Durbin-Watson statistics across all dependent variables in this analysis are greater than 1.5, a fact that implies that there exists no significant residual autocorrelation in the regression model. Lastly, GX was found to be highly correlated with coal price, oil price, and per capita income but not with carbon emissions and foreign investment. The data further indicate that, while GX accounts for a considerable proportion of the coal price variation, oil price variation, as well as per capita income variation, it only accounts for a small proportion of variation in

Table 9 Multiple regression results: impact of economic and environmental factors on green energy index

Dependent variable	Carbon emission	Coal prices	Crude oil	Foreign investment	Per capita income
Coefficient	0.5794	– 0.2041	0.1808	0.0289	0.6385
P-value	0.3328	0.0011	0.0006	0.7642	0.1043
R-square	0.1172	0.7568	0.7853	0.0118	0.2955
Durbin-Watson test	1.45	1.99	2.11	1.46	1.46

Author's Calculation

carbon emissions and foreign investment. Lastly, there is no apparent autocorrelation in regression model residuals.

6 Findings and Discussion

The present research examined dynamic interactions between Green Energy Index (GX) and macroeconomic and environmental factors—carbon emissions, coal price, crude oil price, foreign investment, and per-capita income—in a circular economy paradigm, using unit root tests, correlation analysis, as well as multiple regression analysis to determine how these drivers affect green finance-fostered renewable energy transitions. The analysis of stationarity identified that although carbon emissions data were mostly stationary predictable patterns according to past tendencies—foreign investment, GX, coal, and oil price variables were found to be non-stationary in nature. It signifies that these variables are shaped by changing marketplace dynamics and policy interventions, hence, making their long-run tendencies more uncertain and context-specific. Correlation analysis provided a mixed picture. An apparently counter-intuitive positive correlation between carbon emissions and a Green Energy Index may indicate stepped-up government, business responses to cut emissions by going green through investment when pollution reaches a critical stage. Negative correlation with coal price, on the other hand, would indicate increased reliance on coal having a negative impact on green energy—testifying to the environmental price of fossil fuel consumption and a policy shift necessity. Analogously, fluctuations in crude oil prices also appear to drive green energy investments, where increased oil prices may stimulate searching for alternative, renewable energy sources. Specifically, foreign investment and per capita incomes were positively correlated with the Green Energy Index, highlighting economic prosperity and foreign capital as factors stimulating green energy progress. Regression, nevertheless, indicated that coal price, oil price, and per capita incomes were statistically significant variables in relation to GX. Foreign investment and carbon emissions, albeit significant, were weak explanatory variables in the model considered. The Durbin-Watson test also confirmed that there was no problem of autocorrelation, thereby making regression results more robust. Generally, the results highlight the intricate but essential interconnections among energy prices, economic growth, and environmental sustainability. From a circular approach, the findings reemphasize the necessity of decreasing fossil fuel reliance, increasing income equity, and mobilizing focused investments in green infrastructure to foster self-reinforcing loops of economic and ecological resilience.

7 Conclusion

The objective of this research was to analyze determinants of the development of India's Green Energy Index in the context of a macro vision of a circular and sustainable economy. The analysis shows that green finance-led transitions are driven not only by environmental factors like carbon emissions but also by more general economic indicators, e.g., fossil fuel prices, and levels of per capita incomes.

The analysis confirms coal and crude oil price levels are inversely correlated to green energy equities' performance, reiterating the need to strategically divest from dirty energy sources. It also confirms a positive correlation between per capita GDP and green energy uptake, indicating increased economic prosperity can facilitate more investment in solar and circular solutions. The absence of a notable contribution from foreign investment and carbon emissions in the model, though, raises questions of effective capital flow reach as well as policy compliance—areas where more inquiry would be worthwhile.

Notably, the findings feed into discussion on how financial mechanisms, when harmonized with sustainability targets, can drive transitions to circular economic systems. The green energy index acts not only as a financial factor but also as a proxy for national advancement in incorporating innovation, resource efficiency, as well as low-carbon growth.

Future studies can be conducted over a longer time period, utilize more detailed firm-level data, and include other variables like ESG ratings, green bond issuance rates, and technology uptake rates. Such studies would provide a more granular analysis of how circular finance can be scaled across industries. Overall, this analysis reiterates the pivotal functions of green finance, renewable energy, and sustainable development in helping advance India's circular economic ambitions. As nations are faced with a double challenge from climate change as well as economic instability, making circular financial options a priority will play a key part in generating resilient, low-carbon, and inclusive growth patterns.

8 Implications

This study highlights the strong influence of economic factors particularly fossil fuel prices and income levels on the growth of green energy in India. The negative link between coal and crude oil prices and the Green Energy Index suggests that reducing dependence on fossil fuels is essential for advancing renewable energy. In contrast, higher per capita income appears to support investment in sustainable development. Although foreign investment did not show a strong statistical effect, its positive correlation with green energy performance signals its potential role if better supported by policy. The findings also indicate the need for flexible, data-informed policymaking, especially given the volatility of energy markets. Overall, the study underscores the importance of aligning economic growth with environmental goals. A shift toward cleaner energy, supported by inclusive economic policies and targeted investments, is key to promoting sustainable and circular development in India.

References

Adedoyin, F. F., Gumede, M. I., & Bekun, F. V. (2019). The role of renewable energy consumption and trade in environmental quality improvement: Evidence from a balanced panel of 16 sub-Saharan African countries. *Renewable Energy, 143*, 1194–1201. https://doi.org/10.1016/j.renene.2019.05.087

Agarwalla, S. K., Varma, J. R., & Venkatesh, P. C. (2021). Crude oil price shocks and Indian energy sector: Evidence from BSE energy index. *Energy Economics, 99*, 105285. https://doi.org/10.1016/j.eneco.2021.105285

Aghabalayev, E., Wang, Y., & Liu, Q. (2022). Green technology cooperation, economic policy tools, and environmental outcomes in G7 nations: A dynamic panel study. *Environmental Economics and Policy Studies, 24*, 711–733. https://doi.org/10.1007/s10018-022-00333-y

Ali, E., Anshari, M., Hamdan, M., Ahmad, N., & Surieshtino, Y. (2023). *Green finance for sustainable development: A bibliographic analysis*. https://doi.org/10.1109/sibf60067.2023.10379915

Ali, Q., Sule, A., & Zhang, J. (2022). Mapping the intellectual landscape of green finance: A bibliometric review of 363 publications (1976–2021). *Journal of Cleaner Production, 336*, 130386. https://doi.org/10.1016/j.jclepro.2022.130386

Alkathery, S., Hammoudeh, S., & Shahbaz, M. (2021). Volatility spillovers and co-movements among global clean energy, oil, and GCC energy stocks. *Energy Economics, 96*, 105155. https://doi.org/10.1016/j.eneco.2021.105155

Alola, A. A. (2019). The trilemma of migration, environment and health: Evidence from the United States using flexible ARDL approach. *Science of the Total Environment, 651*, 849–857. https://doi.org/10.1016/j.scitotenv.2018.09.212

Andric, I., Silva, C. A., & Martins, J. (2020). Green roofs and walls as climate adaptation strategies in extreme heat urban areas. *Energy and Buildings, 211*, 109806. https://doi.org/10.1016/j.enbuild.2020.109806

Azam, M., Rafiq, M., & Shafique, M. (2022). Environmental degradation in SAARC countries: The role of globalization, energy use, and urbanization. *Environmental Science and Pollution Research, 29*(3), 3728–3742. https://doi.org/10.1007/s11356-021-15769-2

Barunik, J., & Krehlik, T. (2014). Measuring the frequency dynamics of financial connectedness and systemic risk. *Journal of Financial Econometrics, 16*(2), 271–296. https://doi.org/10.1093/jjfinec/nbx019

Barunik, J., & Krehlik, T. (2018). Disentangling connectedness of economic variables at different frequencies. *Journal of Financial Econometrics, 16*(2), 271–296. https://doi.org/10.1093/jjfinec/nbx019

Bhattacharya, A., & Arora, S. (2022). Green finance and investment in renewable energy in India: The role of stock markets in promoting a circular economy. *Journal of Sustainable Finance & Investment, 12*(2), 115–130.

Bi, X., Qiao, M., & Wang, Z. (2021). Ownership structure, external finance, and firm value: Evidence from green bond-issuing Chinese firms. *China Finance Review International, 11*(2), 217–234. https://doi.org/10.1108/CFRI-05-2020-0061

Bilal, M., & Shaheen, W. A. (2023). *Towards sustainable development: Investigating the effect of green financial indicators on renewable energy via the mediating variable.* https://doi.org/10.2139/ssrn.4534116

Bildirici, M. E., & Kayıkçı, F. (2019). Oil price and clean energy stock returns: Evidence from multivariate Markov-switching VAR model. *Renewable Energy, 138*, 876–887. https://doi.org/10.1016/j.renene.2019.02.017

Bölük, G., Eyuboglu, S., & Yılmaz, M. (2021). The impact of incentive policies on the deployment of renewable energy: A panel data analysis for EU countries and Turkey (2000–2018). *Renewable Energy, 164*, 1555–1564. https://doi.org/10.1016/j.renene.2020.10.066

Busch, T. (2021). Sustainable finance and ESG integration: Theory and practice. *Corporate Governance: The International Journal of Business in Society, 21*(4), 675–690. https://doi.org/10.1108/CG-01-2021-0014

Castro, C., Marques, R., & Ferreira, A. (2021). Do environmental performance and history matter in the stock market? Evidence from global companies. *Journal of Cleaner Production, 314*, 127989. https://doi.org/10.1016/j.jclepro.2021.127989

Cedic, M., Saucic, B., & Markovic, M. (2021). Market uncertainty and the stock returns of energy markets: A sectoral analysis. *Energy Reports, 7*, 5463–5474. https://doi.org/10.1016/j.egyr.2021.08.145

Chai, S., Chu, W., Zhang, Z., Li, Z., & Abedin, M. Z. (2025). Dynamic nonlinear connectedness between the green bonds, clean energy, and stock price: the impact of the COVID-19 pandemic. *Annals of Operations Research, 345*(2), 1137–1164.

Chala, V., & Demidov, B. (2024). Theoretical aspects of green financing for providing sustainable development. *Economic Scope.* https://doi.org/10.30838/ep.193.37-45

Chen, Y., Li, W., & Zhang, Y. (2019). Foreign direct investment and stock performance in China's renewable energy sector. *Renewable Energy, 139*, 1295–1304. https://doi.org/10.1016/j.renene.2019.03.048

Chong, J., & Miffre, J. (2018). The impact of oil prices on clean energy stock returns: The role of investment horizon and industrial maturity. *Energy Economics, 70*, 183–194. https://doi.org/10.1016/j.eneco.2018.01.021

Climate Bonds Initiative. (2023). *India green finance state of the market 2023.* https://www.climatebonds.net/files/reports/india-sotm-2023.pdf

De Grauwe, P. (2019). Green finance and financial stability: Complement or contradiction? Financial Markets, *Institutions & Instruments, 28*(5), 321–333. https://doi.org/10.1111/fmii.12123

Demiral, M., Aydin, R., & Pata, U. K. (2021). Measuring eco-efficiency and eco-productivity in U.S. states: A DEA approach. *Journal of Cleaner Production, 278*, 123740. https://doi.org/10.1016/j.jclepro.2020.123740

Deng, Q., Li, Z., & Wang, Y. (2021). Capital inflows, renewable energy consumption, and environmental quality in BRICS countries. *Renewable Energy, 171*, 1227–1235. https://doi.org/10.1016/j.renene.2021.02.094

Destek, M. A., & Sarkodie, S. A. (2020). Investigation of environmental Kuznets curve for ecological footprint: The role of energy and financial development. *Science of the Total Environment, 721*, 137612. https://doi.org/10.1016/j.scitotenv.2020.137612

Devi, S. (2023). Green finance: A way towards sustainable India. *Asian Journal of Management and Commerce*. https://doi.org/10.22271/27084515.2023.v4.i2c.214

Diebold, F. X., & Yilmaz, K. (2012). Better to give than to receive: Predictive directional measurement of volatility spillovers. *International Journal of Forecasting, 28*(1), 57–66. https://doi.org/10.1016/j.ijforecast.2011.02.006

Dongyang, Z. (2021). Environmental disclosures and financial resilience: Evidence from Chinese listed firms. *Chinese Journal of Environmental Economics, 13*(3), 87–104.

D'Orazio, P. (2021). Mapping the emergence and diffusion of climate-related financial policies. *Ecological Economics, 184*, 106989. https://doi.org/10.1016/j.ecolecon.2021.106989

Driessen, P. P. J. (2021). Understanding the governance of climate finance: Institutional arrangements and legitimacy. *Environmental Policy and Governance, 31*(1), 6–20. https://doi.org/10.1002/eet.1905

Dutta, A. (2019). Impact of silver price on solar energy stock index: Evidence from the GARCH-MIDAS model. *Resources Policy, 62*, 503–508. https://doi.org/10.1016/j.resourpol.2019.04.012

El-Aziz, M. A., Zhao, Y., & Dong, L. (2021). COVID-19 pandemic and oil market crash: Impact assessment and recovery pathways. *Energy Policy, 156*, 112422. https://doi.org/10.1016/j.enpol.2021.112422

Ellen MacArthur Foundation. (2021). *Financing the circular economy: Capturing the opportunity*. https://www.ellenmacarthurfoundation.org/publications

Falcone, P. M., & Sica, E. (2019). Assessing the opportunities and limits of green finance: A review of Italian biomass investment regulations. *Journal of Cleaner Production, 231*, 743–752. https://doi.org/10.1016/j.jclepro.2019.05.283

Ferrer, R., Shahzad, S. J. H., & López, R. (2021). Are green bonds a distinct asset class? An empirical analysis using wavelet coherence and spillover techniques. *Energy Economics, 94*, 105064. https://doi.org/10.1016/j.eneco.2020.105064

Flammer, C. (2020). Green bonds: Effectiveness and implications for public policy. *Journal of Corporate Finance, 61*, 101411. https://doi.org/10.1016/j.jcorpfin.2018.08.008

Frydrych, J. (2021). An overview of the European green bond market: Currency, regional structure, term, sectors, and leading stock exchanges. *Sustainability, 13*(11), 6237. https://doi.org/10.3390/su13116237

Fuentes, M. A., Pérez, C. R., & Morales, L. (2020). Volatility connectedness between clean energy and financial markets: Evidence from dynamic networks. *Energy Economics, 92*, 104950. https://doi.org/10.1016/j.eneco.2020.104950

Gabaix, X., Koijen, R. S. J., & Veldkamp, L. (2020). The dynamics of the stock market. *Brookings Papers on Economic Activity, 2020*(1), 111–191. https://doi.org/10.2139/ssrn.3663105

Gan, C., Hu, B., & Li, Z. (2019). Estimating China's private car ownership: A panel data approach. *Transport Policy, 81*, 17–28. https://doi.org/10.1016/j.tranpol.2019.05.005

Gao, Y., Zhang, D., & Wang, Y. (2019). Crude oil price shocks and solar energy stock returns: Evidence from panel quantile regression. *Energy Economics, 81*, 859–869. https://doi.org/10.1016/j.eneco.2019.05.014

Garcia-Amate, M., De la Flor, M., & Lopez-Menendez, A. J. (2021). A comparative analysis of renewable energy indexes and sustainability stock indexes. *Renewable Energy, 178*, 118–129. https://doi.org/10.1016/j.renene.2021.06.071

Greiner, A., Grüne, T., & Semmler, W. (2013). The global transition to low-carbon economies: A macroeconomic perspective. *Journal of Economic Dynamics and Control, 37*(10), 2062–2083. https://doi.org/10.1016/j.jedc.2013.05.003

Geissdoerfer, M., Savaget, P., Bocken, N. M. P., & Hultink, E. J. (2020). The circular economy—A new sustainability paradigm? *Journal of Cleaner Production, 143*, 757–768.

Gu, H., Li, W., & Zhou, S. (2021). Environmental governance innovation for sustainable development: A Chinese provincial perspective. *Ecological Indicators, 127*, 107695. https://doi.org/10.1016/j.ecolind.2021.107695

Haas, R., Resch, G., & Panzer, C. (2019). The impact of financial system structures on CO_2 emissions: Cross-sectoral and international perspectives. *Energy Policy, 132*, 133–144. https://doi.org/10.1016/j.enpol.2019.05.025

Hawkins, R. (2000). Economic instruments in environmental policy: Lessons from OECD experience and their relevance to developing economies. *Environment and Development Economics, 5*(2), 131–148. https://doi.org/10.1017/S1355770X00000056

Huang, Y., Wang, Y., & Zhu, H. (2020). Tight oil supply shocks and stock return volatility: Evidence from the S&P energy index. *Energy Economics, 89*, 104793. https://doi.org/10.1016/j.eneco.2020.104793

Jackson, T., Drake, F., & Victor, P. (2021). Macro-economic modelling of sustainable transitions: The stock-flow consistent approach. *Ecological Economics, 179,* 106833. https://doi.org/10.1016/j.ecolecon.2020.106833

Jianpeng, L., Jiajun, W., & Xun, Z. (2021). Coordination between industrial growth, pollution reduction, and green innovation: Evidence from Chinese cities. *Journal of Cleaner Production, 278*, 123941. https://doi.org/10.1016/j.jclepro.2020.123941

Jiao, J., Sun, H., & Chen, X. (2021). Regulatory frameworks and the development of green energy stocks in China. *Energy Policy, 156,* 112428. https://doi.org/10.1016/j.enpol.2021.112428

Jobst, A., Oura, H., & Kühn, M. (2019). Greening the financial system: How the World Bank Group is tackling climate-related financial risks. *World Bank Group*. https://documents.worldbank.org/en/publication/documents-reports/documentdetail

Kalantaripor, N., Taghizadeh-Hesary, F., & Yoshino, N. (2021). Green productivity, energy consumption, and spatial spillover in SCO countries: A spatial econometric analysis. *Journal of Cleaner Production, 319*, 128741. https://doi.org/10.1016/j.jclepro.2021.128741

Khairunnessa, M., Islam, M. A., & Akter, R. (2021). Green banking in Bangladesh: A study on financial regulation and policy effectiveness. *Journal of Sustainable Finance & Investment, 11*(3), 234–251. https://doi.org/10.1080/20430795.2020.1739970

Khattak, M. S., Ahmad, M., Khan, Z., & Iqbal, W. (2021). Asymmetric impact of foreign direct investment and trade openness on environmental sustainability: Evidence from BRICS economies. *Environmental Science and Pollution Research, 28*, 26053–26067. https://doi.org/10.1007/s11356-021-12433-0

Khemnar, J., & Pandey, P. (2024). Green finance: Key driver of renewable energy growth in India. *Shanlax International Journal of Arts, Science and Humanities, 12*(S1-i2-Oct), 13–21. https://doi.org/10.34293/sijash.v12is1-i2-oct.8413

Khule, P., Bhilare, M., & Saxena, J. (2024). Revisiting the challenges and opportunities for emerging green finance in india. *ShodhKosh Journal of Visual and Performing Arts, 5*(6). https://doi.org/10.29121/shodhkosh.v5.i6.2024.3868

Kirchherr, J., Reike, D., & Hekkert, M. (2020). Conceptualizing the circular economy: An analysis of 114 definitions. *Resources, Conservation & Recycling, 127*, 221–232.

Kocaarslan, B., Soytas, U., & Yilmaz, M. K. (2019). Asymmetric relationships among oil prices, interest rates and clean energy stock returns. *Energy Economics, 81*, 661–670. https://doi.org/10.1016/j.eneco.2019.05.010

Koengkan, M., Fuinhas, J. A., & Marques, A. C. (2019). The relationship between globalization, private capital stock and renewable energy in Latin American and Caribbean countries. *Renewable Energy, 143*, 1106–1117. https://doi.org/10.1016/j.renene.2019.05.066

Lahiani, A., Shahbaz, M., & Roubaud, D. (2021). Financial development and renewable energy consumption in the USA: New insights from the NARDL approach. *Energy Economics, 94*, 105060. https://doi.org/10.1016/j.eneco.2020.105060

Lei, Y., Zhang, Y., & Wu, Y. (2021). Renewable energy consumption and economic development: New evidence from a nonlinear ARDL approach. *Energy, 235*, 121365. https://doi.org/10.1016/j.energy.2021.121365

Li, J., & Su, M. (2020). Carbon futures in the EU ETS: Characteristics and pricing behavior. *Journal of Cleaner Production, 263*, 121421. https://doi.org/10.1016/j.jclepro.2020.121421

Li, W., Zhou, X., & Liu, Y. (2020). Carbon trading markets and green infrastructure: A study on emissions reduction mechanisms. *Environmental Economics and Policy Studies, 22*(4), 725–743. https://doi.org/10.1007/s10018-020-00270-4

Li, Y., Xu, X., & Zhao, L. (2021). Constraints on the expansion of green energy stocks in China: Financing costs and investment channels. *Energy Reports, 7,* 781–793. https://doi.org/10.1016/j.egyr.2021.01.078

Lin, B., & Chen, Y. (2019). Carbon pricing and green total factor productivity in China's metallurgical industry. *Energy Policy, 132,* 678–689. https://doi.org/10.1016/j.enpol.2019.05.062

Liu, Y., Wei, Y., & Zhang, X. (2019). Climate change and stock returns of renewable energy firms in the United States. *Energy Economics, 81,* 713–725. https://doi.org/10.1016/j.eneco.2019.04.018

Liu, Z. (2024). Assessing the role of green finance in enhancing rural economic resilience and environmental sustainability. *Frontiers in Management Science.* https://doi.org/10.56397/fms.2024.02.09

Liu, L., Wang, Y., & Wei, Y. (2021). COVID-19 and energy stock markets: An insight into volatility connectedness and hedging strategy. *Energy Economics, 99*, 105254. https://doi.org/10.1016/j.eneco.2021.105254

Ma, X., Liu, J., & Wang, S. (2019). Oil price shocks and stock market returns: Evidence from U.S. energy sector using Markov-switching models. *Resources Policy, 64*, 101531. https://doi.org/10.1016/j.resourpol.2019.101531

McKinsey & Company. (2022). *The future of clean energy: How renewables will transform business and economies.* https://www.mckinsey.com/business-functions/sustainability/our-insights/the-future-of-clean-energy-how-renewables-will-transform-business-and-economies

Maghyereh, A. I., Awartani, B., & Bouri, E. (2021). Energy stocks and oil market shocks: Evidence from time–frequency quantile regression. *Energy Economics, 97*, 105180. https://doi.org/10.1016/j.eneco.2021.105180

Mensi, W., Rehman, M. U., & Kang, S. H. (2021). Frequency spillovers between oil futures and stock markets in MENA countries: New evidence from wavelet coherence and rolling window spillover index. *Energy Economics, 95*, 105140. https://doi.org/10.1016/j.eneco.2021.105140

Ministry of New and Renewable Energy. (2023). *India's progress towards renewable energy and the role of green finance in achieving a circular economy.* Government of India. https://mnre.gov.in

Mohanty, B. C. (2024). Opportunities of renewable energy financing in India. *The Management Accountant*, 26–30. https://doi.org/10.33516/maj.v59i6.26-30p

Mundaca, L., Neij, L., Worrell, E., & McNeil, M. (2016). Evaluating energy efficiency policies with energy-economy models. *Energy Policy, 38*(10), 6014–6025. https://doi.org/10.1016/j.enpol.2010.05.053

Mustafayev, S., Ibrahimov, R., & Ahmadov, A. (2022). Policy reform strategies for achieving renewable energy transition goals: A regional perspective. *Energy Reports, 8*, 2345–2356. https://doi.org/10.1016/j.egyr.2022.02.139

Mukhopadhyay, D.P. (2024). Green finance: A study with respect to modern India. *International Journal of Advanced Research in Science, Communication and Technology*. https://doi.org/10.48175/ijarsct-18677

Murray, A., Skene, K., & Haynes, K. (2021). The circular economy: An interdisciplinary exploration of the concept and application in a global context. *Journal of Business Ethics, 140*(3), 369–380.

Naeem, M. A., Bouri, E., Peng, Z., & Shahzad, S. J. H. (2020). Time and frequency dynamics of connectedness between oil, energy, and carbon markets. *Energy Economics, 91*, 104914. https://doi.org/10.1016/j.eneco.2020.104914

Nassary, M. A. (2019). Fintech for green finance: Blockchain-based innovations in renewable energy and climate markets. *Journal of Sustainable Finance & Investment, 9*(4), 312–327. https://doi.org/10.1080/20430795.2019.1628876

Németh-Durkó, E. (2020). Legal regulation of green finance in the EU: An analysis of current directions and implications. *Lex et Scientia International Journal, 27*(1), 85–94.
Nosheen, S., Aziz, N., & Waheed, R. (2021). Climate change mitigation through green technology: Evidence from European economies. *Technological Forecasting and Social Change, 166*, 120635. https://doi.org/10.1016/j.techfore.2021.120635
Nguyen, C. P., Schinckus, C., & Su, T. D. (2021). Does climate change affect renewable energy firms? Evidence from Europe. *Energy Reports, 7*, 648–658. https://doi.org/10.1016/j.egyr.2021.01.040
Oliveira, M., Ferreira, J. J. M., & Fernandes, C. I. (2023). Green energy transition and circular economy: An integrative model and agenda for future research. *Journal of Cleaner Production, 331*, Article 129949.
Pauliuk, S., Heeren, N., Hasan, M. M., & Wang, T. (2023). Circular economy strategies—Implementing resource efficiency and climate mitigation goals. *Resources, Conservation and Recycling, 172*, Article 105663.
Pedroni, P. (2001). Purchasing power parity tests in cointegrated panels. *Review of Economics and Statistics, 83*(4), 727–731.
Paramati, S. R., Mo, D., & Gupta, R. (2017). The effects of stock market growth and renewable energy use on CO_2 emissions: Evidence from G20 countries. *Environmental Science and Pollution Research, 24*(20), 17354–17362. https://doi.org/10.1007/s11356-017-9356-8
Petrukha, S., Tikhonova, N., & Shokhina, Y. (2020). The role of the banking sector in implementing the European Green Deal: Financial sustainability, foreign exchange operations, and national resilience. *Financial Journal, 12*(6), 10–21.
Purkayastha, D., Tripathy, S., & Patel, R. (2021). Designing effective blended finance models for climate-smart infrastructure: Evidence from Asia. *Development Finance Review, 5*(1), 45–63.
Punzi, M. T. (2019). Oil price shocks and the macroeconomy in small open economies. *Energy Economics, 80*, 204–220. https://doi.org/10.1016/j.eneco.2019.01.010
Pham, L. (2019). Do oil prices affect clean energy stock performance? Evidence from a continuous wavelet analysis. *Energy Economics, 80*, 632–645. https://doi.org/10.1016/j.eneco.2019.01.006
Psy, R. (2020). Eco-productivity and eco-efficiency of U.S. states: A Data Envelopment Analysis (DEA) approach. *Journal of Environmental Management, 261*, 110213. https://doi.org/10.1016/j.jenvman.2020.110213
Quatrini, S. (2020). Decision-support tools for green finance: A macro-prudential framework for resilience and recovery. *Global Policy, 11*(3), 345–355. https://doi.org/10.1111/1758-5899.12763
Reddy, S., & Ghosh, A. (2022). Green finance and its role in promoting circular economy in India. *International Journal of Economics and Financial Issues, 12*(3), 45–53.
Sachs, J. D., Woo, W. T., Yoshino, N., & Taghizadeh-Hesary, F. (2023). *Financing green transitions in Asia: Circular economy and sustainable development*. Springer.
Sahu, A., & Khatri, S. (2024). *Green finance as propeller to sustainable development in India.* https://doi.org/10.53361/dmejm.v5i01.02
Sachs, J. D., Schmidt-Traub, G., & Lafortune, G. (2019). Financing sustainable development: Implementing the SDGs through effective capital allocation. *Sustainable Development Report 2019.* Cambridge University Press.
Schumacher, K., Chenet, H., & Volz, U. (2020). Sustainable finance in Japan: Risks and opportunities for climate-aligned development. *Journal of Sustainable Finance & Investment, 10*(2), 213–238. https://doi.org/10.1080/20430795.2020.1711506
Septina, L. (2020). The impact of declining profitability on stock market performance: A case study of ISAT. *International Journal of Business, Economics and Law, 21*(5), 33–39.
Shafiullah, M., Alam, K., & Clark, C. (2021). Economic policy uncertainty and renewable energy consumption in the USA: Evidence from a nonlinear ARDL approach. *Renewable Energy, 164*, 911–923. https://doi.org/10.1016/j.renene.2020.09.070

Sharma, R., Jain, V., & Yadav, R. (2021). Determinants of carbon intensity in South Asian countries: A panel data analysis. *Environmental Science and Pollution Research, 28*(1), 1120–1135. https://doi.org/10.1007/s11356-020-10567-6

Shobande, O. A., & Alege, P. O. (2019). Evaluating the role of energy policy in mitigating carbon emissions in Nigeria, China, and the United States. *Environmental Science and Pollution Research, 26*(9), 8657–8671. https://doi.org/10.1007/s11356-018-1902-z

Sirohi, S., Kapoor, S., & Narula, S. A. (2020). Impact of climate change regulations on Indian renewable energy firms' financial performance. *Renewable and Sustainable Energy Reviews, 124*, 109783. https://doi.org/10.1016/j.rser.2020.109783

Song, D., Yoon, S. M., & Kang, S. H. (2019). Spillover effect between fossil energy, renewable energy and investor sentiment: A network connectedness approach. *Energy Economics, 81*, 455–470. https://doi.org/10.1016/j.eneco.2019.04.018

Song, M., Zhang, Y., & Wang, S. (2022). Linking innovation, green finance and carbon emissions in China: A threshold analysis. *Technological Forecasting and Social Change, 178*, 121586. https://doi.org/10.1016/j.techfore.2022.121586

Su, B., Heshmati, A., Geng, Y., & Yu, X. (2022). A review of the circular economy in China: Moving from rhetoric to implementation. *Journal of Cleaner Production, 173*, 329–339.

Sule, A. K., Eyo-Udo, N. L., Onukwulu, E. C., Agho, M. O., & Azubuike, C. (2024). *Green finance solutions for banking to combat climate change and promote sustainability*. https://doi.org/10.51594/gjabr.v2i6.54

Sun, X., Zhou, M., & Tang, Y. (2019). The impact of fossil fuel price changes on clean energy stock returns: Evidence from China. *Energy Economics, 84*, 104551. https://doi.org/10.1016/j.eneco.2019.104551

Tereshina, N. P., Smirnova, N. V., & Kireev, V. V. (2017). Green economy and sustainable development: Problems and prospects. *Journal of Environmental Management and Tourism, 8(3),* 546–555. https://doi.org/10.14505/jemt.v8.3(19).09

Torvanger, A., Lund, E., & Eskeland, G. S. (2021). The adoption and implementation of green bonds in the Nordic region: A comparative policy study. *Environmental Innovation and Societal Transitions, 41*, 1–12. https://doi.org/10.1016/j.eist.2021.10.005

Ucal, M., Eser, Ö., & Koç, A. A. (2020). Techno-economic analysis of climate change and energy resources under the green economy paradigm. *Journal of Cleaner Production, 276*, 123305. https://doi.org/10.1016/j.jclepro.2020.123305

Ullah, S., Ozturk, I., & Ali, S. (2021). Determinants of sustainable development in Belt and Road Initiative (BRI) countries: Empirical evidence using panel quantile regression. *Sustainability, 13*(2), 739. https://doi.org/10.3390/su13020739

Ulucak, R. (2020). The role of renewable energy, globalization, and technological innovation on environmental degradation: Evidence from the panel data of emerging economies. *Renewable Energy, 146,* 685–696. https://doi.org/10.1016/j.renene.2019.07.078

Verma, M., Sharma, G. D., & Sharma, S. (2021). Stock market reaction to green bond issuance announcement: Evidence from emerging market economies. *Environmental Science and Pollution Research, 28*, 59355–59370. https://doi.org/10.1007/s11356-021-14707-6

Vukovic, M., Jovanovic, D., & Radjenovic, T. (2019). The local green economy: Concept, evaluation and development planning. *Ekonomika, 65*(1), 1–13. https://doi.org/10.5937/ekonomika1901001V

Wang, Q., Su, M., & Li, R. (2021). The Paris Agreement and its impact on China's renewable energy stock market: Evidence from event study. *Journal of Cleaner Production, 280*, 124375. https://doi.org/10.1016/j.jclepro.2020.124375

Winkler, H., Boyd, A., Torres Gunfaus, M., & Climate Policy Team. (2021). Policies, governance and finance for climate-resilient development. *Climate Policy, 21*(5), 603–617. https://doi.org/10.1080/14693062.2021.1903894

Xia, T. (2022). Green technology innovation, renewable energy investment, and carbon neutrality in Chinese provinces. *Sustainability, 14(2),* 876. https://doi.org/10.3390/su14020876

Yahya, M., Wang, Z., & Naseem, M. A. (2019). Dynamic connectedness between oil prices and agricultural commodities: Time and frequency analysis. *Resources Policy, 62*, 605–617. https://doi.org/10.1016/j.resourpol.2019.04.002

Yaoteng, L., Jie, H., & Ming, Z. (2021). Research on green finance strategy to promote sustainable development under big data environment. *Sustainability, 13*(5), 2703. https://doi.org/10.3390/su13052703

You, D., Sorrell, S., & Jin, L. (2021). The role of international collaboration in low-carbon technology development: Evidence from carbon emissions and innovation linkages. *Energy Policy, 152*, 112163. https://doi.org/10.1016/j.enpol.2021.112163

Younis, M., Ali, S., & Khan, M. A. (2021). The role of stock market and renewable energy in environmental degradation: Evidence from BRICS countries. *Environmental Science and Pollution Research, 28*(2), 1711–1724. https://doi.org/10.1007/s11356-020-10570-x

Zahoor, Z., Wang, Y., & Ahmed, M. (2021). Renewable energy investment, financial development, and environmental sustainability in China: Moderating role of urbanization. *Environmental Science and Pollution Research, 28*, 1768–1784. https://doi.org/10.1007/s11356-020-10523-9

Zhang, D., Mohsin, M., Rasheed, A. K., Chang, Y., & Taghizadeh-Hesary, F. (2021). Public spending and green economic growth in BRI region: Mediating role of green finance. *Energy Policy, 153*, Article 112256.

Zhang, X., Shen, L., Tam, V. W. Y., & Lee, W. W. Y. (2019). Barriers to implement extensive green roof systems: A stakeholder perspective. *Science of the Total Environment, 652*, 1205–1215. https://doi.org/10.1016/j.scitotenv.2018.10.182

Zhang, D., Wang, J., & Liu, H. (2019). The impact of foreign institutional investment on India's renewable energy sector stock returns. *Energy Economics*, 81, 610–620

Zhou, Y., Wu, M., & Yang, W. (2019). Does outward foreign direct investment really enhance green total factor productivity at home? Evidence from China. *Journal of Cleaner Production, 233*, 993–1005. https://doi.org/10.1016/j.jclepro.2019.06.077

Zhao, L. (2020). External oil shocks and clean energy stock markets: Evidence from wavelet analysis and quantile regression. *Energy Reports, 6*, 1234–1245. https://doi.org/10.1016/j.egyr.2020.04.013

Zioo, W., Li, F., & Fang, C. (2019). Green financial products and banking network development: A sustainability pathway. *Finance Research Letters, 30*, 385–392. https://doi.org/10.1016/j.frl.2019.03.006

Zuo, Q., Wang, S., & Li, J. (2020). Resource policies, regional growth, and the resource curse in China. *Resources Policy, 68,* 101744. https://doi.org/10.1016/j.resourpol.2020.101744

Circular Practices in Agriculture Sector: Current Challenges and Opportunities

Lipsa Dehal, Abhay Punia, and Nalini Singh Chauhan

Abstract In the agricultural industry, circular techniques seek to minimize waste, improve resource efficiency, and foster ecosystem health in order to develop sustainable and regenerative systems. Although they have great promise, a number of obstacles prevent their widespread use. A biorefinery concept can increase the potential of agricultural value chains; however, there are several obstacles to overcome, including the lack of advanced technology for biorefining to produce value-added products, investment costs, restricted access to advanced technologies, inadequate market development, regulatory barriers, and gaps in farmer knowledge and skills. Significant challenges are also posed by resource degradation and climate change. Nonetheless, there are a lot of advantages to circular agriculture. These consist of lower environmental impact, increased soil health, cost savings, increased resource efficiency, and new revenue streams. Technological developments, cooperative initiatives, and supportive legislation can all help to speed up the shift. Three tools that cover all the sustainability components are part of the proposed framework for sustainability evaluation: life cycle assessment, social life cycle assessment, and cost-benefit analysis. In order to take into account resource prioritization, ecological capacity constraint, and ecosystem protection, respectively, these techniques are further supplemented by Nature-based Solutions, material flow analysis, and ecological footprint. Unlocking the long-term benefits of circular practices and removing obstacles require strategic investments and knowledge-sharing networks, which will help create a more resilient and sustainable agriculture industry.

Keywords Sustainable agriculture · Bio-based economy · Circular agriculture · Opportunities

L. Dehal
Department of Zoology, Punjabi University, Patiala, Punjab, India

A. Punia
Department of Zoology, DAV University, Jalandhar, Punjab, India

N. S. Chauhan (✉)
Department of Zoology, Kanya Maha Vidyalya, Jalandhar, Punjab, India
e-mail: nalini.lloyd@gmail.com

P. Singh et al. (eds.), *Circular Economy and Sustainable Innovation*,
https://doi.org/10.1007/978-3-032-00437-6_6

1 Introduction

In order to satisfy the demands of an increasingly populous world, the agriculture industry faces enormous obstacles in implementing circular processes. Increasing food production is necessary to feed more people, yet doing so strains agricultural ecosystems severely (Bhat, 2021; Singh et al., 2022). According to Bruinsma (2017), there would be a 5.1 billion-ton increase in food production required by 2050. Because traditional farming methods use a lot of water and energy and hurt the environment, this presents an opportunity for circular practices (Aznar-Sánchez et al., 2018). Agriculture is currently responsible for almost 90% of the environmental impact associated with land use (Kusumastuti et al., 2016), underscoring the need for more sustainable methods. Circular agriculture has two advantages: it reduces waste and increases resource efficiency. Two strategies to lessen this industry's environmental impact are to reuse agricultural byproducts and use less energy and water. In 2019, agriculture contributed significantly to material consumption (21.3 billion tonnes) and carbon emissions (10 billion tonnes of CO_2 equivalent), according to Rodríguez-Espinosa et al. (2023). An effective solution to the issues facing the agriculture sector, such as resource waste, land degradation, rising food demand, and monocropping, is the use of circular approaches. Traditional agricultural methods currently generate two significant environmental harms: greenhouse gas emissions and the degradation of natural resources (Farmery et al., 2022). Both the long-term viability of food production and the depletion of significant ecosystems are at risk if these activities are being carried out (Table 1). While maintaining manufacturing, circular techniques offer a chance to reduce adverse externalities.

A shift toward circular agriculture, which prioritizes resource reuse, minimized waste, and the creation of more sustainable food systems, is addressing these problems (Table 1). In an attempt to reduce environmental effects and close local resource loops, governments in the global North, including the Netherlands, are beginning to implement circular agriculture regulations, which is why this approach is growing in popularity there (Hoogstra et al., 2024). Circular agriculture offers opportunities to solve issues like hunger, deforestation, and biodiversity loss, as well as a more environmentally friendly structure to meet requirements for food in the face of urbanization and climate change (Bianchi et al., 2020).

Circular farming techniques, however, are not entirely new. Historically, many low- and middle-income countries (LMICs) have recycled and reused resources using circular methods due to necessity (Skene, 2022). Despite the fact that these techniques are usually disregarded or little documented, they provide valuable information for enhancing modern agricultural systems. Restoring these circular processes could help address inefficiencies in the present food system and reduce its negative environmental effects, particularly in regions where intensive, resource-heavy agriculture has been prevalent (Kongnso 2024).

From input suppliers to final consumers, there are opportunities and challenges in applying circular practices across the agro-food chain. Even while agricultural and food processing remain the most polluting and resource-intensive phases, this burden

Table 1 Comparison of circular and linear (traditional) agriculture practices

Aspect	Circular agriculture	Linear agriculture	References
Definition	A system designed to minimize waste and maximize resource use by reusing, recycling, and reducing inputs	A traditional system where resources are used in a straight path from production to waste, with little focus on reuse	Skene (2022), Farmery et al. (2022), Preston and Lehne (2017)
Resource use	Emphasizes resource efficiency through closed-loop cycles, using waste as a resource (e.g., composting, manure)	Resources like water, fertilizers, and energy are used intensively, with waste often discarded	Bhat (2021), Aznar-Sánchez et al. (2018), Cucagna and Goldsmith (2018)
Waste management	Focuses on reducing waste through reuse, recycling of by-products, and nutrient cycling (e.g., manure, crop residues)	Generates significant waste, with little reuse or recycling, often leading to environmental pollution	Rodríguez-Espinosa et al. (2023), Aznar-Sánchez et al. (2019)
Environmental impact	Designed to have minimal environmental impact by conserving resources and reducing emissions	High environmental impact, contributing to deforestation, soil degradation, and greenhouse gas emissions	Kusumastuti et al. (2016), Hoogstra et al. (2024)
Energy use	Aims to use renewable energy sources (solar, wind, bioenergy) to minimize carbon footprint	Often relies heavily on non-renewable energy sources like fossil fuels	Muscio and Sisto (2020), Preston and Lehne (2017)
Water usage	Emphasizes efficient water use and recycling to reduce water footprint and address water scarcity	High water consumption with inefficient use, often leading to water scarcity and contamination	Bruinsma (2017), Aznar-Sánchez et al. (2018)
Biodiversity	Promotes biodiversity by integrating crop rotation, polyculture, and organic farming	Monocropping and intensive farming practices reduce biodiversity and soil health	Bianchi et al. (2020), Hoogstra et al. (2024)
Economic sustainability	Long-term focus on reducing costs through resource efficiency and waste reduction, benefiting small-scale farmers	Short-term profit focus, with high input costs for fertilizers, pesticides, and energy	Farmery et al. (2022), Skene (2022)

(continued)

Table 1 (continued)

Aspect	Circular agriculture	Linear agriculture	References
Farmer dependency	Reduces dependence on external inputs (fertilizers, pesticides) by using natural cycles (e.g., compost, manure)	High dependency on external inputs, often leading to increased costs and reduced sustainability	Bhat (2021), Preston and Lehne (2017)
Soil health	Enhances soil fertility through organic matter recycling and sustainable practices	Declining soil fertility due to overuse of chemical fertilizers and monocropping	Cucagna and Goldsmith (2018), Rodríguez-Espinosa et al. (2023)
Example practices	Composting, manure recycling, crop rotation, use of renewable energy, agroecology	Monocropping, heavy pesticide and fertilizer use, over-irrigation, intensive livestock farming	Preston and Lehne (2017)
Economic model	Circular economy model where waste is minimized, and by-products are reintegrated into the production cycle	Linear economy model with a 'take, make, dispose' approach, leading to high resource depletion and waste	Kongnso (2024), Skene (2022)
Sustainability	Designed to be sustainable by minimizing environmental damage and promoting long-term viability	Unsustainable in the long run due to high resource consumption and environmental degradation	Hoogstra et al. (2024), Bianchi et al. (2020)
Adoption and policy	Gaining traction in regions like the EU and the Netherlands, with policy support for sustainable practices	Common in most parts of the world, especially in industrialized agricultural systems, but facing growing criticism	Hoogstra et al. (2024), Farmery et al. (2022)

can be reduced by using circular solutions such increasing recycling and reducing resource waste. This transition to circular agriculture may cause the farming sector to evolve in the future, becoming more robust and sustainable for coming generations.

2 Core Principles of Circular Practices in Agriculture

"Circular agriculture" aims to make the most efficient use of all biomass. Closing the material and resource loop aims to reduce environmental effect and resource consumption (vanBerkum & Dengerink, 2019). Importantly, these goals are interconnected, which means that certain actions may concurrently benefit many sectors. Three fundamental ideas form the foundation of circular agriculture, and they are.

2.1 *Regenerative Agriculture: Maintaining and Improving Natural Resources*

The first principle discusses how important it is to enhance ecosystem services and organic processes while avoiding harmful or non-renewable inputs. To do this, one must stay away from materials, chemicals, and other objects that are hazardous or difficult to recycle or reuse (Akinsemolu, 2020). Agricultural intensification focuses on large-scale monocultures of high-yield varieties with the primary objective of increasing crop yields. This approach often uses intense soil management and large amounts of chemicals to control weeds, pests, and diseases. Even while farming practices differ across and within countries and depend on the type of crops grown—from massive industrial farms to small organic ones—agricultural intensification has become the most common farming model in recent decades (Tittonell et al., 2016). Stronger agroecosystems that are inherently resistant to weeds, pests, diseases, and harsh weather conditions depend on maintaining the health of the soil. This approach is more consistent with the circular agriculture paradigm since it uses fewer external inputs (Zimmermann et al., 2021). Strategies like introducing beneficial species and using less disruptive methods for soil management can be used to increase resource utilization and close nutrient loops within the system. Regenerative circular agriculture's primary objectives are resource reuse, waste reduction, and ecosystem restoration. It draws attention to the constant circulation of biomass (Fig. 1), which transforms waste from a particular procedure into an input for another and supports the sustainability of agricultural systems.

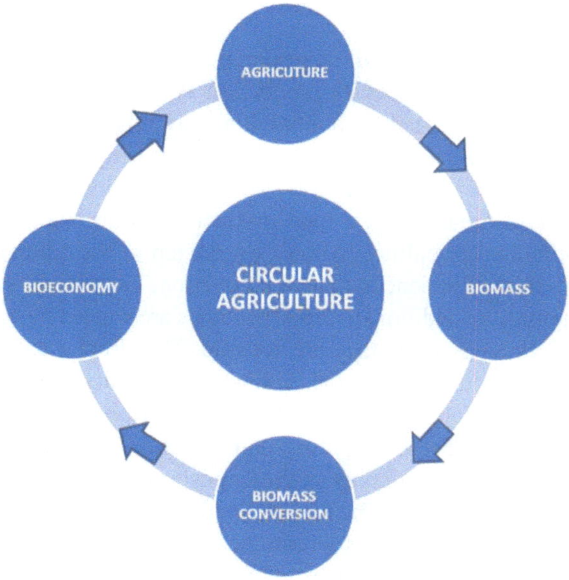

Fig. 1 Regenerative circular agriculture

2.2 Efficient Use of Resources: Closing Nutrient Loops

The second principle highlights the importance of "circular metabolism" in agriculture. Resource efficiency can be improved by optimizing feedback loops and reusing resources within production systems (Redlingshofer, 2022). Natural ecosystems are known for their efficient cycling of energy, water, and nutrients. For example, diverse species with different root systems or beneficial interactions, such as plants forming symbiosis with mycorrhizal fungi to improve nutrient uptake, play a key role in this process. By incorporating species that efficiently capture resources or benefit from interspecies interactions, agricultural systems can mimic nature's efficiency in circular agriculture (Domeignoz-Horta et al., 2024). Resource use efficiency can be enhanced on three levels. The initial and most fundamental stage of production development is achieved through the use of conventional agricultural production methods, whose efficiency in using resources is maximized through careful management and breeding.

In the second stage, species with comparatively poor resource use efficiency are replaced by those with better resource use efficiencies. This doesn't always necessitate a total overhaul of the agricultural production system; it just entails changing out a specific part of it, like switching out one crop type for another (Gosai et al., 2024).

2.3 Multi-purpose Use and Recovering Value: From Waste to Value

The third level at which resource use efficiency can be achieved involves designing circular, multi-species production systems. In these systems, waste streams from one species, such as animal dung, are repurposed as inputs for another, like using manure as a high-quality fertilizer, or by utilizing crop residues as animal feed (Chary et al., 2024). Reducing food loss by utilizing waste streams and converting them into useful inputs for the food production cycle is also involved in the third tenet of circular agriculture. Almost every step of the value chain in the mainstream food system experiences losses: cultivation, harvesting, storage, processing, and transportation; additionally, food products are wasted by consumers and are thrown away by retailers when they are not sold. The circular economy, also known as the "economic counterpart of the ecological circularity concept, is similar to circular agriculture (Bianchi et al., 2020). It is characterized by three economic activities known as the "3Rs": reuse, recycling, and reduction of existing (used) materials and products. The circular economy is the opposite of the take-produce-consume-discard linear economic model. The maxim "reuse and recycle" refers to converting waste streams into useful resources within the context of circular agriculture. Sorting waste into water bodies that can be enhanced and those that cannot, establishing processing facilities that can produce upgraded goods, creating a market for these products, and arranging trade logistics are some of the steps that may be required to establish waste valorization chains (Zheng & Zhou, 2023).

3 Opportunities in Circular Agriculture

Circular agriculture offers a number of chances to create farming systems that are more robust, sustainable, and lucrative. Adopting circular concepts can benefit communities, businesses, and farmers in a variety of ways, such as seeing trash as a precious resource, reducing waste from efficient input usage, and increasing resource efficiency. It also enhances soil health and productivity through nutrient recycling and carbon sequestration (Asadu et al., 2024). Furthermore, value-added products like organic fertilizers, biofuels, and biochemicals that can be sold both domestically and internationally are promoted by circular agriculture (Toplicean & Datcu, 2024), which reduced pollution and greenhouse gas emissions, preserved natural resources, and contributed to the creation of more sustainable farming systems and biodiversity by using polyculture, agroforestry, and varied crop rotations. Farmers can also earn extra money by selling products like compost, bioenergy, and bioplastics that are made from agricultural waste (Kumar et al., 2022). Furthermore, this diversity strengthens the financial resiliency of farms. Water-saving techniques including rainwater collection, effective irrigation systems, and water recycling are encouraged by circular agriculture since they lessen the demand on freshwater resources (Sabir et al., 2024). It strengthens local economies by creating jobs in composting facilities, biorefineries, and markets for organic and sustainable products. Governments and organizations may offer financial assistance, subsidies, or exemptions from taxes as incentives to promote sustainable practices in recognition of the benefits of circular agriculture. The Sustainable Development Goals (SDGs) of the UN are in line with circular agriculture, especially those that deal with land-based living, climate action, and responsible consumption and production (Ortiz-de-Montellano et al., 2023). Circular agriculture has the potential to enhance sustainability, protect the environment, increase farm revenue, and promote global food security in a number of ways. Here's an example of how the agriculture sector could adopt the three principles of the circular economy.

3.1 Enhancing Efficiency with Precision Technology in Circular Agriculture

When paired with regenerative methods, precision farming has demonstrated good potential for improving the efficiency of traditional agricultural systems. Crop yields can be optimized to increase farmer income while lowering environmental externalities through the use of agricultural IT, remote sensing, and real-time environmental data (Sishodia et al., 2020). By applying geospatial variability in soils, microclimate, and other pertinent husbandry aspects to supply the correct amount, at the right time, and at the right place, precision agriculture improves control over inputs including energy, water, fertilizers, and agrochemicals. Precision farming conserves the land's potential for various uses and encourages appropriate stewardship of the land for future generations (Grover et al., 2024). One such instance is Fruit look, a program run by the Western Cape Department of Agriculture (WCDoA) that provides cutting edge information technology to assist growers of grapes and deciduous fruit in being climate-smart and water-efficient. With the use of geographical data from remote sensing, Fruitlook enables these farmers to increase the efficiency with which they use water. The data gives semireal-time information on crop growth, evapotranspiration deficiencies, and nitrogen status for irrigation blocks. Currently, only vineyards and orchards in the Western Cape's primary growing regions have access to this technology (Mandondo & Troskie, 2021).

An important factor in farming systems that are effective is agricultural mechanization. The sharing economy can be used to develop business models for mechanization hire services that provide access to cutting-edge machinery and boost production, especially for small-scale farmers (Diao et al., 2023). Through the usage of product-as-service, the agriculture industry is witnessing a global transition from an owner-focused value proposition to a customer-centered one. Service providers for farmer groups are now included in this. By pooling their resources, a group of farmers can enhance access to services related to agricultural mechanization in this paradigm. Services include every step of the agricultural value chain, including tillage, postharvest tasks, transportation, and processing. The advantages include cheaper rates for the farmer; optimized use of heavy machinery (such as tractors); expertly driven or operated machinery; timely equipment access; accessibility for vulnerable groups, such as women; and flexible payment options. More circularity in South Africa's food systems may be made possible by emerging technological advancements, for instance through improved citizen nutrition, increased farm output, and more efficient supply chains (El Bilali & Allahyari, 2018).

Examples include:

- Chemical leasing, where the focus of the payment is on the advantages of the chemical rather than the chemical itself. The emphasis switches from growing the volume of chemicals sold to a value-added strategy (Seyoum, 2024).
- Digital or mobile platforms that improve the relationship between customers and producers. Digitally-driven business models that facilitate on-demand services or equipment sharing (Rong & Luo, 2023).
- Agroprocessing and packaging innovations to lower food losses and postharvest losses (Urugo et al., 2024).

3.2 *Implementing Circular Practices to Preserve Agricultural Resources and Products*

From farm to table, there are significant food losses and wastage in supply networks. This is mostly because of inadequate cold chains, unsafe transportation, inappropriate handling and packaging, inadequate standardization and grading, inadequate market infrastructure, and incorrect processing. In accordance with the Sustainable Development Goals, South Africa has pledged to cut food losses throughout the production and supply chains, including post-harvest losses, and to halve per capita food waste at the retail and consumer levels by 2030. This has been implemented by the Consumer Goods Council of South Africa (CGCSA) through a voluntary agreement on food loss and waste (Hobson, 2023). If food waste and losses are unavoidable, they can be economically recycled back into the system via a number of technological approaches, such as anaerobic digestion to create bioenergy or composting to create organic manure. The average consumption of chemical fertilizers in sub-Saharan Africa is less than three times that in South Africa (Dimkpa et al., 2023). There is a chance to utilize more organic fertilizers in order to stop nutrient loops and decrease the need for mineral fertilizers. Adding organic fertilizers to agricultural lands will also aid in replenishing the topsoil and soil holding ability. Agriculture has been using animal manure from livestock farms for generations as a source of fertilizers and soil conditioners. Packed with macronutrients including potassium, phosphorus, and nitrogen, they enhance the soil's appearance and enrichment. By using biorefinery technology, organic waste streams can also be utilized to recover high-value resources, such as different platform chemicals (Chatterjee & Mohan, 2022). The CSIR-hosted Biorefinery Industry Development Facility (BIDF) is collaborating with multiple agriculture and agro-processing companies to identify suitable, high-value alternatives for diverse organic waste streams that are conventionally intended for land disposal. These pathways have the potential to process waste biomass across the entire value chain when used effectively. By eliminating the need to dispose of waste by burning, landfilling, stockpiling, or discharge into the sea, this reduces the amount of greenhouse gas emissions that are produced. Food security, water shortages, declining arable land, growing urbanization, and high population growth are already problems facing South Africa. Adoption of sustainable farming techniques, like aquaculture, vertical agriculture, hydroponics, aquafeed made from agricultural waste, and aquaponics, in urban and peri-urban areas could have the following advantages:

- The year-round production of fresh, wholesome, nutrient-dense, and pest-free food, independent of weather conditions.
- Significant water consumption reduction: vertical farming recycles and consumes 96% less water than other agricultural practices.
- The application of green energy.
- Production and land may be used efficiently in industrial, urban, and rural areas when they are close to markets and customers. This lowers transportation costs and, as a result, carbon emissions.

- Higher yields per cultivated acre and no loss of harvest because to unfavourable weather conditions, drought, or pests.

Increased produce shelf life and lower input costs (Al-Kodmany, 2020).Essential components of circular agriculture are the utilization of treated wastewater, rainwater collection, and return-flow systems, which send excess irrigation and drainage backward. Additionally, industrial symbiosis offers ways to maintain resources in use (Goyal & Sivanappan, 2017). By-products and waste materials from one company can be used as input into another through the purposeful clustering of mutually beneficial resource-converting industries. The creation of integrated agro-industrial parks (IAIP) across Africa is being assisted by UNIDO. The three main goals of IAIPs (Integraed Agro-Industrial Parks) are to lower rural poverty, promote structural economic transformation, and improve the climate for greater investment in the agro-food and related industries (Jenane et al., 2022). A strict foundation of specialized services, based on the idea of the agri-business value chain, is implemented to target the weaker links in the chain. These services include agricultural mechanization, contemporary processing technologies, perishable product packaging, the promotion of food safety in the processing and regulatory environment, and interventions to increase productivity and competitiveness. The IAIP is a geographical cluster of autonomous businesses brought together to take advantage of economies of scale and enhance the value of the agriculture sector's output, boosting productivity and competitiveness. This involves offering shared services for trash beneficiation and reuse.

3.3 Rejuvenate Ecosystems

In South Africa, regenerative agriculture is nothing new, nor is indigenous knowledge-based agriculture. The goal of regenerative agriculture is to protect the natural system's integrity. Crop rotation, intercropping, cover crops, minimal tillage, agroforestry (planting trees alongside crops or pastures), ecological aquaculture, food waste application in gardens, mixed farming (combining livestock and crop production to create additional nutrient loops), and minimum tillage are examples of traditional practices that can support circular economy goals towards an optimized agricultural system. This strategy can facilitate numerous long-term advantages, such as improved nutrition, soil protection, water conservation, carbon capture, genetic biodiversity preservation, reduced external inputs, and overall ecosystem improvement (Mishra et al., 2024). This stands in sharp contrast to linear agriculture, which prioritizes short-term production increases and productivity at the expense of the environment and long-term farm health by using synthetic inputs and intensive farming techniques. Adopting crop biodiversity, integrating traditional agro-ecological knowledge, and leveraging appropriate technology and scientific insights on ecosystems, soil health, nutrition, and resilience while adapting to the specific context and geography are key elements of successful regenerative agriculture practices (Seymour &

Connelly, 2023).It is evident that the South African agriculture industry is not unfamiliar with the tenets of a circular economy. To enhance food security and open up new economic prospects, the industry must shift from a linear to a circular one. But for the industry to produce technologically advanced, sustainable, and innovative processes, goods, and services, new knowledge must be developed and applied (Ejaz, 2024). Numerous global trends, many of which directly affect the agriculture industry, have the potential to accelerate and scale the circular economy. Astute farming, biorefinery and biofuels; bioinformatics; robotics; aquaculture; autonomous micro-robots; vertical agriculture; bioinformatics; genetics; information technology; synthetic biology; 3D and 4D printing; protein transition; food design; robotics; aquaculture; autonomous micro-robots; bioinformatics; conservation technology; renewable energy; and transportation technology (De Wilde, 2016).

4 Biorefineries and Nature-Based Solutions: Driving Sustainability in Circular Agriculture

Similar to how a conventional oil refinery processes crude oil into various products like petrol, diesel, and petrochemicals, a biorefinery is a facility that processes biomass into a number of products (Mailaram et al., 2021). High-value chemicals, energy, and fuels are produced by a biorefinery. The idea is comparable to that of an oil refinery, which generates a variety of petroleum products and fuels. But in order to create biofuels (such as ethanol, biodiesel, and biogas) (Azeez & Al-Zuhairi, 2020), biochemicals (such as bioplastics, bio-based solvents, and other chemicals typically derived from petroleum) (Kumar & Verma, 2021), and bioproducts (such as food additives, fertilizers, animal feed, and other materials) (Amin & Bari, 2023), biorefineries concentrate on renewable resources, mainly agricultural and forestry residues. A sugarcane farm could serve as an illustration of how biorefinery integration fits into circular agriculture:

- **Bioethanol Production**: Bioethanol is a fuel that is produced through fermentation of sugarcane juice (Gupta et al., 2024).
- **Utilization of Bagasse**: The remaining bagasse, or fibrous waste, is either used to generate electricity through cogeneration or is converted into biochar and returned to the fields as a soil amendment (Awasthi et al., 2024).
- **Recycling of Vinasse**: Vinasse, a byproduct of the production of ethanol, is prepared and spread as a liquid fertilizer that is high in organic matter and nutrients (Silva & Kardos, 2024).

5 Nature-Based Solutions for Sustainable Agriculture

"Nature-based" solutions (NBS) are those that make use of ecosystems and natural processes to address a variety of environmental, social, and economic issues (Davies et al., 2021). In order to improve production, sustainability, and resilience, NBS includes the preservation and restoration of natural ecosystems as well as the development of natural processes in artificial or modified ecosystems. "Nature-based solutions" are agricultural practices that bolster soil health, increase biodiversity, and lessen dependency on artificial inputs. Crucial NbS farming recommendations include:

- **Enhancing Ecological Resources**: Encouraging methods that aid with water regulation, natural insect control, and pollination.
- **Restoring Degraded Land**: Using reforestation, agroforestry, or wetland restoration to rehabilitate degraded agricultural land.
- **Promoting Biodiversity**: Encouraging a diversity of crops and integrating native species to enhance ecological balance and resilience.
- **Climate Resilience**: Implementing practices that help farms adapt to and mitigate climate change, such as through carbon sequestration in soils.

One example of a nature-based approach to circular farming is agroforestry, which incorporates trees and shrubs into agricultural systems to improve soil fertility, offer wildlife habitat, and yield additional crops including fruits and nuts (Ballinas et al., 2022). Rainwater harvesting, polyculture Systems (growing multiple crops together to improve biodiversity, improve pest control naturally, and maintain soil health), and composting. Reusing and harvesting rainwater for gardening reduces the need for external water sources and promotes water conservation. Additionally, biogas production from animal waste generates energy, while the remaining byproduct serves as a nutrient-rich fertilizer).

6 Challenges to Circular agriculture

Circular agriculture confronts numerous fundamental issues, including the complexity of successfully managing resources like water, energy, and nutrients in a closed-loop system. Adoption may be hampered by high upfront expenditures and the requirement for cutting-edge equipment, especially for small-scale farmers. Barriers to changing conventional farming practices include a lack of understanding and resistance, and the shift is made more difficult by insufficient policies and problems with market integration. Furthermore, it is crucial but difficult to scale up breakthroughs like trash recycling and renewable energy, necessitating concerted efforts by industries, governments, and farming communities. Here are some of the key challenges in circular agriculture (Table 2):

Table 2 Challenges to circular agriculture

Challenge	Description	References
Environmental impact of residue management	Focuses on the environmental aspects such as soil degradation, water and air pollution, and resource overuse	Ballinas et al. (2022), Lampridi et al. (2022)
Upgrading biogas and biofertilizer technology	Involves the technological challenges related to biogas production, biofertilizer, and renewable energy advancements	Ardolino and Arena (2019), Saini et al. (2023)
Innovation in biomaterial production	Highlights the innovation in creating biomaterials from agricultural residues and the need for technological advancements	Tardy et al. (2021), Banu et al. (2021), Harmsen et al. (2014)
Promoting agricultural residue trade	Emphasizes the importance of promoting agricultural residue trade to enhance resource efficiency	Kowalski et al. (2023), Krom (2020)
Stakeholder collaboration	Addresses the need for collaboration between stakeholders in managing agricultural residue and exchanging information	Iqbal et al. (2022)

Challenge I: Analyzing and anticipating the environmental effects of agricultural residue management strategies.

Significant environmental costs are linked to agricultural practices, which impact all environmental components. These costs include deteriorating the quality of the soil, air, and water; using scarce resources (such as land, water, energy, etc.) directly and indirectly; and, most obviously, producing numerous, diverse waste streams that are not effectively used (Ballinas et al., 2022).

The agro-food industry and modern agriculture are both putting a lot of strain on soils because of production intensification, which causes soil erosion and quality degradation, imbalanced nutrient cycles, uneven distribution of organic matter, and pollution from the overuse of chemical pesticides, fertilizers, and fertilizers, among other things (Lampridi et al., 2022).

Challenge II: Upgrading technologies is necessary to convert agricultural leftovers into biogas and biofertilizer.

The most developed and widely used agricultural residue conversion process is anaerobic digestion (AD). The European Biogas Association estimates that there are currently 17,500 AD plants operating in Europe, the majority of which are farm-based, with a total installed capacity of 9.98 GWe (Ardolino & Arena, 2019). Over 65 TWh of power is projected to be produced from biogas annually. Regarding the production of biomethane, around 500 plants with a capacity of 17,264 GWh are now operating in Europe (Cucchiella et al., 2019). Despite its strength, this method still has a number of drawbacks, including low conversion yields of lignin-rich organic material with little economic value and problems with feedstock availability, digestate processing, and storage (Saini et al., 2023).

Challenge III: Creating novel building blocks, compounds, and materials from agricultural waste: Overcoming major obstacles.

The potential for creating bioproducts from agricultural residues is well known, but from a technological perspective, biomass conversion processes require advancements and a deeper understanding of their possibilities in order to be sustainable in terms of the environment, the economy, and society. Despite serious mounting worries about environmental implications (e.g., global warming, the seventh plastic continent), the petrochemical industry continues to offer the most widely used chemicals, versatile plastic materials, and energy in the global economy by utilizing established technology. Chemically and biotechnologically speaking, almost all polymers and their constituent compounds can be produced from renewable basic materials (Tardy et al., 2021). Nevertheless, not every procedure is efficient and commercially viable. Moreover, goods frequently exhibit inadequate purity or are overpriced (Harmsen et al., 2014). Nowadays, platform chemicals serve as the primary feedstock for the production of secondary chemicals, chemical intermediates, and final products, whereas only small percentages of chemicals are bio-based (Banu et al., 2021). The literature has reported on the biological synthesis of platform chemicals from biomass waste, agro-resources, and food processing leftovers (Ezealigo, 2022; Mahato et al., 2023). However, only lab-scale validation has been provided for valorization approaches. It is necessary to conduct more optimization and technique integration studies. Thus, in order to maximize the potential for the valorization of waste, byproducts, and co-products and to promote both environmental and economic benefits for the farming sector (e.g., the development of new products and processes), it is imperative to develop breakthrough technologies in agricultural residue conversion and bio-refinery (Ufitikirezi et al., 2024).

Challenge IV: Encouraging the sale of agricultural residues in the framework of the circular bioeconomy.

Some of the first people to propose the idea of industrial ecology in eco-industrial parks were Frosch and Gallopoulos (1989). Under such a system, "industry's environmental impact can be reduced by using wastes from one industrial process as the raw material for another" An eco-industrial park is a collection of manufacturing and service companies aiming to improve their economic and environmental performance by working together to manage resource-related challenges, such as energy, water, and materials (Kowalski et al., 2023). The goal of the business community's collaboration is to achieve a collective benefit that exceeds the total of the gains that each company would have realized if it had maximized its own interests. At the moment, the concept of an eco-industrial park refers to business-to-business cooperation with the goal of maximizing resource efficiency a practice known as industrial symbiosis (Krom, 2020).

Challenge V: Bringing interested parties together and exchanging information regarding agricultural residue management.

Many European Unioncountries' stalled agricultural development has resulted in environmental and socioeconomic problems that are exacerbated at particular supply chain/regional intersections. As was previously said (Challenge I), while soil nutrient depletion is a common occurrence in arable areas devoid of livestock, livestock rearing zones experience problems with surplus nutrients, odorous compounds, methane emissions, and disease transmission. However, a huge amount of food is imported into metropolitan areas, and the organic nutrient load from wastewater and sewage sludge has a significant detrimental ecological impact (Iqbal et al., 2022). While certain agricultural residue conversion processes have been put into place, most of them are now not adequately satisfying local environmental and economic aspirations, which are impeding their development.

7 Policies and Recommendations

Circular agriculture is emerging as a key approach to achieving sustainability in global food systems, addressing challenges such as resource depletion, environmental degradation, and climate change. Governments around the world are passing legislation to transform traditional agriculture into more productive, environmentally friendly systems that improve biodiversity, reduce waste, and complete resource loops. One of the main goals of these programs is to change the linear "take-make-dispose" model to a circular one, in which resources are recycled, regenerated, and reused.

The European Union has developed initiatives such as the Farm to Fork Strategy and the Green Deal to reduce food waste, enhance nutrient recycling, and promote the use of organic and sustainable farming practices (Wrzaszcz & Prandecki, 2020). With an emphasis on greater biodiversity, less environmental effect, and less chemical inputs, these policies aim to create a climate-neutral food system by 2050. Similarly, China's Circular Agriculture Development Plan and Zero Growth of Fertilizer Use policies promote organic alternatives and agricultural waste recycling as ways to reduce the use of synthetic fertilizers (Chen et al., 2023). These rules reduce environmental contamination and improve soil health. Countries in other regions are adopting similar goals. India's National Mission on Sustainable Agriculture promotes organic farming and the recycling of agricultural residues with the primary goal of enhancing climate change resistance (Bhargavi, 2024). Japan's Biomass Japan Strategy highlights the need of recycling trash, particularly in the context of converting agricultural waste into biomass energy (Swastika et al., 2024). The United States promotes sustainable farming through programs such as Climate-Smart Agriculture and SARE (Sustainable Agriculture Research and Education), which incentivize practices that enhance soil health, boost water use efficiency, and reduce greenhouse gas emissions (Shedekar, 2018). While Brazil's ABC (Low-Carbon Agriculture) Plan promotes the incorporation of organic waste into farming systems to create sustainable, low-emission agricultural practices, the Netherlands' Circular Agriculture Vision focuses on precision farming and nutrient reuse to reduce the environmental impact of food production (Vogel, 2022).

To improve circular agriculture, several significant recommendations could be made. The most crucial is encouraging waste-to-resource initiatives, such as composting agricultural waste, producing bioenergy from organic matter, and using crop wastes and manure as natural fertilizers. By enhancing soil quality and reducing the need for artificial inputs, these methods reduce waste output. Precision farming methods, which use advanced technology like sensors, GPS, and data analytics, can also greatly reduce waste and boost efficiency by optimizing the utilization of resources like water, electricity, and fertilizers (Ogbu & Okey, 2023).

Crop rotation and other regenerative farming practices, like agroforestry, are essential for improving soil health and biodiversity. While agroforestry, which includes trees in farming systems, can improve soil fertility, store carbon, and provide animal habitats, crop rotation prevents soil degradation and promotes natural pest control (Sekhar et al., 2024). These methods aid in the development of more adaptable and robust agricultural systems that can cope with shifting climatic circumstances. To encourage innovation in circular agriculture, governments are also recommended to fund research, educate farmers, and offer monetary incentives like tax breaks or subsidies (Gemtou et al., 2024). This could make it easier for farmers to implement innovative techniques and equipment that promote sustainability. Integrating renewable energy sources, such as solar and biogas, into farming systems can help close resource loops, reduce reliance on fossil fuels, and reduce greenhouse gas emissions, all of which contribute to a more sustainable agricultural sector.

As a sustainable way to meet the rising needs of food production while reducing its negative environmental effects, circular agriculture is becoming more and more popular. Through waste recycling, input reduction, and the growth of regenerative activities, it places a strong emphasis on resource efficiency. The following case studies (Table 3) provide effective implementations in a range of industries and domains to enhance comprehension of the use and results of circular agriculture in various situations. These illustrations show how focused regulation and creative problem-solving may contribute to the development of a more robust and sustainable farm sector.

8 Conclusion

In conclusion, circular agriculture provides a ground-breaking path toward developing sustainable food systems by promoting resource efficiency, reducing waste, and restoring ecological balance. This idea has several benefits, including increased biodiversity, improved soil health, less reliance on chemical fertilizers, and a smaller carbon impact. To create more robust and sustainable systems that increase long-term profitability while lessening their impact on the environment, close resource loops

Table 3 Case studies highlighting circular agriculture policies and implementation

	Location	Focus	Details	Key outcome	Circular practice	References
AgroCycle: valorizing agricultural waste	European Union	Recycling agricultural waste into bio-products	Converts agricultural waste (crop residues, livestock waste) into biofuels, bioplastics, and fertilizers	Reduced environmental impact, waste management	Closing nutrient loops, energy recovery from waste	Varvaringos et al. (2023)
TerraCycle's regenerative agriculture	Brazil	Closed-loop agriculture using composting	Uses organic waste to create compost for crop growth, reducing chemical fertilizer usage	Improved soil fertility, reduced chemical inputs	Composting, nutrient recycling	Ócsai and Ócsai (2021)
Kringlooplandbouw (circular agriculture)	Netherlands	Integration of livestock and crop farming	Utilizes livestock manure as fertilizer for crops and crop residues as feed for livestock	Higher resource efficiency, lower emissions	Resource-efficient nutrient cycling	Hoogstraet al. (2024)
Rice straw management	Punjab, India	Circular use of rice straw waste	Converts rice straw into bioenergy, compost, and animal feed, reducing burning and air pollution	Air pollution reduction, enhanced soil quality	Waste-to-energy, organic soil amendments	Mahalet al. (2019)
Closed-loop aquaponics system	USA	Integration of fish farming and plant cultivation	Combines fish farming (aquaculture) and plant growing (hydroponics) in a closed-loop system	Efficient water use, nutrient recycling	Closed-loop nutrient and water cycles	Goddek and Korner (2019)
Zero-waste farm	New Zealand	Whole-farm circular system	A farm uses animal manure, crop residues, and green waste to create biogas and fertilizers for crops	Reduced fossil fuel use, improved energy efficiency	Waste-to-energy, nutrient cycling	Diprose et al. (2023)

(continued)

Table 3 (continued)

	Location	Focus	Details	Key outcome	Circular practice	References
Sustainable viticulture (wine production)	Spain	Organic viticulture and waste recycling	Organic wine production using recycled organic waste for compost and energy recovery from grape residues	Improved soil health, reduced waste	Organic farming, waste-to-compost and energy	Falco et al. (2023)
Circular dairy farm	Denmark	Integration of dairy farming and crop production	Uses manure from dairy cattle to fertilize crops, and crop waste is used as feed for cattle	Reduced input costs, improved resource efficiency	Nutrient recycling, closed-loop livestock and crops	Khoshnevisan et al. (2023)

and repurpose agricultural waste. Circular methods, such as regenerative agriculture, nutrient recycling, and converting waste into valuable inputs, provide practical solutions to pressing global problems including resource scarcity, global warming, and food poverty. However, the transition to circular agriculture is hampered by a number of factors. Due to their high upfront costs and ignorance, many small-scale farmers still have no access to technical advancements, despite their significance. Making the switch from conventional to circular systems also requires significant investments in infrastructure, training, and technology—all of which could be unaffordable without outside help. The absence of robust regulatory frameworks and economic incentives that may otherwise speed adoption exacerbates this. Cultural opposition makes the widespread implementation of circular agricultural practices much more difficult, especially in areas where traditional farming methods are deeply ingrained. To get over these challenges, the farming community, researchers, and policymakers will need to collaborate. Only by making investments in education, infrastructure, and enabling legislation can this transition be feasible. Governments may play a vital role in encouraging farmers to adopt circular farming practices by providing training programs, subsidies, and incentives. If given the right support, circular agriculture has the potential to completely transform food production by improving its sustainability, profitability, and advantages for both the environment and the general public. By addressing these problems, circular agriculture offers a useful and necessary way to increase global food security and decrease the adverse environmental effects of current farming methods.

References

Akinsemolu, A. A. (2020). *The principles of green and sustainability science*. Springer.

Al-Kodmany, K. (2020). The vertical farm: Exploring applications for peri-urban areas. In *Smart village technology: Concepts and developments* (pp. 203–232).

Amin, M. S., & Bari, A. B. M. (2023). *Production and combustion analysis of microalgae-based biodiesel* (Doctoral dissertation, Department of Mechanical and Production Engineering (MPE), Islamic University of Technology (IUT), Board Bazar, Gazipur-1704, Bangladesh).

Ardolino, F., & Arena, U. (2019). Biowaste-to-biomethane: An LCA study on biogas and syngas roads. *Waste Management, 87*, 441–453.

Asadu, C. O., Ezema, C. A., Ekwueme, B. N., Onu, C. E., Onoh, I. M., Adejoh, T., Ezeorba, T. P. C., Ogbonna, C. C., Otuh, P. I., Okoye, J. O., & Emmanuel, U. O. (2024). Enhanced efficiency fertilizers: Overview of production methods, materials used, nutrients release mechanisms, benefits and considerations. *Environmental Pollution and Management*.

Ashokkumar, V., Flora, G., Venkatkarthick, R., SenthilKannan, K., Kuppam, C., Stephy, G. M., Kamyab, H., Chen, W. H., Thomas, J., & Ngamcharussrivichai, C. (2022). Advanced technologies on the sustainable approaches for conversion of organic waste to valuable bioproducts: Emerging circular bioeconomy perspective. *Fuel, 324*, Article 124313.

Awasthi, S., Mishra, A., & Pal, D. B. (2024). Energy production from sugarcane bagasse and rice husk. In *Sustainable clean energy production using waste biomass: Sustainable energy production and utilization* (pp. 157–181). Springer Nature Singapore.

Azeez, R. A., & Al-Zuhairi, F. K. (2020). Biofuels (bioethanol, biodiesel, and biogas) from lignocellulosic biomass: A review. *Journal of University of Babylon for Engineering Sciences*, 202–215.

Aznar-Sánchez, J. A., Piquer-Rodríguez, M., Velasco-Muñoz, J. F., & Manzano-Agugliaro, F. (2019). Worldwide research trends on sustainable land use in agriculture. *Land Use Policy, 87*, Article 104069.

Aznar-Sánchez, J. A., Velasco-Muñoz, J. F., García-Arca, D., & López-Felices, B. (2020). Identification of opportunities for applying the circular economy to intensive agriculture in Almería (South-East Spain). *Agronomy, 10*(10), 1499.

Ballinas, M. B. P., Arumugam, P., Verstand, D., Siegmund-Schultze, M. G. R., Keesstra, S. D., Chavez, L. G., de Boer, R., van Eldik, Z. C. S., & Voskamp, I. M. (2022). Discussion paper: Nature based solutions in food systems: Review of nature-based solutions towards more sustainable agriculture and food production.

Banu, J. R., Kavitha, S., Tyagi, V. K., Gunasekaran, M., Karthikeyan, O. P., & Kumar, G. (2021). Lignocellulosic biomass based biorefinery: A successful platform towards circular bioeconomy. *Fuel, 302*, Article 121086.

Bhargavi, P. (2024). Exploring the viability of organic farming for sustainable agriculture in India. *International Research Journal on Advanced Engineering and Management (IRJAEM), 2*(03), 267–275.

Bhat, R. (2021). *Future foods: Global trends, opportunities, and sustainability challenges.* Academic Press.

Bianchi, F., van Beek, C., de Winter, D., & Lammers, E. (2020). Opportunities and barriers of circular agriculture. *Insights from a synthesis study of the Food & Business Research Programme.*

Bruinsma, J. (2017). *World agriculture: Towards 2015/2030: An FAO study.* Routledge.

Carlson, K. M., Gerber, J. S., Mueller, N. D., Herrero, M., MacDonald, G. K., Brauman, K. A., Havlik, P., O'Connell, C. S., Johnson, J. A., Saatchi, S., & West, P. C. (2017). Greenhouse gas emissions intensity of global croplands. *Nature Climate Change, 7*(1), 63–68.

Chary, K., van Riel, A. J., Muscat, A., Wilfart, A., Harchaoui, S., Verdegem, M., Filgueira, R., Troell, M., Henriksson, P. J., de Boer, I. J., & Wiegertjes, G. F. (2024). Transforming sustainable aquaculture by applying circularity principles. *Reviews in Aquaculture, 16*(2), 656–673.

Chatterjee, S., & Mohan, S. V. (2022). Fungal biorefinery for sustainable resource recovery from waste. *Bioresource Technology, 345*, Article 126443.

Chen, F., Yin, X., & Jiang, S. (2023). *Climate-smart agriculture in China.* Food & Agriculture Org.

Cucagna, M. E., & Goldsmith, P. D. (2018). Value adding in the agri-food value chain. *International Food and Agribusiness Management Review, 21*(3), 293–316.

Cucchiella, F., D'Adamo, I., & Gastaldi, M. (2019). An economic analysis of biogas-biomethane chain from animal residues in Italy. *Journal of Cleaner Production, 230*, 888–897.

Davies, C., Chen, W. Y., Sanesi, G., & Lafortezza, R. (2021). The European Union roadmap for implementing nature-based solutions: A review. *Environmental Science & Policy, 121*, 49–67.

De Wilde, S. (2016). *The future of technology in agriculture.* StichtingToekomstbeeld der Techniek.

Diao, X., Reardon, T., Kennedy, A., DeFries, R. S., Koo, J., Minten, B., Takeshima, H., & Thornton, P. (2023). The future of small farms: Innovations for inclusive transformation. *Science and innovations for food systems transformation* (pp. 191–205). Springer.

Dimkpa, C., Adzawla, W., Pandey, R., Atakora, W. K., Kouame, A. K., Jemo, M., & Bindraban, P. S. (2023). Fertilizers for food and nutrition security in sub-Saharan Africa: An overview of soil health implications. *Frontiers in Soil Science, 3*, 1123931.

Diprose, G., Dombroski, K., Sharp, E., Yates, A., Peryman, B., & Barnes, M. (2023). Emerging transitions in organic waste infrastructure in Aotearoa New Zealand. *New Zealand Geographer, 79*(1), 15–26.

Domeignoz-Horta, L. A., Cappelli, S. L., Shrestha, R., Gerin, S., Lohila, A. K., Heinonsalo, J., Nelson, D. B., Kahmen, A., Duan, P., Sebag, D., & Verrecchia, E. (2024). Plant diversity drives positive microbial associations in the rhizosphere enhancing carbon use efficiency in agricultural soils. *Nature Communications, 15*(1), 8065.

Ejaz, M. R. (2024). Smart manufacturing as a management strategy to achieve sustainable competitiveness. *Journal of the Knowledge Economy, 15*(1), 682–705.

El Bilali, H., & Allahyari, M. S. (2018). Transition towards sustainability in agriculture and food systems: Role of information and communication technologies. *Information Processing in Agriculture, 5*(4), 456–464.

Ezealigo, U. (2022). *Biomass valorization: Assessment and characterization of biomass waste for valuable products* (Doctoral dissertation, AUST and PAMI).

Falco, J. M., Marco-Lajara, B., Sánchez-García, E., &GilinskyJr, A. (2023). The economic, social, and environmental value of the Spanish wine industry. In *Handbook of research on sustainability challenges in the wine industry* (pp. 169–193). IGI Global.

Farmery, A. K., Alexander, K., Anderson, K., Blanchard, J. L., Carter, C. G., Evans, K., Fischer, M., Fleming, A., Frusher, S., Fulton, E. A., & Haas, B. (2022). Food for all: Designing sustainable and secure future seafood systems. *Reviews in Fish Biology and Fisheries, 32*(1), 101–121.

Frosch, R. A., & Gallopoulos, N. E. (1989). Strategies for manufacturing. *Scientific American, 261*(3), 94–102.

Gemtou, M., Kakkavou, K., Anastasiou, E., Fountas, S., Pedersen, S. M., Isakhanyan, G., Erekalo, K. T., & Pazos-Vidal, S. (2024). Farmers' transition to climate-smart agriculture: A systematic review of the decision-making factors affecting adoption. *Sustainability, 16*(7), 2828.

Goddek, S., & Körner, O. (2019). A fully integrated simulation model of multi-loop aquaponics: A case study for system sizing in different environments. *Agricultural Systems, 171*, 143–154.

Gosai, H. G., Sharma, A., &Mankodi, P. (2024). Climate change's impact on agricultural food production. In *Food security in a developing world: Status, challenges, and opportunities* (pp. 117–132). Springer Nature Switzerland.

Goyal, M. R., &Sivanappan, R. K. (Eds.). (2017). *Engineering practices for agricultural production and water conservation: An interdisciplinary approach* (p. 400).

Grover, D., Mishra, A. K., Rani, P., Kalonia, N., Chaudhary, A., & Sharma, S. (2024). Soil Management in sustainable agriculture: Principles and techniques. *Technological approaches for climate smart agriculture* (pp. 41–77). Springer.

Gupta, A., Kashyap, L., Sanghera, G. S., Bhushan, K., & Kochar, G. S. (2024). Analyzing the potential of sugarcane juice for bio-ethanol production. *Sugar Tech, 26*(4), 1179–1193.

Harmsen, P. F., Hackmann, M. M., & Bos, H. L. (2014). Green building blocks for bio-based plastics. *Biofuels, Bioproducts and Biorefining, 8*(3), 306–324.

Hobson, S. (2023). *Optimising food loss and waste measurement practices in the South African grocery retail sector*.

Hoogstra, A. G., Silvius, J., de Olde, E. M., Candel, J. J. L., Termeer, C. J. A. M., van Ittersum, M. K., & de Boer, I. J. M. (2024). The transformative potential of circular agriculture initiatives in the North of the Netherlands. *Agricultural Systems, 214*, Article 103833.

Iqbal, A., Zan, F., Siddiqui, M. A., Nizamuddin, S., & Chen, G. (2022). Integrated treatment of food waste with wastewater and sewage sludge: Energy and carbon footprint analysis with economic implications. *Science of the Total Environment, 825*, Article 154052.

Jenane, C., Ulimwengu, J. M., Ono, T., & Kirui, O. K. (2022). *Agro-parks as drivers of the African food processing sector: Review of conditions for success*.

Khoshnevisan, B., Fog, E., Baladi, S., Chan, S. W. S., & Birkved, M. (2023). Using the product environmental footprint to strengthen the green market for sustainable feed ingredients; Lessons from a green biomass biorefinery in Denmark. *Science of the Total Environment, 877*, Article 162858.

Kongnso, M. E. (2024). Practice of compost use in urban farming: Opportunities and constraints in the West Region, Cameroon. *International Journal of Environment, Agriculture and Biotechnology, 9*(5), 144–156.

Kowalski, Z., Kulczycka, J., Makara, A., Mondello, G., & Salomone, R. (2023). Industrial symbiosis for sustainable management of meat waste: The case of Śmiłowo Eco-Industrial Park, Poland. *International Journal of Environmental Research and Public Health, 20*(6), 5162.

Krom, P. T. (2020). *Digital platforms for industrial symbiosis. An exploration of industrial parks in Norway* (Master's thesis).
Kumar, B., & Verma, P. (2021). Biomass-based biorefineries: An important architype towards a circular economy. *Fuel, 288*, Article 119622.
Kusumastuti, R. D., Van Donk, D. P., & Teunter, R. (2016). Crop-related harvesting and processing planning: A review. *International Journal of Production Economics, 174*, 76–92.
Lampridi, M., Angelopoulou, T., Tagarakis, A. C., & Bochtis, D. D. (2022). A circular precision farming system towards the optimization of dairy value-chains. *Information and communication technologies for agriculture—Theme IV: actions* (pp. 77–93). Springer.
Mahal, J. S., Manes, G. S., Singh, A., Kaur, S., & Singh, M. (2019). Complementing solutions and strategies for managing rice straw and their impact in the state of Punjab. *Agricultural Research Journal, 56*(3).
Mahato, R. P., Kumar, S., & Singh, P. (2023). Production of polyhydroxyalkanoates from renewable resources: A review on prospects, challenges and applications. *Archives of Microbiology, 205*(5), 172.
Mailaram, S., Kumar, P., Kunamalla, A., Saklecha, P., & Maity, S. K. (2021). Biomass, biorefinery, and biofuels. In *Sustainable fuel technologies handbook* (pp. 51–87). Academic Press.
Mandondo, M. S., & Troskie, D. (2021). *Western Cape Department of Agriculture.*
Mishra, N., Pradhan, S. K., Patra, S. K., & Ghosh, S. (2024). Preservation of biodiversity and sustainability of ecosystem. *Asian Journal of Agricultural Extension, Economics & Sociology, 42*(9), 140–150.
Muscio, A., & Sisto, R. (2020). Are agri-food systems really switching to a circular economy model? Implications for European research and innovation policy. *Sustainability, 12*(14), 5554.
Ócsai, A., & Ócsai, A. (2021). Ecologically conscious business organizations and their value orientations. In *Ecologically conscious organizations: new business practices based on ecological commitment* (pp. 93–150).
Ogbu, C. C., & Okey, S. N. (2023). Agro-industrial waste management: The circular and bioeconomic perspective. In *Agricultural waste-new insights.*
Ortiz-de-Montellano, C. G. S., Samani, P., & van der Meer, Y. (2023). How can the circular economy support the advancement of the sustainable development goals (SDGs)? A comprehensive analysis. *Sustainable Production and Consumption, 40*, 352–362.
Preston, F., & Lehne, J. (2017). *A wider circle? The circular economy in developing countries.*
Redlingshöfer, B. (2022). Food waste in cities: an urban metabolism approach applied to Paris and Île-de-France.
Rodríguez-Espinosa, T., Papamichael, I., Voukkali, I., Gimeno, A. P., Candel, M. B. A., Navarro-Pedreño, J., Zorpas, A. A., & Lucas, I. G. (2023). Nitrogen management in farming systems under the use of agricultural wastes and circular economy. *Science of the Total Environment, 876*, Article 162666.
Rong, K., & Luo, Y. (2023). Toward born sharing: The sharing economy evolution enabled by the digital ecosystems. *Technological Forecasting and Social Change, 196*, Article 122776.
Sabir, R. M., Sarwar, A., Shoaib, M., Saleem, A., Alhousain, M. H., Wajid, S. A., Rasul, F., Shahid, M. A., Anjum, L., Safdar, M., Muhammad, N., Waqas, R., Zafar, U., & Raza, A. (2024). Managing water resources for sustainable agricultural production. In *Transforming agricultural management for a sustainable future: Climate change and machine learning perspectives* (pp. 47–74). Springer Nature Switzerland.
Saini, R., Kaur, A., Saini, J. K., Patel, A. K., Varjani, S., Chen, C. W., Singhania, R. R., & Dong, C. D. (2023). Trends in lignin biotransformations for bio-based products and energy applications. *BioEnergy Research, 16*(1), 88–104.
Sekhar, M., Lallawmkimi, M. C., Rani, D. V., Upadhyay, L., Tiwari, P., Verma, U., & Rajput, A. (2024). Agroforestry practices and their impact on soil health and fertility: A review. *Journal of Experimental Agriculture International, 46*(9), 511–528.

Seymour, M., & Connelly, S. (2023). Regenerative agriculture and a more-than-human ethic of care: A relational approach to understanding transformation. *Agriculture and Human Values, 40*(1), 231–244.

Seyoum, B. (2024). Industrialization and the role of foreign direct investment. In S. Fragility (Ed.), *Business, and economic performance: An Ethiopian perspective* (pp. 179–224). Springer.

Shedekar, V. (2018). Decision tools for climate-smart agriculture. In D. Verma, V. Shedekar, A. Murumkar, R. Sharma, A. Kumar, V. Rani, F. Yusuf, & Brajendra (Eds.), Climate Smart Agriculture: Training Manual, Certified Training Workshop During 2nd International Conference on Food & Agriculture (pp. 23–34).

Silva, L. R. B., & Kardos, L. (2024). Composting of distillery spent wash. *Journal of Environmental Geography, 17*(1–4), 15–28.

Singh, S., Singh, P., Sharma, A., & Choudhury, M. (Eds.). (2022). *Agriculture waste management and bioresource: The circular economy perspective.* Wiley.

Sishodia, R. P., Ray, R. L., & Singh, S. K. (2020). Applications of remote sensing in precision agriculture: A review. *Remote Sensing, 12*(19), 3136.

Skene, K. R. (2022). The circular economy: A critique of the concept. *Towards a circular economy: Transdisciplinary approach for business* (pp. 99–116). Springer.

Swastika, D. K. S., Priyanti, A., Hasibuan, A. M., Sahara, D., Arya, N. N., Malik, A., Ilham, N., Sayekti, A. L., Triastono, J., Asnawi, R., & Sugandi, D. (2024). Pursuing circular economics through the integrated crop-livestock systems: An integrative review on practices, strategies and challenges post green revolution in Indonesia. *Journal of Agriculture and Food Research, 1*, Article 101269.

Tardy, B. L., Mattos, B. D., Otoni, C. G., Beaumont, M., Majoinen, J., Kämäräinen, T., & Rojas, O. J. (2021). Deconstruction and reassembly of renewable polymers and biocolloids into next generation structured materials. *Chemical Reviews, 121*(22), 14088–14188.

Tittonell, P., Klerkx, L., Baudron, F., Félix, G. F., Ruggia, A., Van Apeldoorn, D., Dogliotti, S., Mapfumo, P., & Rossing, W. A. (2016). Ecological intensification: Local innovation to address global challenges. *Sustainable Agriculture Reviews, 19*, 1–34.

Toplicean, I. M., & Datcu, A. D. (2024). An overview on bioeconomy in agricultural sector, biomass production, recycling methods, and circular economy considerations. *Agriculture, 14*(7), 1143.

Ufitikirezi, J. D. D. M., Filip, M., Ghorbani, M., Zoubek, T., Olšan, P., Bumbálek, R., Strob, M., Bartoš, P., Umurungi, S. N., Murindangabo, Y. T., & Heřmánek, A. (2024). Agricultural waste valorization: Exploring environmentally friendly approaches to bioenergy conversion. *Sustainability, 16*(9), 3617.

Urugo, M. M., Yohannis, E., Teka, T. A., Gemede, H. F., Tola, Y. B., Forsido, S. F., Tessema, A., Suraj, M., & Abdu, J. (2024). Addressing post-harvest losses through agro-processing for sustainable development in Ethiopia. *Journal of Agriculture and Food Research, 1*, Article 101316.

vanBerkum, S., & Dengerink, J. (2019). *Transition to sustainable food systems: the Dutch circular approach providing solutions to global challenges* (No. 2019-082). Wageningen Economic Research.

Varvaringos, I., Skourtanioti, E., Letsos, G., Rizoudi, E., Makras, E., Panagiotopoulou, M., Papadaki, S., & Valta, K. (2023). *Promoting sustainable agricultural waste management and industrial symbiosis through an innovative web platform.*

Vogel, E. (2022). *Assessing the economic and environmental performance of dairy and paddy field agricultural systems in Brazil.*

Wrzaszcz, W., & Prandecki, K. (2020). Agriculture and the European green deal. *ZagadnieniaEkonomikiRolnej/Problems of Agricultural Economics*, 4.

Zheng, F., & Zhou, X. (2023). Sustainable model of agricultural product logistics integration based on intelligent blockchain technology. *Sustainable Energy Technologies and Assessments, 57*, Article 103258.

Zimmermann, B., Claß-Mahler, I., von Cossel, M., Lewandowski, I., Weik, J., Spiller, A., Nitzko, S., Lippert, C., Krimly, T., Pergner, I., & Zörb, C. (2021). Mineral-ecological cropping systems—a new approach to improve ecosystem services by farming without chemical synthetic plant protection. *Agronomy, 11*(9), 1710.

Harmonization Harvests: Circular Practices in Agriculture Supporting Global Food System

Daniel Balsalobre-Lorente, Syed Asif Ali Naqvi, Syed Ale Raza Shah, and Naila Abbas

Abstract Zero hunger has become a top priority across the nations, and global leaders have presented the Sustainable Development Goals (SDGs), which centered on socio-economic and environmental development. Particularly, this study focuses on "SDG 2: Zero Hunger" and tries to show the importance of the food system in dealing with those challenges. Under the food system, the circular economy (CE) has received considerable attention from policymakers and OECD nations to transition from a linear to a circular economy. Under the shadow of CE—food system nexus, the present study ponders digitalization (internet use and mobile subscription), green energy, financial development, farm machinery, and agricultural labor force as determinants of food supply for OECD nations covering the period of 2005–2023. The Quantile Generalized Method of Moments is applied to explore the study objectives. The enquiry results show that digitalization, green energy, circular economy, and farm machinery increase the food supply. On the other hand, financial development and the agricultural labor force cause a decline in food supply. This study offers ecological implications to strengthen the specified factors' response to food supply.

D. Balsalobre-Lorente
Department of Applied Economics I, University of Castilla La Mancha, Ciudad Real, Spain

UNEC Research Methods Application Center, Azerbaijan State University of Economics (UNEC), Baku 1001, Azerbaijan

Economic Research Center (WCERC), Western Caspian University, Baku, Azerbaijan

D. Balsalobre-Lorente
e-mail: Daniel.balsalobre@uclm.es

S. A. A. Naqvi · S. A. R. Shah (✉)
Department of Economics, Government College University, Faisalabad 38000, Pakistan
e-mail: syedaaleraza56@gmail.com

S. A. A. Naqvi
e-mail: syedasif_1@yahoo.com

N. Abbas
School of Economics and Finance, Xi'an Jiao Tong University, Xian 710061, China
e-mail: nailaabbaskml@gmail.com

P. Singh et al. (eds.), *Circular Economy and Sustainable Innovation*,
https://doi.org/10.1007/978-3-032-00437-6_7

Keywords Circular economy · Food supply · Green energy · Farm machinery · Labor force

1 Introduction

Recently, the world has faced many challenges, including increasing hunger due to food shortages (Lowe, 2021). Practitioners have thrived in maximizing food supply to reach the food shortage threshold, but the problem remains (Sporchia et al., 2024). In these efforts, the paramount importance of Sustainable Development Goals (SDGs) cannot be ignored, especially SDG 2: "Zero Hunger" (Velazquez et al., 2023). Economies should review their agricultural development plans to meet the growing food demand and maximize the sustainable food supply (Narwane et al., 2022). The main question remains: is a sustainable food supply sufficient for economies? The consultants consistently respond that this remains a considerable challenge, as economies may be unable to maximize food production along with population growth (Hemathilake & Gunathilake, 2022). To address these issues, exponents must strive to understand the chief roots and smart alternatives to maximize food production and ensure the threshold level. From this perspective, the overall food production still faces ***two distinct problems***. The first major obstacle most farmers face is financial constraint: without adequate financing, they cannot engage all inputs and maximize the output level (Hong, 2021). Even though economies have encouraged financial institutions to provide agricultural-related loans, farmers may still have a gap in access to sufficient loans to maximize productivity (Muntaka et al., 2024). Labor force performance in agriculture may also be another foremost problem for farmers in augmenting their productivity and efficiently contributing to the world food supply (Karan & Asgari, 2021). Most of the labor force has fewer skills and may face problems managing pesticide usage, insect-appropriate solutions, the number of seeds, etc., indirectly increasing production costs. In compile words, due to inefficient allocation of labor services, farmers always face scale diseconomies (Harimaya & Kagitani, 2022).

Besides addressing the major challenges, this study offers ***concrete solutions*** implemented globally to maximize food production. Among these solutions, the role of agricultural machinery is essential (Rehman et al., 2023). In recent years, the world has moved toward technological progress and is strained to replace outdated methods with modern ones, which may not be possible without advanced technologies. Certain technologies increase food productivity and effectively reduce costs (Emami et al., 2018). The energy division is indispensable for the agricultural sector but indirectly creates unbearable environmental stress (Koondhar et al., 2021). That's why most nations have transformed their energy sector from conventional to modern and tried to maximize sustainable food production (Capurso et al., 2022). Green energy and digitalization may help deal with agricultural problems such as agrarian disease and its solution, advanced seeds, and sustainable methods (Shen et al., 2022). Digital transformation has reduced the obstacles of asymmetric information regarding farmers' market situation during agricultural product supply (Khanna, 2020). The globe is acquainted with it to manage productivity issues within a few jiffies. Finally, the circular economy has received immense attention from policy analysts to address food scarcity (Tamasiga et al., 2022). Recently, food waste has become a leading challenge for most nations, and they need to recycle all waste into useable food products (Pannila et al., 2022). This method is made known for two purposes, i.e., protecting the environmental quality and increasing food supply. These ***core problems and solutions*** encourage proponents to consider an empirical study and evaluate their response to the food supply.

The current work focuses on 32 OECD economies covering the period of 2005–2023 and underwrites the prevailing literature in the following ways. Firstly, digital transformation is a determinant of food supply because it has been adapted across the nations to facilitate humankind. Its green role is specific to managing all human and economic doings and can also be utilized in agricultural activities. Two different measures for digital transformation are adopted here, i.e., mobile use and internet use, to explore the digital role in food supply. This kind of inclusion in the food supply model would help to understand whether digital adoption assists in handling the overall increase in food supply. Obtaining outcomes would help exponents suggest appropriate policies to maximize agricultural output and sustain it shortly. Secondly, green energy is measured as another food supply element due to its ecological behavior and sustainability. Over time, economies have used conventional energy sources to facilitate the agricultural sector at the cost of environmental deterioration. International initiatives like the Paris Agreement, the Kyoto Protocol, and SDGs have been obliged to reduce their conventional pattern and conversion to modern ones for sustainable agri-output. But no systematic analysis has been observed evaluating its role in food supply for the mentioned economies. This study tries to portray its actual picture and may help to suggest green implications to increase the food supply. Thirdly, the financial role in all human and economic actions has become an important instrument to improve overall performance; this study uses it as a food supply factor to assess its performance in agricultural activities. Some countries made special sections where the farmer can take loans to purchase efficient technology, green energy sources, seeds, agricultural machinery, cattle, and skilled labor to maximize their productivity level. But still, there is an ambiguity as to whether its adoption would benefit food supply in specified nations. This study considers it to answer such a question and may submit core implications to maximize the overall output level.

Fourthly, the circular economy is an efficient tool to tackle all kinds of wastage into useable products, and its significant performance in food supply cannot be ignored. The present study considers food supply an element of food supply because food wastage has become a common problem across nations. In this regard, the circular economy has received significant importance in dealing with food wastage and increasing food supply. This inclusion in the food supply model would guide officials regarding its true response and may help to give efficient guidelines to maximize food productivity. Fifthly, farm machinery is another instrument that can help increase productivity. In the modern world, this strategy has been approved to finish the cultivation and harvesting process without any delay. This study includes it in the food model to depict its response to food supply across the specified regions. Sixthly, the labor force in the agriculture sector operates as another determinant of food supply. It is accepted that without factors of production, especially labor, the desired production may not be possible because the ability of agricultural labor to boost the overall food supply remains uncertain. To answer this question, this research attempts to analyze an entire nexus between labor and food supply, which could help suggest relevant implications for sustainable food supply. Finally, this research uses an advanced series of data reliability tests, such as cross-sectional dependence, data integration, and cointegration. The generalized method of moment-quantile regression (GMM-QR) was also used to analyze the long-term association.

2 Literature Review on Circular Economy-Food Supply Nexus

The current section of the present study précises core revisions directly linked to circular economy (CE) and food supply across the different regions. For example, Ojha et al. (2020) inspected the green role of CE in preventing massive food quantity from becoming waste. They concisely show the significant importance of CE in the food system and the different methods employed to secure food from the garbage. They introduced insect-based bioconversion as a green solution under the CE's theme. From this approach, they recommended that the economy attain a significant rise in food supply over the conversion of thousands of tons of human food, animal feed, fertilizers, and industrial material. The specified strategy not only helps to maximize food production but also secures the biodiversity level. Coderoni and Perito (2020) focused on the Italian case study to gauge the response of circular economy in food prevention. They collected data from 477 Italian customers to assess their response, which might help in the impact of emissions on agricultural production. They employed the binary regression estimator and confirmed the key interest of consumers in new food items rather than recycled ones. In other words, only 56% of respondents favored the theme of waste-to-value. Cicculo et al. (2021) demonstrated the effective role of CE in agri-food supply with advanced technologies. They collected only 34 semi-structured interviews from technology providers and companies related to agri-food supply. They applied certain technologies to prevent waste and validated that advanced technologies may help each economy secure food from wastage. Interestingly, they have not used any estimator to obtain valuable contributions to the existing literature in numerical form, and they only considered descriptive themes to show the role of technology in sustainable food supply. Kharola et al. (2022) potted the role of CE in viable food supply by having Indian economy. Over time, 25 different criteria were established to reduce food waste and increase food supply in the agriculture sector. They used the Best-Worst method to determine CE's part in food hoard. In concluding remarks, they illustrated that preventing food products earlier was preferable to treatment at later stages. Also, their skills and training were the leading themes for dealing with food wastage and its better usage in the overall food supply. Saha (2023) discussed the study of China to appraise CE's role in the food industry. They give the key logic behind the food industry selection: massive waste generation and resource exploitation. They mainly concentrated on two perspectives, i.e., circular economy strategies (waste reduction, recycling, and resource recovery) for a green environment and economic benefits from the food industry. Under the second perspective, they debated the challenges to a circular economy, such as consumer uncertain behavior, market demand, and regulation. Then, they proved that the circular economy was the best strategy for dealing with waste and managing the food industry. Rabbi and Bin Amin (2024) conducted the green role of CE in the food industry in view of the bibliometric analysis of 1000 studies. They validated the ecological role of CE in business activities, and most studies centered on food waste reduction, climate policy, and climate action strategy. Some say a circular economy may amplify the food industry through efficient resource usage, green waste management practices, and transforming food waste into usable foods. Banerjee et al. (2024) focused on linear to circular economy that may emphasize waste management, efficient resource allocation, and sustainable practices. They conducted a bibliometric analysis of 364 studies to clarify their performance on global growth activities and tried to offer insight regarding authors, regions, and keywords, but they did not offer any solid message concerning role of CE and food supply. Chen and Eweje (2025) deliberated the case study of New Zealand to assess an important role in food manufacturing industries. They concluded that, CE was played an essential character to identify the institutional drivers. They discussed about the multiple barrier that faced circular economy theme such as regulations, social reasoning, and supply chain. This research study offers bibliometric graph thru co-occurrence analysis and doughnut graph to show the circular economy across the different regions (See Fig. 1).

The subsisting literature has endeavored to weigh up the role of a circular economy in the food supply, but still, some flaws must be scrutinized by forthcoming studies. First, no study has been seen that considers secondary data sources and evaluates the circular economy's part in food production. This study finds 32 OECD economies and gives the most attractive food supply factor under the prism of the circular economy. Most studies have researched using other study's review themes and made their arguments regarding CE and food supply. In order to minimize these flaws, this study chooses empirical themes demonstrating the direct role of CE and other factors in food production across the selected nations. Finally, most sum-up studies have used general information to provide a sound message, but this study considers a proper estimator scheme to reveal the core response of selected variables to food production.

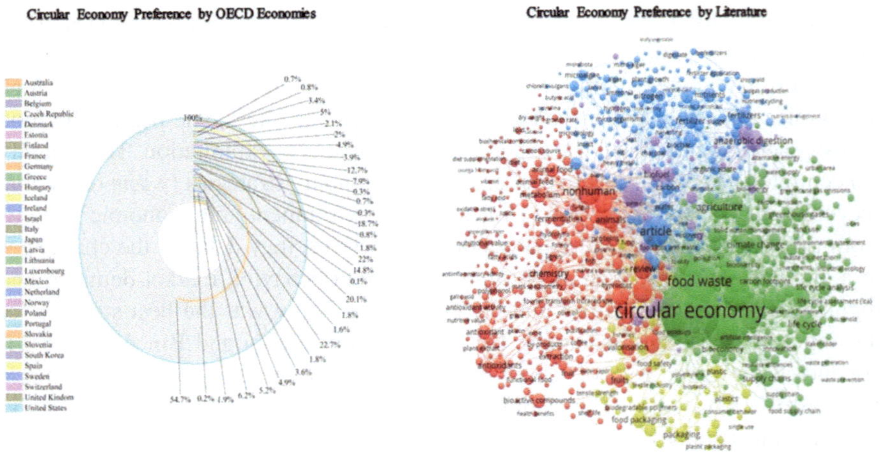

Fig. 1 Circular economy preference by literature and OECD economies

3 Data and Methods

This empirical study concentrates on 32 OECD economies to validate the response of circular economy (CE) in food supply covering the period of 2005–2023. Under the CE-food nexus, this study chooses digital transformation, green energy, financial development, farm machinery, and agricultural labor force as indicators of food supply. The data on given variables are collected from different sources, and information is presented in Table 1.

Table 1 Variable description

Symbol	Description	Unit	Source
FS	Food supply	Food Production index	WDI
DT	Digital transformation	Individuals using the internet (% of population)	WDI
–	–	Mobile cellular subscriptions (per 100 people)	WDI
GE	Green energy	Renewable energy consumption (% of total final energy consumption)	WDI
FD	Financial development	Domestic credit to private sector by banks (% of GDP)	WDI
CE	Circular economy	Municipal waste recycled (Tonnes)	OWD
FM	Farm machinery	Machinery per unit of agricultural land	OWD
LF	Labor force	Share of the labor force employed in agriculture	OWD

The initial screening tests, such as descriptive statistics, correlation, and VIF, contemplate here to evaluate the data reliability. Table 2 expresses the results of the descriptive statistics and validates no chance of an outlier in specified data.

Table 3 shows the outcomes of correlation and VIF analysis, which tells us that no single variable has a more than 0.80% correlation. Thus, the obtained outcomes strongly validate that there is no chance of multicollinearity in the selected panel. Figure 2 labels the parallel plot to understand the progress of selected nations towards food production.

Table 2 Descriptive statistics

	LNFP	LNIU	LNGE	LNFIN	LNCE	LNFM	LNLFA
Mean	1.9876	1.8821	1.1643	1.9104	6.2484	0.3798	0.5464
Median	1.9912	1.9151	1.1818	1.9507	6.2919	0.3597	0.5611
Maximum	2.1074	2.0107	1.9410	2.4836	7.8166	1.2630	1.2400
Minimum	1.8231	1.2355	– 0.0457	1.1241	4.3802	– 0.6528	– 0.2057

Table 3 Correlation analysis

Correlation	LNFP	LNIU	LNGE	LNFIN	LNCE	LNFM	LNLFA	VIF	1/VIF
LNFP	1.000							–	–
LNIU	0.2811	1.000						1.72	0.5808
	0.000	–						–	–
LNGE	0.0949	0.2334	1.000					1.38	0.7252
	0.019	0.000	–					–	–
LNFIN	– 0.0609	0.3516	0.0458	1.000				1.29	0.7772
	0.133	0.000	0.259	–				–	–
LNCE	0.1957	0.1209	– 0.3692	0.1926	1.000			1.29	0.7749
	0.000	0.002	0.000	0.000	–			–	–
LNFM	0.1395	0.1592	– 0.0785	0.2865	0.0857	1.000		1.22	0.8214
	0.000	0.000	0.052	0.000	0.034	–		–	–
LNLFA	– 0.1145	– 0.5051	0.1318	– 0.2806	– 0.3011	0.0958	1.000	1.69	0.5927
	0.004	0.000	0.001	0.000	0.000	0.018	–	–	–
Mean VIF								1.43	

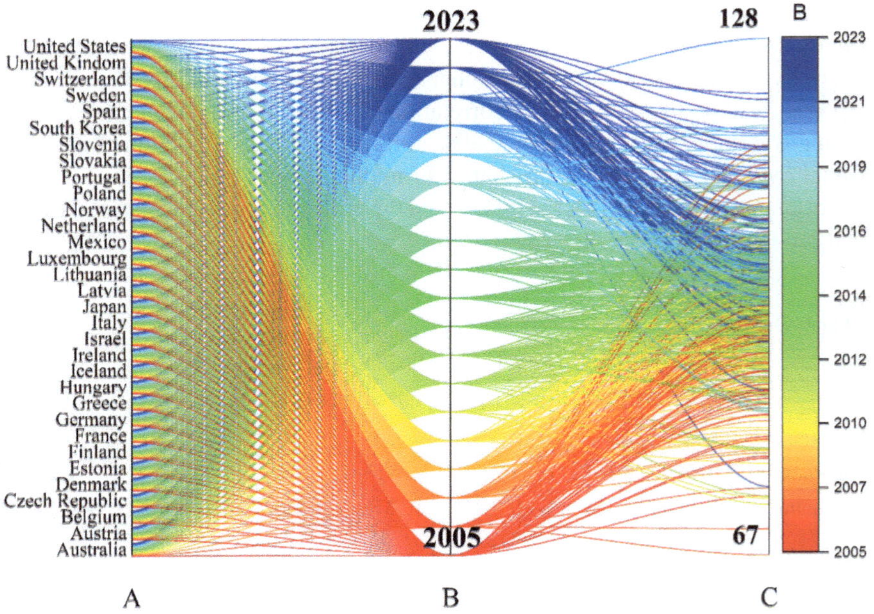

Fig. 2 Parallel plot for food supply

3.1 Empirical Model Construction

It is essential to choose appropriate studies to construct an empirical model, thus, the present study followed some recent studies [refers: (Brenya et al., 2024; Lee & Song, 2024; Marczewska, 2023; Tong et al., 2024; Zou & Mishra, 2024)]. The proposed variables can be expressed in a function,

$$FS_{i,t} = \mathrm{f}\Big(\beta_0, DT_{i,t}^{\beta_1}, GE_{i,t}^{\beta_2}, FD_{i,t}^{\beta_3}, CE_{i,t}^{\beta_4}, FM_{i,t}^{\beta_5}, LFA_{i,t}^{\beta_6}, \mu_{i,t}\Big) \tag{1}$$

Equation (1) defines the non-linear function of food supply and its determinants: digital transformation, green energy, financial development, circular economy, farm machinery, and labor force in the agriculture sector. In order to transform into the log–log model, this study takes logs on both sides,

$$\begin{aligned}\ln FS_{i,t} &= \beta_0 + \beta_1 \ln DTa, b_{i,t} + \beta_2 \ln GE_{i,t} + \beta_3 \ln FD_{i,t} \\ &\quad + \beta_4 \ln CE_{i,t} + \beta_5 \ln FM_{i,t} + \beta_6 \ln LFA_{i,t} + \mu_{i,t}\end{aligned} \tag{2}$$

Equation (2), gives the natural log form of food supply, digital transformation, green energy, financial development, circular economy, farm machinery, and labor force in the agriculture sector. This research work has few hypotheses that must be discussed here. Firstly, in the modern world, digital transformation has become an important instrument, and it facilitates all human beings at the domestic level with its green facilities. To validate its role in food supply, this study uses two distinct measures: internet use and mobile subscription. Due to their green role, it is understood that both factors may help to facilitate food supply with innovative and efficient methods at the perfect time and reduce asymmetric information. Due to green spillover effects, both factors contribute to food supply [hint: DT_a (internet use) and DT_b (mobile subscription) > 0]. Green energy has a significant role in environmental correction and may contribute to overall development plans. Its inclusion in the food supply encourages well-organized methods and helps in sustainable production (GE > 0). Financial development is being used as another food supply determinant, and most nations have tried to convince their financial sector to provide effective loans for development ventures. Farmers cannot allocate these loans in real time for specified purposes that may reduce the overall food supply (FD < 0). The circular economy takes time, and countries have adopted it to convert all waste into usable products. Resultantly, its green role in the overall food supply can't be ignored, and this study considers that it can enhance food supply through the coherent use of food resources (CE > 0). In addition, the ecological role of farm machinery in agri-food activities is considerable, and it condenses all impediments related to production, cultivation, and harvesting. Its supportive behavior induces this study to reflect its positive role in food supply (FM > 0). Finally, the labor force in the agriculture sector is an essential tool for dealing with all production affairs, either from the industrial, services, or agriculture sectors, because efficient labor matters more in this era to attain a significant rise in productivity. It is understood that most laborers in the agriculture sector have fewer skills and agree to work at lower wages. Due to excessive labor force, most economies face diminishing returns to scale, and its slope in terms of food supply would be negative (LFA < 0).

3.2 Estimation Strategy

Due to globalization, economies are interconnected, and there is a chance of cross-sectional dependence (CSD). Then to confirm CSD, this study deliberates the four most reliable CSD tests proposed by (Frees, 1995; Friedman, 1937; Pesaran, 2014; Pesaran & Smith, 2004). Under the CSD, the next step is to use the slope of homogeneity to validate further data properties (Hashem Pesaran & Yamagata, 2008).

To escape from the trap of inappropriate outcomes from conventional data integration tests, the present work uses the advanced series of data integration such as the CADF and CIPS and these advanced tests proposed by Pesaran (2007). Without long-term cointegration among selected variables, this study uses another unconventional test projected by Westerlund (2007).

In the current situation, this empirical research has the number of cross-sections is greater than the time; therefore, the existing literature strongly supports applying the Generalized Method of Moments (GMM) as anticipated by Arellano and Bond (1991) and Blundell and Bond (2000). In late, its updated form was explained by

Machado and Santos Silva (2019) and known as the Quantile Regression GMM estimator. Hence, the current study employs the Quantile GMM and the Panel quantile regression for the robustness and heterogeneity analysis. Its (Q-GMM) general form is given as,

$$\min \sum_{i:u\theta\geq 0}^{n} \theta \times |u\theta| + \sum_{i:u\theta\geq 0}^{n} 1-\theta \times |u\theta i| = \sum_{i:yi-x'i\,.\,\beta\theta\geq 0}^{n} \theta \times \left|yi - x'i.\beta\theta\right| + \sum_{i:yi-x'i\,.\,\beta\theta\leq 0}^{n} (1-\theta)\times \left|yi - x'i.\beta\theta\right| \quad (3)$$

4 Results and Discussion

Before further empirical analysis, this study gives additional tests to validate the robustness of selected data. Table 4 describes the results of CSD and slope of homogeneity that validate the prior expectations.

Table 4 CSD and slope of homogeneity test outcomes

Statistics	Pesaran (2004)	Pesaran (2014)	Frees (1995)	Friedman (1937)
lnFS	26.486***	26.486***	8.410***	167.532***
lnMB	45.322***	45.322***	6.462***	187.265***
lnIU	89.544***	89.544***	25.515***	530.675***
lnGE	75.499***	75.499***	22.364***	454.923***
lnCE	38.433***	38.433***	14.468***	245.385***
lnFD	22.052***	22.052***	11.545***	37.089***
lnFM	14.321***	14.321***	18.212***	106.873***
lnLFA	64.370	64.370***	19.137***	380.276***
Slope of homogeneity				
	Value		*P-value*	
Δ	5.414		0.000	
Δ Adj	6.925		0.000	
Bias-adjusted LM test				
LM	697.9		0.000	
LM adj*	3.522		0.000	
LM CD*	9.263		0.000	

Note *** shows levels of significance at 1%

Table 5 Data integration results

Variable	CADF unit root		CIPS unit root	
	Level	1st difference	Level	1st difference
lnFS	–	– 5.197***	–	– 5.197***
lnMB	–	– 3.780***	–	– 3.780***
lnIU	– 2.899***	–	– 2.899***	–
lnGE	– 2.675***	–	– 2.675***	–
lnCE	– 2.525***	–	– 2.525***	–
lnFD	–	– 3.116***	–	– 3.116***
lnFM	–	– 4.527***	–	– 4.527***
lnLFA	– 2.630***	–	– 2.630***	–

Note *** shows levels of significance at 1%

Table 6 Cointegration test outcomes

Statistics	Value	Z-value	P-value	Robust P-value
Gt	– 3.205	– 2.172	0.015	0.000
Ga	– 4.103	8.245	1.000	0.000
Pt	– 18.680	– 3.813	0.000	0.000
Pa	– 4.681	5.427	1.000	0.000

Table 5 gives us fascinating upshots about data integration, and it shows all variables are integrated at first difference except for internet use, green energy, circular economy, and labor force in the agriculture sector.

Table 6 exhibits cointegration conclusions and strongly validates the alternative hypothesis of cointegration among the designated variables.

4.1 Long-Run Results by Q-GMM Estimator

The present study uses the Q-GMM estimator to demonstrate the study objectives and find the most reliable results (See Table 7). For example, technology in internet use is being used as a first food supply determinant, and it shows any significant change in internet facility would significantly raise food supply across all quantiles. Such a connection is crucial for the specified nations and can be explained with some actual logic. Under the old age, most nations have faced core challenges of market asymmetric information and lack of knowledge of prices. Digitalization has helped to escape this trap and become more informed about the market situation. As a result, internet use helps attain all basic information, and producers can maximize their output and subsidize the overall food supply. In parallel, the harmful role of insects on overall growth is also substantial, and without perfect information, farmers may not secure their agricultural products. Internet use has transformed overall economies into global villages, and farmers can get help managing their Agri-products through innovative methods. The well-organized storage has remained another key problem that can be resolved when farmers have modern technology and methods to preserve their agri products for a long time. To address this problem, economies have availed digital services to preserve food items for a long time and ensure the overall food supply meets the zero-hunger target. This target aligns with the findings of Liu et al. (2025).

Green energy (GE) is employed as another food supply element and finds a positive role in explaining the variable. Since the Industrial Revolution, economies have been highly dependent on conventional energy resources that they used to accelerate their overall growth. At later stages, these nations have felt a rise in environmental deterioration due to excessive conventional energy use and obligated those unsustainable productions to rethink their energy policies and transition to green ones. Green energy has leading features that help supply food and abate environmental costs. GE has become a symbol of green activities due to its efficient behavior; the selected economies may urge their farmer to practice green energy and ensure sustainable food production. In recent years, agricultural technology has been upgraded, and most economies offer green energy installation projects at subsidized rates to multiply their share. Due to cheap energy, the farmers may increase their plowing and manage all stores through updated technology to preserve food items and supply them at market time. Therefore, an efficient energy source helps in the sustainable production and consumption of food items. This factor is highly accepted in cropping and can be verified via the results (Ermiş & Güven, 2022).

The given coefficient value of circular economy infers that any significant change in this factor would cause a rise in food supply across all quantiles. In recent years, this strategy has become a leading contribution by economies to environmental sustainability. Under this race, the core role of CE in food supply cannot be ignored from the contest. Agriculture can work at different stages to preserve food items and expand

Table 7 Q-GMM estimator outcomes

Quantile	lnIU	lnGE	lnCE	lnFD	lnFM	lnLFA	Cons
Q.10	0.1190***	0.0182***	0.0257***	− 0.0349***	0.0298***	− 0.0091*	1.6004***
Q.20	0.1080***	0.0167***	0.0205***	− 0.0382***	0.0237***	− 0.0058*	1.6788***
Q.30	0.1009***	0.0158***	0.0171***	− 0.0402***	0.0199***	− 0.0036*	1.7288***
Q.40	0.0967***	0.0152***	0.0151***	− 0.0415***	0.0176***	− 0.0023*	1.7585***
Q.50	0.0916***	0.0145***	0.0127***	− 0.0430***	0.0148***	− 0.0008*	1.7947***
Q.60	0.0867***	0.0139***	0.0103***	− 0.0444***	0.0120***	0.0006*	1.8300***
Q.70	0.0813***	0.0132***	0.0078***	− 0.0460***	0.0091***	0.0023*	1.8683***
Q.80	0.0758***	0.0125***	0.0052**	− 0.0476***	0.0061***	0.0039*	1.9069***
Q.90	0.0656***	0.0111***	0.0003*	− 0.0506***	0.0005*	0.0070*	1.9793***

Note ***, **, and * show significance levels at 1%, 5% and 10% respectively the supply. Firstly, during the harvesting process, there was a massive loss of garbage that could not be operated for final products. In the past, most of the produce was wasted as garbage, and farmers may have faced a decline in production. To escape this trap, the concept of the circular economy was introduced. It allows farmers to reuse their garbage during harvesting to make useable products. Such action not only preserves the environmental quality but also improves the food supply. Most stockholders face product waste issues at various storage points, even if they keep them. Also, stockholders strive to maintain the quality of their products but mostly waste lower-quality items, such as grains, that can be used to produce certain foods. Therefore, stockholders may use CE services to reuse their waste and convert it into usable products. Finally, the transportation and processing of agri-products at the industrial level also results in waste that the specified economies must reuse and increase their supply to achieve the desired level of zero hunger. This outcome is consistent with recent findings by Kumar et al. (2022).

Financial resources and their role in all human and economic activities have become most prominent. Over time, there has been a variation in financial interest, but the leading aim is to intensify social well-being. In the framework of social well-being, zero hunger is a leading dream, and economies are jointly working to deal with it. Most nations have paid attention to agriculture because it can enhance the food supply to meet the desired level (zero-hunger). For these reasons, economies have duty-bound financial institutions to offer loans cheaply and encourage farmers to maximize their agri-related activities. Financial development is a food component in the present case but shows an inverse connection from lower to upper quantiles. This connection explains that financial loans do not effectively work for food supply, and it can be described for several reasons. Firstly, most loans are allocated for agricultural activities such as purchasing seeds, agricultural machinery, green energy installation projects, and cattle farms, but farmers may not distribute these resources for definite purposes. Secondly, the financial institutions' performance may not be satisfactory because these institutions provide loans, but the overall process is too lengthy. Due to those processes, most farmers may not share financial resources; consequently, the overall production level decreases. Finally, the financial institution may not have strict checks and balances to audit and confirm whether created loans are owed to stated projects. These are a few rationales that can help to understand the valid reasons behind the scenes, and this outcome can be tested by the results (Arshad, 2022).

Farm machinery is essential in increasing agricultural productivity in the agriculture sector. The current empirical research has taken this factor as a food supply determinant and confirms a positive and significant role across all quantiles. Farm machinery has numerous advantages over conventional methods in the real world, which may enhance overall productivity. For example, before the agricultural revolution, most economies hired outdated methods to cultivate their agricultural land, which may face limited area access, time shortage, and other relevant problems. Those issues have been resolved proficiently by adding agrarian technology to manage irrigation, time, and land use, increasing food supply. In case of preservation from wastage, modern agri technology tries to manage and ensure its contribution to the overall food supply. From the harvesting perspective, most farmers have matters related to protecting themselves from damages and harvesting without delay. To solve this problem, nations have adopted advanced machinery directly supporting food production. Finally, to deal with the harmful role of insects in the agriculture sector, many economies have appraised using agricultural machinery to spray green fertilizers that kill insects and boost overall productivity. These could be a few lucidities to clarify the positive connection of farm machinery with food supply, and this effect can be substantiated by the results of Xu et al. (2025).

Finally, the labor force in the agriculture sector also performs well in dealing with all agricultural needs. The foremost economist and their growth theories directly support the performance of the labor force in output. The scenario describes an inverse and insignificant role in food supply across all quantiles. It is scandalous news for specified nations, but it can be elucidated in some comments. Over time, there has been a rising trend in population growth, indirectly increasing the labor force. Due to the gigantic rise in the labor force, employment in different sectors may bring diminishing returns to scale. A massive share of labor effectively reduces their overall productivity, and a decline in the food supply is a leading example in the current study. It is a common belief due to the huge labor share; economies may not provide better facilities to upgrade their knowledge and skills; that being so, more labor force employed may bring a decline in the food supply. Finally, lower wage rates in the agriculture sector may be another primary cause of the inverse association because workers may not efficiently provide their services to manage all agricultural obstacles. This sequel can tally with the conclusions of Kuiper et al. (2020).

Figure 3 expresses the graphical presentation of Q-GMM estimator outcomes.

4.2 Response of Mobile Use in Food Supply

Keeping all things constant, this study uses "mobile subscription" as an alternative proxy for technology and finds any significant change in this factor would cause an increase in the food supply in the specified economies (see Table 8). Generally, specified technology forms have been familiarized with three distinct perspectives: efficiency, transparency, and sustainability under the food supply theme. That kind of positive connection can be enlightened using deep understanding, and this study offers a true explanation. Firstly, in the modern world, all the investors use mobile technology for real-time communication, exact locations, and tracking food orders. Mobile technology has seen a rising trend in sustaining the production and consumption of goods because it has efficiently reduced traveling, time, and transaction costs, which not only helps in food supply but also protects the environmental quality. In addition, mobile technology has been used over the years to get perfect information relating to weather forecasts, market price fluctuations, advanced learning methods, and agricultural management. Such timely information helps in food supply and protects farmers from losses. Also, the efficient utilization of mobile technology in

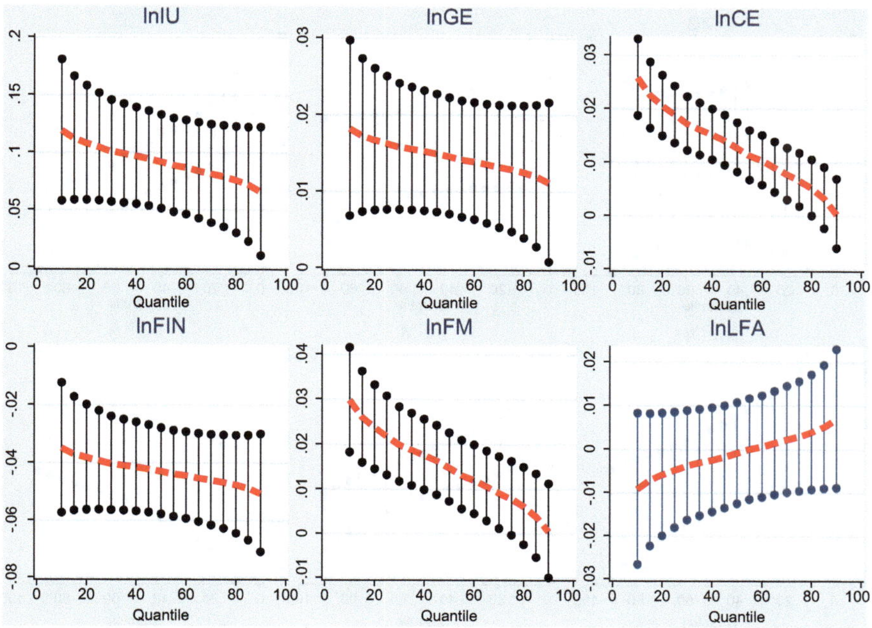

Fig. 3 Q-GMM outcomes' graphical presentation

product management is considerable, which helps protect all products from expiration and reduces owners' losses. The results can prove this outcome (Fernando et al., 2025).

Figure 4 conveyed the graphical presentation of Q-GMM estimator outcomes with alternative proxies.

Table 8 Q-GMM estimator outcomes with alternative proxy

Quantile	lnMB	lnGE	lnCE	lnFD	lnFM	lnLFA	Cons
Q.10	0.0724***	0.0260***	0.0271***	– 0.0182*	0.0338***	– 0.0261***	1.6324***
Q.20	0.0700***	0.0241***	0.0221***	– 0.0264***	0.0271***	– 0.0213***	1.7029***
Q.30	0.0687**	0.0230***	0.0192***	– 0.0310***	0.0233***	– 0.0186***	1.7427***
Q.40	0.0677**	0.0221***	0.0169***	– 0.0346***	0.0203***	– 0.0164***	1.7741***
Q.50	0.0665**	0.0211***	0.0143***	– 0.0388***	0.0169***	– 0.0140***	1.8103***
Q.60	0.0653**	0.0201***	0.0116***	– 0.0432***	0.0132***	– 0.0113***	1.8487***
Q.70	0.0641**	0.0191***	0.0091***	– 0.0472***	0.0099**	– 0.0090*	1.8835***
Q.80	0.0630**	0.0182***	0.0066**	– 0.0512***	0.0066*	– 0.0066*	1.9181***
Q.90	0.0610**	0.0165***	0.0022*	– 0.0583***	0.0008*	– 0.0024*	1.9793***

Note ***, **, and * show significance levels at 1%, 5% and 10% respectively

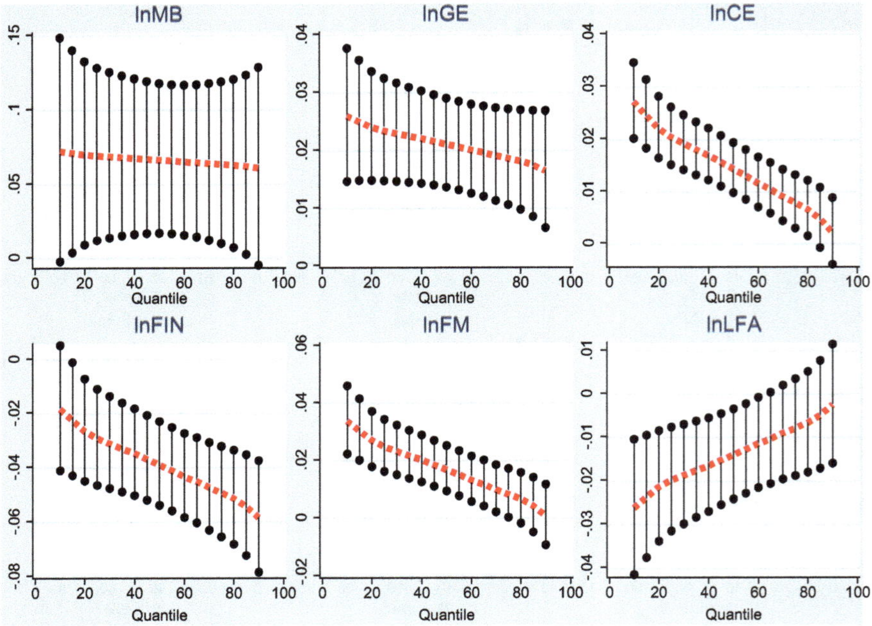

Fig. 4 Q-GMM outcomes' graphical presentation

5 Conclusion and Policy Recommendations

This study emphasizes food supply under the prism of circular economy to show real problems under zero hunger in the case of 32 OECD nations covering the period of 2005–2023. It employed the quantile GMM estimator to illustrate an entire connection of technology (internet use and mobile subscription), green energy, circular economy, financial development, farm machinery, and labor force in agriculture. The explored results narrated how internet and mobile-related technology competently increase the food supply across the selected nations. The circular economy, green energy, and farm machinery increase the food supply, as well as financial development, and the labor force in the agriculture sector also perform well in boosting the food supply.

This study seeks to put forward core implications for boosting the overall food supply globally.

- Under the scenario of technology via mobile and internet use, this study finds significant support for food supply, but still, there is a need to strengthen the overall food supply further to meet the threshold level of zero hunger. Firstly, the telecommunication authorities should try to provide green and cheap facilities to overcome obstacles under the internet services. Sometimes, these services may perform poorly to facilitate the farmers, which may cause them to make wrong decisions. Also, it remains a familiar taste that every economy primarily focuses on urban development plans and voluntarily ignores the agriculture sector to provide internet services for agricultural development. Mobile-use technology should also be offered cheaply and easily. Farmers can efficiently protect their fields from such inclusion, which may help the food supply.

- In the case of green energy, it significantly supports the rise of the food supply. The studied nations have not yet reached their threshold level to maximize their usage in the agriculture sector. These nations should struggle to offer green energy resources at subsidized rates that encourage farmers to augment their investment in quantified energy installations. Higher authorities should also make efforts from their financial institutions and ensure cheap rate loans for agricultural activities, which indirectly support the sustainable production and consumption of food items. The public-private partnership is highly recommended to manage the green energy flaws in the food supply.
- The circular economy capably supports food supply across the selected panel. Few recommendations directly support the role of CE in food supply. Firstly, the specified economies should prevent food wastage using efficient technologies and ensure minimum production, processing, distribution, and final consumption loss. Secondly, these nations should strain to reutilize waste and convert it into useable products such as animal feeds, bioenergy, etc. Also, governing bodies should consider adding digital technologies to the circular economy theme to manage all challenges and directly support the food supply.
- Currently, most farmers face financial constraints, significantly declining overall productivity. In order to meet their agri-expenditure, the government should undertake public–private partnerships and boost agri-finance. Secondly, quantified economies' institutions should be independent and have audits strictly for allocated loans. Such action would promote the consumption of selected projects. And government should endeavor to impose heavy plenty when any flaws in loan allocation are observed.
- Farm machinery is essential to manage all obstacles and expand the food supply. Under this theme, stated economies should introduce a specific sector that provides green farm machinery loans to increase the food supply. Farmers should allocate their revenue to purchase energy-efficient machinery that helps in food production and reduces over-dependence on conventional energy sources. The private sector must try to be involved in agricultural activities and offer services to maximize productivity.
- Labor force participation in agriculture doesn't perform well in managing the food supply; therefore, specified nations should attempt to focus on these suggestions. Firstly, the government should introduce a universal law for labor wages in the agriculture sector and strictly implement it to encourage labor force performance. After that, starting training programs through skilled labor guides is necessary to utilize available products competently and use the food supply best. Another leading suggestion is that legislators should provide the best opportunities for the agri labor force where they get training and boost their capabilities.

The present empirical study is not free from future limitations. Firstly, this study only uses the selected 32 OECD economies because data was missing for all nations; therefore, forthcoming studies should make an exertion to consider some alternative variables when choosing the complete panel. Another major restraint is that this study attempts to utilize the core role of the circular economy along with digitalization, but imminent works must deliberate the role of artificial intelligence in the food supply. Finally, this study only uses a quantile GMM estimator, but impending searches can use alternative methodologies to validate the study outcomes.

References

Arellano, M., & Bond, S. (1991). Some tests of specification for panel data: Monte Carlo evidence and an application to employment equations. *The Review of Economic Studies, 58*(2), 277–297.

Arshad, A. (2022). Impact of financial inclusion on food security: Evidence from developing countries. *International Journal of Social Economics, 49*(3), 336–355. https://doi.org/10.1108/ijse-08-2021-0462

Banerjee, P., Singh, D., & Kunja, S. R. (2024). Circular economy in agro food supply chain: Bibliometric and network analysis. *Business Strategy and Development, 7*(2), Article e360.

Blundell, R., & Bond, S. (2000). GMM estimation with persistent panel data: An application to production functions. *Econometric Reviews, 19*(3), 321–340.

Brenya, R., Jiang, Y., Sampene, A. K., & Zhu, J. (2024). Food security in sub-Sahara Africa: Exploring the nexus between nutrition, innovation, circular economy, and climate change. *Journal of Cleaner Production, 438*, Article 140805. https://doi.org/10.1016/j.jclepro.2024.140805

Capurso, T., Stefanizzi, M., Torresi, M., & Camporeale, S. M. (2022). Perspective of the role of hydrogen in the 21st century energy transition. *Energy Conversion and Management, 251*, Article 114898. https://doi.org/10.1016/j.enconman.2021.114898

Chen, S. M., & Eweje, G. (2025). An extended institutional theory perspective on the implementation of the circular economy: Empirical insights from New Zealand food manufacturing firms. *Business Strategy and the Environment, 34*(3), 2964–2992. https://doi.org/10.1002/bse.4117

Ciccullo, F., Cagliano, R., Bartezzaghi, G., & Perego, A. (2021). Implementing the circular economy paradigm in the agri-food supply chain: The role of food waste prevention technologies. *Resources Conservation and Recycling, 164*, Article 105114. https://doi.org/10.1016/j.resconrec.2020.105114

Coderoni, S., & Perito, M. A. (2020). Sustainable consumption in the circular economy. An analysis of consumers' purchase intentions for waste-to-value food. *Journal of Cleaner Production, 252*, 119870. https://doi.org/10.1016/j.jclepro.2019.119870

Emami, M., Almassi, M., Bakhoda, H., & Kalantari, I. (2018). Agricultural mechanization, a key to food security in developing countries: strategy formulating for Iran. *Agriculture & Food Security, 7*(1), 24. https://doi.org/10.1186/s40066-018-0176-2

Ermiş, A. M., & Güven, M. E. (2022). The role of green energy investments in energy supply security. In H. Dinçer & S. Yüksel (Eds.), *Circular economy and the energy market* (pp. 67–77). Springer International Publishing. https://doi.org/10.1007/978-3-031-13146-2_6

Fernando, W. M., Perera, H. N., Ratnayake, R. M. C., & Thibbotuwawa, A. (2025). Storm in a teacup: Implications of mobile phone literacy on sustainable smallholder agri-food supply chains in developing economies. *The International Journal of Logistics Management, 36*(2), 452–473. https://doi.org/10.1108/IJLM-09-2023-0413

Frees, E. W. (1995). Assessing cross-sectional correlation in panel data. *Journal of Econometrics, 69*(2), 393–414.

Friedman, M. (1937). The use of ranks to avoid the assumption of normality implicit in the analysis of variance. *Journal of the American Statistical Association, 32*(200), 675–701.

Harimaya, K., & Kagitani, K. (2022). Efficiency, and economies of scale and scope in Japanese agricultural cooperatives. *Journal of Economic Structures, 11*(1), 21. https://doi.org/10.1186/s40008-022-00282-8

Hashem Pesaran, M., & Yamagata, T. (2008). Testing slope homogeneity in large panels. *Journal of Econometrics, 142*(1), 50–93. https://doi.org/10.1016/j.jeconom.2007.05.010

Hemathilake, D. M. K. S., & Gunathilake, D. M. C. C. (2022). Agricultural productivity and food supply to meet increased demands. In R. Bhat (Ed.), *Future foods* (pp. 539–553). Academic Press. https://doi.org/10.1016/b978-0-323-91001-9.00016-5

Hong, Y. (2021). New model of food supply chain finance based on the internet of things and blockchain. *Mobile Information Systems, 2021*(1), 7589964. https://doi.org/10.1155/2021/7589964

Karan, E., & Asgari, S. (2021). Resilience of food, energy, and water systems to a sudden labor shortage. *Environment Systems and Decisions, 41*(1), 63–81. https://doi.org/10.1007/s10669-020-09793-w

Khanna, M. (2020). Digital transformation of the agricultural sector: Pathways, drivers and policy implications. *Applied Economic Perspectives and Policy, 43*(4), 1221–1242. https://doi.org/10.1002/aepp.13103

Kharola, S., Ram, M., Mangla, S. K., Goyal, N., Nautiyal, O. P., Pant, D., & Kazancoglu, Y. (2022). Exploring the green waste management problem in food supply chains: A circular economy context. *Journal of Cleaner Production, 351*, Article 131355. https://doi.org/10.1016/j.jclepro.2022.131355

Koondhar, M. A., Tan, Z., Alam, G. M., Khan, Z. A., Wang, L., & Kong, R. (2021). Bioenergy consumption, carbon emissions, and agricultural bioeconomic growth: A systematic approach to carbon neutrality in China. *Journal of Environmental Management, 296*, Article 113242. https://doi.org/10.1016/j.jenvman.2021.113242

Kuiper, M., Shutes, L., van Meijl, H., Oudendag, D., & Tabeau, A. (2020). Labor supply assumptions—A missing link in food security projections. *Global Food Security, 25*, Article 100328. https://doi.org/10.1016/j.gfs.2019.100328

Kumar, M., Raut, R. D., Jagtap, S., & Choubey, V. K. (2022). Circular economy adoption challenges in the food supply chain for sustainable development. *Business Strategy and the Environment, 32*(4), 1334–1356. https://doi.org/10.1002/bse.3191

Lee, C.-C., & Song, T. (2024). Green finance and food production: Evidence from cities in China. *Journal of Cleaner Production, 458*, Article 142423. https://doi.org/10.1016/j.jclepro.2024.142423

Liu, J., Ren, Y., Hong, Y., Glauben, T., & Li, Q. (2025). Does Internet use help to achieve sustainable food consumption? Evidence from rural China. *Sustainable Futures, 9*, Article 100466. https://doi.org/10.1016/j.sftr.2025.100466

Lowe, N. M. (2021). The global challenge of hidden hunger: Perspectives from the field. *The Proceedings of the Nutrition Society, 80*(3), 283–289. https://doi.org/10.1017/S0029665121000902

Machado, J. A. F., & Santos Silva, J. M. C. (2019). Quantiles via moments. *Journal of Econometrics, 213*(1), 145–173. https://doi.org/10.1016/j.jeconom.2019.04.009

Marczewska, M. (2023). Digital transformation: A challenging opportunity for the food industry companies. *British Food Journal, 126*(5), 2027–2040. https://doi.org/10.1108/bfj-01-2023-0065

Muntaka, A. S., Apike, I. A., & Antwi, D. (2024). Antecedents of sustainable agricultural supply chain financing for achieving food security in resource-constrained developing economies: A complex adaptive systems theory perspective. *Australian Journal of Agricultural and Resource Economics, 68*(4), 868–890. https://doi.org/10.1111/1467-8489.12579

Narwane, V. S., Gunasekaran, A., & Gardas, B. B. (2022). Unlocking adoption challenges of IoT in Indian agricultural and food supply chain. *Smart Agricultural Technology, 2*, Article 100035. https://doi.org/10.1016/j.atech.2022.100035

Ojha, S., Bussler, S., & Schluter, O. K. (2020). Food waste valorisation and circular economy concepts in insect production and processing. *Waste Management, 118*, 600–609. https://doi.org/10.1016/j.wasman.2020.09.010

Pannila, N., Jayalath, M. M., Thibbotuwawa, A., Nielsen, I., & Uthpala, T. G. G. (2022). Challenges in applying circular economy concepts to food supply chains. *Sustainability, 14*(24).

Pesaran. (2004). General diagnostic tests for cross section dependence in panels. Cambridge working papers. *Economics, 1240*, 1.

Pesaran, M. H. (2007). A simple panel unit root test in the presence of cross-section dependence. *Journal of Applied Econometrics, 22*(2), 265–312. https://doi.org/10.1002/jae.951

Pesaran, M. H. (2014). Testing weak cross-sectional dependence in large panels. *Econometric Reviews, 34*(6–10), 1089–1117. https://doi.org/10.1080/07474938.2014.956623

Pesaran, M., & Smith, R. (2004). General diagnostic test for cross sectional independence in panel. *Journal of Econometrics, 68*(1), 79–113.

Rabbi, M. F., & Bin Amin, M. (2024). Circular economy and sustainable practices in the food industry: A comprehensive bibliometric analysis. *Cleaner and Responsible Consumption, 14*, Article 100206. https://doi.org/10.1016/j.clrc.2024.100206

Rehman, A., Ma, H., Oláh, J., Alvarado, R., & Işık, C. (2023). Agricultural machinery, irrigation systems and food grains: A symmetric novel analysis. *Cogent Food & Agriculture, 9*(2), 2279714. https://doi.org/10.1080/23311932.2023.2279714

Saha, A. (2023). Circular economy strategies for sustainable waste management in the food industry. *Journal of Recycling Economy & Sustainability Policy, 2*(2), 1–16. https://respjournal.com/index.php/pub/article/view/17

Shen, Z., Wang, S., Boussemart, J.-P., & Hao, Y. (2022). Digital transition and green growth in Chinese agriculture. *Technological Forecasting and Social Change, 181*, Article 121742. https://doi.org/10.1016/j.techfore.2022.121742

Sporchia, F., Antonelli, M., Aguilar-Martínez, A., Bach-Faig, A., Caro, D., Davis, K. F., Sonnino, R., & Galli, A. (2024). Zero hunger: Future challenges and the way forward towards the achievement of sustainable development goal 2. *Sustainable Earth Reviews, 7*(1), 10. https://doi.org/10.1186/s42055-024-00078-7

Tamasiga, P., Miri, T., Onyeaka, H., & Hart, A. (2022). Food waste and circular economy: Challenges and opportunities. *Sustainability, 14*(16).

Tong, T., Ye, F., Zhang, Q., Liao, W. M., Ding, Y. F., Liu, Y., & Li, G. C. (2024). The impact of labor force aging on agricultural total factor productivity of farmers in China: Implications for food sustainability [Original Research]. *Frontiers in Sustainable Food Systems, 8*, 1434604. https://doi.org/10.3389/fsufs.2024.1434604

Velazquez, L., Munguia, N., Alvarez-Alvarez, D., Cuamea-Cruz, G., Anaya-Eredias, C., & Martinez-Castañeda, F. (2023). Residential waste segregation: The interconnection with SDG 2 zero hunger. *Environmental Challenges, 10*, Article 100675. https://doi.org/10.1016/j.envc.2022.100675

Westerlund, J. (2007). Testing for error correction in panel data. *Oxford Bulletin of Economics and Statistics, 69*(6), 709–748. https://doi.org/10.1111/j.1468-0084.2007.00477.x

Xu, Y., Lyu, J., Yuan, D., Yin, G., & Zhang, J. (2025). The impact of agricultural machinery services on food loss at the producer level: Evidence from China. *Agriculture, 15*(3).

Zou, B. L., & Mishra, A. K. (2024). Modernizing smallholder agriculture and achieving food security: An exploration in machinery services and labor reallocation in China. *Applied Economic Perspectives and Policy, 46*(4), 1662–1691. https://doi.org/10.1002/aepp.13433

Building Sustainable Business Models in the Textile Industry with a Circular Economy Approach

Avi Sharma, Shivani Raheja, Vandana Mishra, and Radhey Shyam Sharma

Abstract The textile industry, contributing nearly 20% of global industrial water pollution, faces increasing scrutiny for its environmental footprint. This study presents a novel circular economy-driven business model utilizing mesquite peroxidase enzymes, extracted from the invasive plant *Prosopis juliflora*, as a sustainable solution for textile wastewater treatment. By valorizing an ecologically harmful species into a bioresource, this model epitomizes circularity—transforming environmental threats into scalable industrial solutions. The business model features two enzyme product lines—Orbie Balls for small-scale dyers and Gummy Bears for large-scale applications in textile industry—enabling tailored adoption across diverse manufacturing scales. The enzymatic approach offers cost-effective compliance with environmental regulations, reduces health risks, and contributes to ecosystem restoration. Strategic alliances with national stakeholders, including the Ministry of Textiles and NITI Aayog, would further support the model's expansion and policy integration. Positioned within the rapidly growing $20.4 billion global industrial enzyme market, this innovation exemplifies how circular economy principles can drive both

A. Sharma
Department of Psychology, Aryabhatta College, University of Delhi South Campus, New Delhi, India

S. Raheja
Department of Commerce, Aryabhatta College, University of Delhi South Campus, New Delhi, India

V. Mishra · R. S. Sharma (✉)
Bioresources and Environmental Biotechnology Laboratory, Department of Environmental Studies, University of Delhi, Delhi, India
e-mail: rads26@hotmail.com

V. Mishra
Centre for Interdisciplinary Studies on Mountain & Hill Environment (CISMHE), University of Delhi, Delhi, India

DU-DDA Biodiversity Parks Programme, CEMDE, University of Delhi, Delhi, India

R. S. Sharma
Delhi School of Climate Change & Sustainability, Institute of Eminence, University of Delhi, Delhi, India

P. Singh et al. (eds.), *Circular Economy and Sustainable Innovation*,
https://doi.org/10.1007/978-3-032-00437-6_8

ecological resilience and economic opportunity within high-impact sectors such as textile manufacturing.

Keywords Circular economy · Sustainable business models · Textile manufacturing · Environmental sustainability · Resource efficiency · Wastewater management

1 Introduction

The Need for Sustainable Business Practices

Sustainable business practices are no longer optional but integral to achieving enduring economic resilience while simultaneously confronting escalating environmental and social challenges. The notion of sustainability in business encompasses the strategic integration of resource efficiency, environmental stewardship, and social equity, ultimately driving cost reductions and minimizing waste (MacArthur, 2013). Amid intensifying global environmental crises, the onus is on businesses to transition towards models that maintain profitability while aligning with the principles of sustainability and corporate social responsibility. According to the World Economic Forum (2020), embedding sustainability into business operations is vital for fostering long-term economic growth and systemic resilience. Innovative business models rooted in sustainability can significantly reduce environmental footprints, conserve natural resources, and improves brand reputation (Table 1). For instance, integrating circular economy principles enables enterprises to realize substantial cost efficiencies, curtail waste, and unlock new revenue opportunities (MacArthur, 2017).

The Strategic Importance of the Textile Industry in the Global Economy

The textile industry remains a foundational pillar of the global economy, contributing significantly to GDP and employment, particularly in emerging economies. As reported by the World Bank Group (2019), the textile and apparel sector accounts for approximately 2.3% of global GDP and provides livelihoods to over 75 million individuals worldwide. In India, this sector contributes around 5% to the national GDP and directly employs more than 45 million people, ranking as the second-largest employment sector after agriculture (Ministry of Textiles, 2020). Yet, the remarkable expansion and profitability of this industry have incurred substantial environmental and social costs, thereby necessitating a paradigm shift towards more sustainable and innovative business practices (Mishra et al., 2025).

Fast Fashion: An Economic Powerhouse with Environmental Costs

Fast fashion epitomizes a lucrative and swiftly adaptive business model within the global fashion industry, marked by the rapid production of affordable, trend-responsive garments. Trends are often shaped by celebrity endorsements, high-fashion runway influences, and the pervasive impact of social media. Notably, this model facilitates the design, manufacture, and distribution of new collections within mere weeks—a stark departure from traditional production cycles (Joy et al.,

Table 1 Key Components of sustainable business thoughts

Component	Description	Example	Benefit	Source
Resource efficiency	Efficient use of materials and resources	Recycling programs	Cost savings, reduced waste	MacArthur (2013, 2017)
Waste reduction	Minimizing waste production	Zero-waste manufacturing	Environmental conservation	World Economic Forum (2020)
Renewable energy	Utilizing renewable energy sources	Solar panels in factories	Reduced carbon footprint	UN Environment Programme (2019)
Social responsibility	Commitment to ethical practices	Fair labor practices	Enhanced corporate reputation	Porter and Kramer (2006)
Economic resilience	Building economic systems that withstand disruptions	Diversified supply chains	Long-term growth and stability	MarketsandMarkets (2024)

2012). The affordability of these garments democratizes fashion, making it accessible to a broad consumer base, while frequent inventory turnovers encourage repeat purchasing behavior (Bick et al., 2018).

However, this model is inherently predicated on short product lifespans, with garments often designed for limited use before being discarded, thereby generating considerable environmental and public health burdens (Table 2). The fashion industry ranks among the top ten contributors to industrial water pollution, largely due to dyeing and textile finishing operations (Rawat et al., 2016; UNEP, 2019). Moreover, the extensive global supply chain supporting fast fashion frequently relies on labor in developing nations working under exploitative and unsafe conditions (Caniato et al., 2012; Niinimäki et al., 2020). Consequently, sustainability experts have raised concerns of environmental justice and ecological integrity, demanding efforts for ethical and sustainable transformation across the industry (Mishra et al., 2020; Vijeyarasa & Liu, 2022).

1.1 Problem Statement—Major Sustainability Concerns in the Fashion Industry

Water Pollution from Dyeing Processes

Release of partially treated or untreated industrial dye effluent in the environment has emerged as one of the most acute environmental threats associated with fast fashion. Most dyeing industries are concentrated in the global south,

Table 2 Environmental impact of fast fashion industry

Aspect	Description	Statistics (Year)	Consequence	Source
Water usage	High consumption of water resources	20% of global wastewater	Depletion of freshwater resources	UN Environment Programme (2019)
Carbon emissions	Emissions from production processes	High global emissions	Climate change	UN Environment Programme (2019)
Waste generation	Large amounts of textile waste	Millions of tons annually	Environmental pollution	Niinimäki et al. (2020)
Labor practices	Poor working conditions and wages	Prevalent in developing countries	Social inequality	Niinimäki et al. (2020)
Synthetic materials	Use of non-biodegradable fibers	Extensive use of polyester	Long-term environmental damage	MacArthur (2013, 2017)

comprising clusters of small-scale to mid-scale to large-scale enterprises, where environmental regulations are either relaxed or not strictly complied (Rawat et al., 2018, 2022; Mukherjee et al., 2024a). Therefore, inadequately treated effluents containing hazardous dyes and metabolites challenge the ecological integrity of adjacent water bodies (Garg et al., 2020; Kaur et al., 2022), severely threatening aquatic biodiversity and ecosystem services (Mukherjee et al., 2024b). Besides threatening local ecology, the toxic dye effluent has been suggested to pose severe adverse health impacts on local populations, including dermatological and genetic disorders and other chronic diseases (Holkar et al., 2016; Kumar et al., 2022a). Textile industries rely primarily on synthetic and recalcitrant dyes that resist biodegradation and remain persistent in the environment, causing public health risks (Rawat et al., 2016). Addressing these complex challenges warrants stringent regulatory enforcement and widespread adoption of ecologically sound and environmentally responsible dye effluent treatment practices.

1.2 Economic Constraints and Environmental Costs

Inadequate Wastewater Treatment

Textile dyeing industries across different developing countries face highly competitive environment to remain profitable for meeting global demands while ensuring rapid scalability and cost minimization by compromising with environmental and labour standards. Therefore, emphasizing high-throughput model, characterized by swift production cycles and affordability, leads to inadvertent unsustainable outcomes (Niinimäki et al., 2020). Common economic and environmental issues of fast fashion industries are poorly managed wastewater systems due to widespread use of synthetic fibres and energy-intensive manufacturing processes (Mishra et al., 2023; Rawat et al., 2023; Yadav et al., 2022). Above all, it is estimated that the fashion industry contributes ~ 20% of global water pollution due to release of untreated industrial wastewater in river ecosystems (UN Environment Programme, 2019). These environmentally damaging practices leading to ecological, financial and social burdens besides increasing reputational risks, demand strict regulatory provision. It is further supported by the fact that consumer expectations increasingly favor sustainable brands.

Eonomic Constraints and Regulatory Non-compliance

Although current textile industry models are characterized by considerable inherent operational burdens and regulatory challenges due to widely used conventional wastewater treatment technologies, mostly relying on chemical coagulation and flocculation and often end up generating secondary pollutants (Holkar et al., 2016). Many small and medium-sized enterprises (SMEs) in developing nations resort to non-compliance with environmental mandates due to the heavy cost of chemical treatments (Mukherjee et al., 2024a, 2024b). Furthermore, the fashion sector faces financial volatility due to heavy reliance on non-renewable inputs and inefficient production workflows. This highlights an urgent need of circular business models, promoting regenerative and economically viable alternatives, a major shift from the linear economic model, characterized by substantial resource depletion and economic inefficiencies due to focus on extractive 'take-make-dispose' practices (MacArthur Foundation, 2013, 2017).

1.3 Proposed Solutions and Innovations

Mesquite Peroxidase-Based Wastewater Treatment

A significantly high environmental and operational cost associated with conventional wastewater treatment in the textile industry demands innovative, sustainable, and cost-efficient solutions (Mishra et al., 2025). Microbial methods have been suggested as cost-effective biological methods for treatment of industrial wastewater but the complexity of dyeing effluent threatens the viability of microbes, often due to their population density below the minimum effective level. Among the biological methods, enzymatic treatment has emerged as an innovative approach for treating complex textile effluent, owing to their high stability and specificity under a complex environment. However, identifying a sustainable and effective source of enzyme and developing an economically viable approach for enzyme preparation remain the key for its industrial integration. Application of mesquite peroxidase-based enzymatic treatment represents a promising, effective and eco-friendly methods for detoxifying textile effluents (Garg et al., 2020). Peroxidase from *Prosopis juliflora*, an invasive mesquite species, facilitates the transformation of toxicants into less harmful compounds, thereby significantly reducing the risks to human and ecological health (Singh et al., 2017, 2020). Moreover, creating harvesting stress on invasive species biomass not only contributes to ecological restoration of native ecosystems but also adds value to an otherwise problematic biomass.

Advantages and Novelty of Mesquite Peroxidase

Innovative convergence of environmental biotechnological approaches and ecological principles exemplifies a wastewater treatment system that provides dual benefit: environmental detoxification and the valorization of invasive plant biomass. *P. juliflora* peroxidase demonstrates high efficacy in degradation of synthetic dyes and other recalcitrant pollutants commonly found in textile industry effluents (Garg et al., 2020; Singh et al., 2017, 2020). In contrast with conventional physico-chemical treatment that rely on costly and environmentally hazardous chemical reagents, enzymatic treatment using mesquite peroxidase offers a low-cost, biodegradable, and efficient alternative with minimal secondary pollution.

This business model, using peroxidases from invasive plant species aligns with circular economy objectives, converting ecological stressors into valuable economic assets, transforming an invasive species into a critical input for sustainable and circular industrial practices. Such a strategy fosters both environmental conservation, ecosystem restoration and biodiversity recovery (Shiferaw et al., 2018). Furthermore, it illustrates how nature-based innovations can configure linear economy industrial systems into circular economic systems with greater ecological harmony and economic resilience.

Knowledge Gaps and Study Objectives

Although potential of mesquite peroxidase in textile effluent remediation has been previously demostrated, several critical gaps remain for its successful industrial integration. Some of the challenges for its onsite application include (i) scaling up of enzyme production, (ii) ensuring consistent quality, and activity at commercial levels, and (iii) optimizing extraction and stabilization processes. Additionally, robust techno-economic models and operational frameworks are required for its integration into existing wastewater infrastructure, especially for small- and medium-sized industries.

Therefore, the current study proposes a business strategy that advocates adoption of mesquite peroxidase in textile industrial wastewater treatment by exploring avenues for (i) cost-effective enzyme production, (ii) enzymatic process integration, and (iii) supply chain design for enzyme. The perspective developed will help facilitate ecological entrepreneurs to innovate and develop bioformulations to catalyze the transition of textile industry toward a sustainable circular bioeconomy.

2 Methodology

This study employed an ecological innovation approach for designing a sustainable business model. The enzymatic treatment was anchored in the extraction and characterization of mesquite peroxidase from the invasive plant *Prosopis juliflora*, and using it as a cost-effective biocatalyst for textile wastewater remediation (Garg et al., 2020; Singh et al., 2017, 2020). Technical feasibility was evaluated based on enzyme properties relevant to its real-world application in industries, including efficacy and stability in a wide range of thermal and pH regimes, besides the cost compared to conventional chemical treatments. Given the invasive nature of *P. juliflora*, perspective centered around ecological restoration potential, and socioeconomic feasibility (Table 3) by analyzing biomass availability, ecological impacts, and integration potential for small and medium enterprises (SMEs), as well as attention to public health outcomes in developing nations affected by inadequate wastewater treatment (CPCB, 2017; Pichot & Delarue, 2019).

Building on scientific validation, a dual-product strategy was developed to fit in the need of different market segments: (i) the Orbie Ball model, proposing Orbie balls soaked in enzyme solution, dried, and used as a low-cost option suitable for small micro-enterprises; and, (ii) the Gummy Bear model, proposing dry gummy bears soaked in a concentrated enzyme solution and supplied in moist form, requiring refrigeration to fit in the needs of medium to large enterprises. This two-tiered approach aims to serve both grassroots entrepreneurial ventures and large-scale industrial manufacturers, ensuring fair access and broad market reach (Porter & Kramer, 2006). A comparative cost–benefit analysis was conducted to assess the commercial viability against conventional treatments, focusing on return on investment (ROI), capital and operational expenditure, and long-term profitability (MarketsandMarkets, 2024). The implementation plan included pilot trials, informed by prior enzymatic studies, stakeholder consultations, and the development of strategic partnerships with textile industries and policy bodies.

Table 3 Advantages of mesquite peroxidase-based business model

Advantage	Description	Example	Benefit
Cost-Effective	Low-cost enzymatic treatment	Reduced use of chemical reagents	Cost savings, less secondary pollution
Sustainable	Eco-friendly wastewater treatment	Degradation of synthetic dyes	Reduced environmental impact
Resource Utilization	Use of invasive plant species	*Prosopis juliflora* as enzyme source	Biodiversity restoration
Scalability	Applicable to various industries	Paper and pulp, food and beverage	Revenue generation, market expansion
Public Health	Reduces toxic pollutants in water	Improved water quality	Reduced health risks, better ecosystems

3 Results and Discussion

3.1 Product Innovation and Business Model Design

The proposed business model is an innovation grounded in circular economy principles of using invasive species biomass for textile industrial sustainability, addressing ecotoxic challenges imposed by textile wastewater to ecosystems and human society. At its core is the enzymatic biotreatment of industrial effluents using the world's worst invasive mesquite (*P. juliflora*) causing ecological degradation across semi-arid landscapes (Ellsworth et al., 2018; Shiferaw et al., 2018). This innovative approach offers an ecologically sound alternative to conventional chemical treatments by involving valorization of an unwanted biomass, aligning ecological restoration with environmental compliance and cost efficiency (Singh et al., 2020).

Two product lines designed to cater to varied industrial scales is proposed (i) Orbie Ball for small-scale dyers (Fig. 1), and (ii) Gummy Bear for medium and large-scale enterprises (Fig. 2), by encapsulating mesquite peroxidase to enhance stability and shelve life, for efficient degradation of synthetic dyes to minimize or avoid formation of toxic aromatic amines in untreated wastewater. To enhance coverage across the textile industry by reducing environmental and health burdens, two proposed strategies differ in formulation, target markets, storage requirements, and pricing strategies (Table 4).

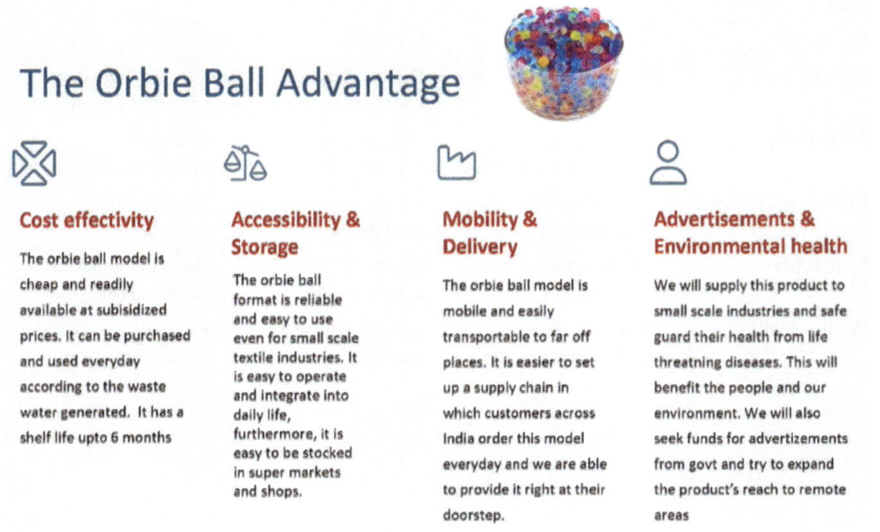

Fig. 1 Outline of advantages of Orbie Ball business model for wastewater treatment and textile industry sustainability

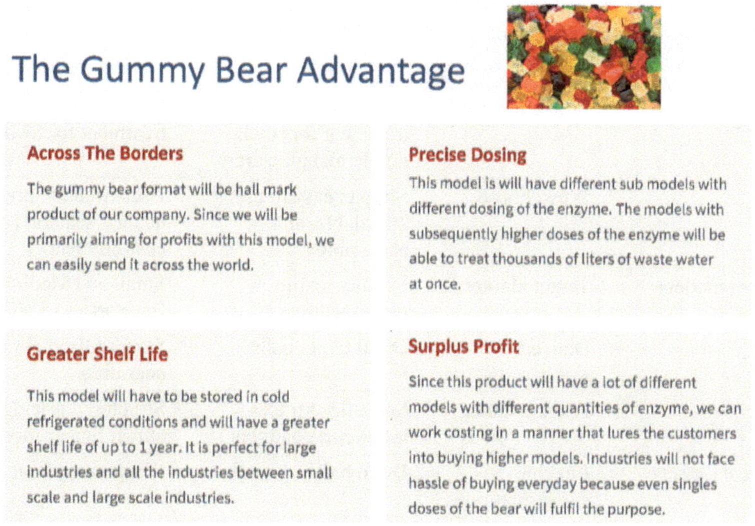

Fig. 2 Outline of advantages of Gummy Bear business model for wastewater treatment and textile industry sustainability

The Orbie Ball Model: Low-Cost Accessibility for Decentralized Units

The Orbie Ball model, as a dry-form enzyme capsule with a six-month shelf life, is tailored for small-scale, decentralized, or informal dyeing units because it requires no refrigeration and supports easy distribution. Not only is it low cost, but also easy to use under non-industrial conditions. Orbie ball enables microenterprises to comply with environmental regulations without formal wastewater infrastructure. Integration of Orbie business model with government schemes like the Viksit Bharat mission can facilitate subsidized distribution, awareness campaigns, and community engagement around safe dyeing practices, creating pathways for public funding and health-based interventions.

Under unregulated settings, the Orbie Ball model has the potential to reduce the untreated discharge, and addresses the challenges of pollution of aquatic ecosystems (Rawat et al., 2016). Thus, it would serve dual purposes: (i) Preventing pollution of water bodies and associated environmental and health challenges, (ii) A step towards social equity through employment generation in rural and peri-urban settings by collection of *P. juliflora* leaves from the field.

The Gummy Bear Model: Precision and Integration for Industrial Users

The Gummy Bear model targets medium to large-scale industrial users, and provides a high-performance, gel-encapsulated enzyme formulation. This model can be easily integrated with the existing Effluent Treatment Plants (ETPs), which are mandatory for the industries involved in the dying business. Due to low temperature storage, they would have a shelf life of up to one year. Also, the potential for the precision

Table 4 Comparison of the Orbie and the Gummy Bear business models based on various aspects relevant to their business and technical features

Category	Aspect	Orbie model	Gummy bear model
Value proposition	Problem solved	Affordable wastewater treatment for small textile manufacturers	Efficient wastewater treatment for SMEs
	Why choose	Simple, easy to use, affordable, and accessible	User-friendly, precise dosing, scalable, and cost-efficient
Business model	Target market	Smaller textile manufacturers	Small and Medium-sized Enterprises (SMEs)
	Delivery mechanism	Dried Orbei balls	Standardized gummy bear units
	Pricing strategy	Subsidized prices, long-term contracts	Structured pricing to attract higher models
Technical aspects	Enzyme encapsulation	Dried balls	Wet gummy bears
	Stability and shelf life	Stable for up to six months in dry form	Higher stability, up to two years when stored properly
	Dosage and efficiency	Tailored to daily wastewater generation	Multiple dosing options for higher efficiency
Environmental and economic impact	Production cost	Low production cost due to stability of mesquite peroxidase in crude form	Slightly higher due to packaging and dosing
	Environmental benefits	Reduces water pollution, improves effluent quality	Reduces water pollution, improves effluent quality
	Economic benefits	Cost savings for small textile manufacturers, job creation in local areas	Cost savings for SMEs, potential revenue from enzyme sales
Scalability and market reach	Production scale	Small scale, focusing on local and remote areas	Large scale, aiming for global distribution
	Marketing strategy	Government funds for advertising, partnerships with small industries	Strategic marketing campaigns, collaborations with industry associations
	Distribution strategy	Establishing supply chains for timely delivery	Easy distribution through standardized units
Collaborative efforts	Government support	Seeking government funds for advertising and expansion into remote regions	Engaging with government bodies for support and capacity building

(continued)

Table 4 (continued)

Category	Aspect	Orbie model	Gummy bear model
	Industry partnerships	Partnerships with small industries, promoting adoption	Collaborations with industry associations and large-scale manufacturers
	Community involvement	Involving community for environmental conservation and socioeconomic benefits	Broadening market reach through partnerships with dyers' associations
Technical support and training	Support mechanism	Providing technical support and customized implementation strategies	Capacity-building programs and training sessions
	Capacity building	Training programs to motivate small retail dyers	Workshops and training sessions for industry personnel and policymakers
Unique selling proposition (USP)	Source of enzyme	Derived from *Prosopis juliflora*, an unwanted invasive plant	Derived from *Prosopis juliflora*, an unwanted invasive plant
	Circular economy	Converts waste into a resource, promotes environmental conservation and community involvement	Converts waste into a resource, aligns with global sustainability initiatives
	Competitive advantage	Low production cost, easy to use, accessible	Low production cost, scalable, precise dosing

dosing based on effluent volume and composition, supports scalability and export readiness of the gummy bears for adoption in regulated sectors and international markets.

There are several advantages of Gummy Bear model over the Orbie one, including but not limited to dose efficiency and process optimization, as fewer units are needed for the same treatment effect, maximizing efficacy. In fact, it is also characterized by tailoring or customization across diverse operational scales and industries.

Comparative Analysis: Complementary Models for Inclusive Impact

The two models fulfil complementary roles, as the Orbie Ball Model prioritizes accessibility to informal sectors, enabling environmental justice. On the other hand, the Gummy Bear targets formal sectors, ensuring high-end precision, and scalability. Ultimately, both target reductions in dye toxicity and improve compliance with effluent discharge norms while taking advantage of the ecological value of *P. juliflora* as a raw material. Importantly, our interactions with both informal small-to-medium scale dyers and formal large-scale operators at Panipat—India's textile dyeing hub—revealed their strong interest in adopting such treatment technologies as long as they are affordable. Regulatory incentives can further boost the adoption of such low cost technologies by the dyers of every scale.

Product Validation through Comparable Technologies

Enzyme-based commercial formulations, such as Biogel (laccase-embedded beads) and Bioclean™ (urease gel units) serve as scientific validity and credibility and market feasibility for gel-encapsulated enzymatic preparations (Aslam et al., 2022; Darwesh et al., 2019; Kabra et al., 2013). The mesquite peroxidase model differentiates itself due to its ubiquitous presence and low-technology extraction of the enzyme preparation. Further, it exemplifies the principle of ecosystem conservation by utilization of invasive species. This positioning enhances its sustainability value proposition and makes it a lucrative business model.

Further, the enzyme exhibits high catalytic efficiency even in crude form, and under diverse environmental conditions, viz. pH, temperature, contaminant concentration etc. reducing processing complexity and cost (Singh et al., 2017). Community-based harvesting of *P. juliflora* from wild serves as rural employment opportunities integrating social entrepreneurship into the business model. A similar community-based harvesting model proposing the use of *Sacchrum ravennae* species for restoration of barren fly ash dumps has received community appreciation and adoption (Rau et al., 2009). Hence this approach not only aligns with the core principles of ecosystem restoration (Karmakar et al., 2024) but also contributes to community-based resource management.

Economic Feasibility: Disruptive Cost Advantage

A lower cost of harvesting *P. juliflora* and producing MPx makes it economically viable. Singh et al. (2020) have estimated that for 1,000 enzyme units, the total cost is \$0.18 (raw material collection: \$0.014; chemical reagents: \$0.14; and processing energy: \$0.032). In contrast, other commercially available peroxidases, such as horseradish peroxidase (HRP), cost ~ \$7.45 for the same quantity (Singh et al., 2020). This 40-fold low price of MPx shows its high potential in enzyme markets, particularly in textile manufacturing sector, where even marginal cost increases result in loss of international orders to other countries with relatively less stringent policy legislation and even poorer implementation (pers. comm.). This cost difference provides a competitive advantage for entering into the market while encouraging public–private cost sharing and CSR funding. As the global industrial enzyme market is projected to grow from USD 14.0 billion in 2024 to USD 20.4 billion by 2029 (MarketsandMarkets, 2024), MPx has a fairly high market acceptability, besides the environmental and ethical appeal.

The dual-product strategy—Orbie Balls for decentralized small-scale industry and Gummy Bears for formal medium to large-scale industry aligns well with India's textile sector and its regulatory provisions. By balancing affordability, scalability, and ecological regeneration, the mesquite peroxidase model offers a rare combination of innovation, inclusion, and environmental management. Future research should focus on enzyme formulation optimization under variable climatic and effluent conditions associated with variation in textile characteristics. Further, a digital monitoring system for treatment efficacy, and startups for enzyme manufacturing would enhance local ownership and adoption of this innovative business. With these refinements, the model can serve as a scalable, equitable framework for sustainable transformation in even other pollution-intensive industries.

3.2 Ecosystem Integration and Implementation Strategy

Both scientific validity and supporting institutional, policy, and market ecosystems remain a prerequisite for the success of any environmental innovative technology. Therefore, mesquite peroxidase-based wastewater treatment model can be a success, if the efforts integrate government support, stakeholder engagement, market infrastructure, revenue viability, and milestone-driven execution. To evaluate the potential of proposed buisness model in India's textile sector, it is important to examine necessary ecosystem that ensures its integration through inclusive innovation and systemic partnerships.

Government Support and Institutional Enablement

Large-scale adoption of clean technologies cannot be a reality unless Governmental serves as a catalyst, particularly in pollution-intensive sectors like textiles. Proposed mesquite peroxidase-based wastewater treatment model aligns with the Charter for the Textile Industry (CPCB, 2017), and circular economy frameworks advocated by NITI Aayog, which is conducive for its institutionalization, particularly in mobilizing public funding and regulatory incentives.

Among the Indian Government institutes and initiatives, Technology Mission of the Department of Science and Technology (DST, 2021) and state-level MSME departments help enable capacity-building and innovation adoption. Programmed collaboration may include: (i) grant-based pilot funding for decentralized enzyme production, (ii) technical training workshops for dyers, and (iii) dissemination of best practices across textile clusters (Fig. 3). Notably, capacity-building is not limited to technical training but also involves sensitization of factory owners and workers to the long-term ecological and health implications of untreated dye effluents (Kumar et al., 2022a; Rawat et al., 2016). Without such interventions, even promising technologies may fail to gain ground in entrenched operational cultures.

Scalability, Production Strategy, and Market Reach

Scalability is embedded into the business model through a bifurcated production and delivery strategy—dry-form Orbie Balls for decentralized enterprises and refrigerated gel-form Gummy Bears for regulated industrial units. These delivery systems address divergent operational capacities and wastewater treatment needs. Scalability is also facilitated through the modularity of the Gummy Bear formulation, which can be tailored to various effluent volumes, thereby ensuring adaptability to small, medium, and large-scale production units.

To enable this scale-up, the strategy incorporates pilot-to-commercial pipeline development—beginning with enzyme yield optimization at laboratory and pilot

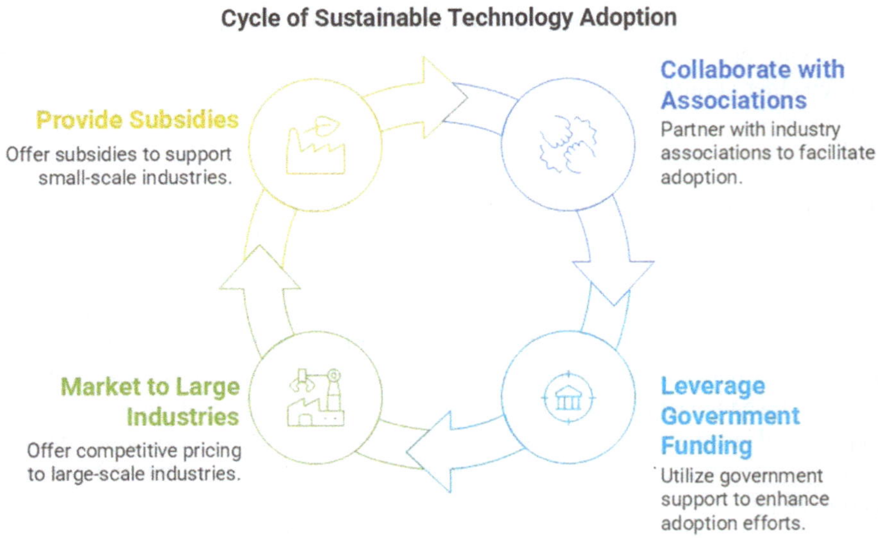

Fig. 3 Strategy for reaching a broader market through outsourcing, symbiotic partnerships, and revenue generation

scales, followed by industrial extraction protocols for batch consistency and quality assurance. The availability of *P. juliflora* biomass across semi-arid regions further ensures raw material abundance, while the crude enzyme's catalytic stability reduces downstream purification costs (Garg et al., 2020; Singh et al., 2020). These attributes make mass production technically feasible and economically viable, particularly for deployment in dyeing hubs such as Panipat, Ludhiana, and Surat.

Reaching the Broader Market: A Multi-Actor Dissemination Model

Effective dissemination requires an ecosystem approach encompassing government bodies, textile industry associations, civil society organizations, and media platforms. Outreach will be operationalized through demonstration pilots, localized awareness campaigns, and regional policy dialogues. Collaborative linkages with dyers' associations offer an efficient mechanism for reaching a fragmented base of informal operators, many of whom remain unaware of sustainable alternatives or lack the financial capacity to implement them.

In line with inclusive innovation models, this approach emphasizes co-creation, where knowledge transfer is not top-down but iterative, culturally contextualized, and capacity-building oriented. Demonstration pilots, particularly in semi-urban textile hubs, can serve as proof-of-concept installations—generating data on efficacy, cost savings, and health outcomes that feed into marketing, regulatory endorsement, and investor interest (Fig. 4). Such system-level approaches mirror those used in environmental quality assessments of river systems (Kumar et al., 2022b). Public–private partnerships (PPPs) and CSR-backed initiatives can support subsidized distribution of Orbie Balls in low-income settings. Meanwhile, strategic branding and differentiated packaging of the Gummy Bear line will target industrial clients, both domestic and international. The dual-channel outreach strategy thus ensures the democratization of technology access without compromising on commercial sustainability.

Revenue Potential and Market Traction

The economic foundation of the business model is reinforced by favourable macroeconomic trends in the industrial enzyme sector. With a projected Compound Annual Growth Rate of 7.8%, the global enzyme market is expected to expand from USD 14.0 billion in 2024 to USD 20.4 billion by 2029 (MarketsandMarkets, 2024). Mesquite peroxidase offers a low-cost, eco-origin alternative in this expanding domain, with applicability extending beyond textiles to paper, leather, and agro-industrial wastewater streams (Kundu et al., 2021).

The revenue model is structured around volume-driven sales, licensing of production technology, and customized service packages (e.g., dosing plans, monitoring protocols). Orbie Balls—priced for accessibility—target mass deployment in the domestic informal sector, while Gummy Bears command premium pricing through performance guarantees, regulatory compliance, and reduced operational overheads. Moreover, enzyme sourcing from an invasive species creates positive externalities, potentially qualifying the business for green tax credits, ecosystem service compensation, and biodiversity-linked incentives.

Early adopters stand to gain from reduced chemical input costs, improved effluent quality, and enhanced brand equity, as the sustainability norms gain traction globally. Importantly, the model introduces circular revenue flows, where value generation

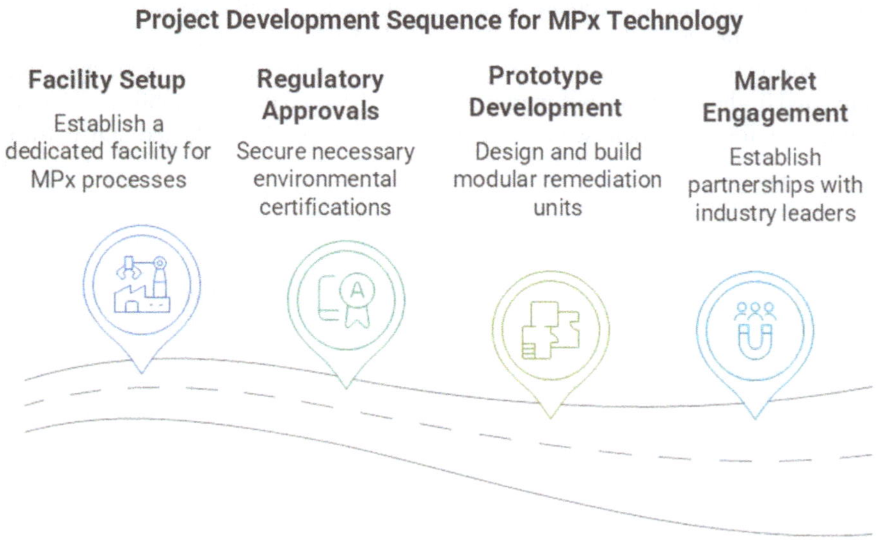

Fig. 4 Key milestones and roadmap for the proposed business model

supports not only manufacturers but also local communities engaged in biomass harvesting and preliminary processing.

Collaborative Frameworks and Stakeholder Alliances

System-wide transformation in the textile industry necessitates coalition-building across institutional and industrial actors. Academic institutions and research laboratories provide the scientific foundation, while NGOs and community organizations facilitate grassroots acceptance. Industry consortia, including the Confederation of Indian Textile Industry (CITI) and state-level textile development corporations, serve as key partners in mainstreaming adoption.

Formal alliances with large dyeing corporations enable scaling through co-branding, co-financing, and pilot-to-product transition. These partnerships also allow for joint applications to green innovation funds and climate-linked financing. Meanwhile, government ministries (Textiles, Jal Shakti, MSME, Environment) are strategic in aligning the initiative with national missions on clean water, climate adaptation, and sustainable industrialization (Fig. 5).

These networks are not merely instrumental; they also serve epistemic and normative functions—setting industry benchmarks, embedding environmental accountability, and reinforcing legitimacy in a traditionally polluting sector. Ultimately, the model's success will depend on the robustness of these collaborative arrangements, particularly in ensuring continuity of funding, technology support, and policy coherence.

Key Milestones and Strategic Roadmap

The progression of MPx from laboratory innovation in the form of an enzyme preparation to a full-scale, marketable product is guided by a well-structured, milestone-based implementation framework. The initial phase involves rigorous laboratory experiments for validation of the enzyme's efficacy in degrading complex dye structures, and further ensuring its mineralization which would further establish a scientific foundation for subsequent scaling. Next is pilot-scale trials conducted in real-world

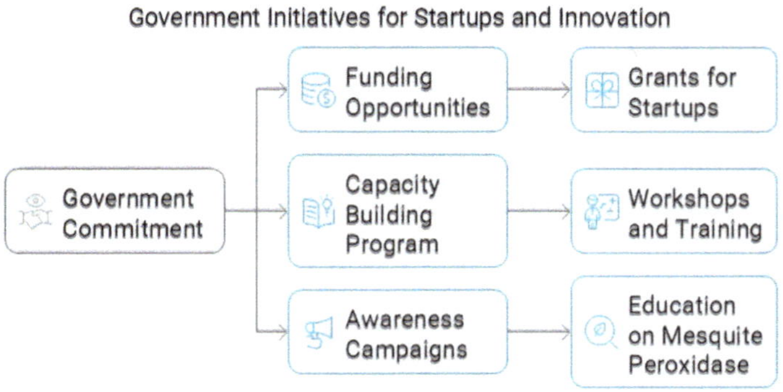

Fig. 5 Government support and capacity-building initiatives to supporting startups for industrial integration of the business model

scenario to evaluate its performance and adaptability involving dyeing clusters characterized with releasing of effluent that varies in composition (Yaseen & Scholz, 2019). Building on these findings, the enzyme preparations can enter into the product launch phase, with Orbie Balls and Gummy Bears—tailored for specific wastewater treatment needs. Efforts would be required to secure regulatory approvals from the state pollution control boards and relevant technology accreditation bodies, for legal and technological compliance. It is proposed to develop a dedicated production-cum-processing facility as a part of a modular remediation unit for seamless extraction, formulation and packaging of enzymes and integration with existing Effluent Treatment Plants (ETPs) and Common Effluent Treatment Plants (CETPs), enhancing deployment flexibility. However, strategic partnerships with textile manufacturers and municipal utilities based on perception analyses will help accelerate early adoption and further tailoring of enzyme preparation. Taken together, to facilitate adaptive scaling across the industries, minimize systemic risk and enhance long-term viability, it is essential that technical innovation, regulatory alignment, and market readiness evolve in parallel.

3.3 Environmental and Social Impact Assessment

The prime objective of the MPx model is to reduce the environmental footprint of the textile industry, besides addressing the associated issues of environmental justice and socio-ecological inequities arising from inadequate wastewater treatment. This section assesses the economic and environmental impacts of the proposed biotechnological intervention in conjunction with public health, aligning with the principles of circular economy and sustainability.

Environmental Restoration and Pollution Mitigation

The textile dyeing industry ranks among the top five contributors to industrial water pollution globally due to the wide use of synthetic dyes characterized by high persistence, toxicity, and resistance to biodegradation (Holkar et al., 2016; Rawat et al., 2016). The adoption of MPx offers an excellent alternative and/or complimentary treatment technology to the conventional treatment protocols for managing overall aquatic toxicity of effluents. Using enzyme-based detoxification of hazardous dye compounds could reduce the chemical oxygen demand (COD) and biological oxygen demand (BOD) of the real industrial effluents (Garg et al., 2020).

Pilot studies indicate the superiority of MPx treatment over conventional chemical treatments by up to 90% pollutant removal (Singh et al., 2020). Besides pollutant removal, enzymatic treatment does not generate harmful sludge or secondary pollutants, common during the application of chemical coagulants and oxidants. Thus enzymatic treatment minimizes the risk of downstream contamination and ecological disruption due to the inefficient and incomplete treatment (Mishra et al., 2025). This directly supports the regeneration of freshwater ecosystems, especially in the global south, intensively engaged in the dyeing industry, where the health of aquatic ecosystems has been severely compromised. Similar nature-based solutions have also been shown to restore industrially degraded terrestrial ecosystems (Sharma et al., 2019a, 2019b).

The environmental benefits are evident through our model's uniqueness in its sourcing of the raw material. Valorizing *P. juliflora*, a widely distributed invasive species, would contribute to biodiversity restoration by utilising invasive species. Globally, and in India, a substantial amount of money is spent annually for the control of *P. juliflora* so that other native species can regenerate. *P juliflora* is an aggressive invasive that does not allow other native species to grow and substantially reduces the biodiversity. Therefore, managing invasive species by removal reduces pressure on habitat and allows the recovery of soil fertility and native flora, and restoration of local ecological balance (Shiferaw et al., 2018). This ecological feedback loop—where pollution mitigation and invasive species management reinforce one another—illustrates the system-level potential of circular bio-innovations.

Public Health Outcomes and Social Equity

The social dimension of untreated/ partially treated textile effluents is largely underrepresented in industral sustainability, despite significant evidence linking dye pollutants to a range of health outcomes, including carcinogenicity, endocrine disruption, respiratory diseases, and dermatological conditions (Kumar et al., 2022a, 2022b). As per the Lancet commission on pollution and health (2017), extremely high global disease burden and the highest number of pollution-associated deaths are seen in some of the Asian countries of Global South. In fact, most of these countries, including India, China, Bangladesh, etc. have a flourishing dying industry. Here, the global burden of environmental pollution associated with the ever-increasing demand of textile, partially due to the adoption of fast fashion has severely affected a handful of south east Asian countries. As a result, although the fashion is enjoyed by the whole world, its negative environmental and social impact are outsourced to just a few countries, raising serious Environmental justice concerns that has received minimal global attention (Narayan et al., 2025). In informal and poorly regulated dyeing sectors, exposure risks extend beyond factory workers to include surrounding communities, particularly women and children.

The use of enzymes in wastewater treatment in the form of biodegradable formulations has the potential to reduce ambient exposure to hazardous compounds by generating non-toxic derivatives. As highlighted in global health assessments, improving water quality alone could prevent 1.4 million deaths and 74 million DALYs (disability-adjusted life years) annually, representing nearly 8% of global health losses in children under five (Wolf et al., 2023). These findings show a critical linkage between environmental innovation and health outcomes, particularly in the Global South, where regulatory enforcement is often non-stringent.

The Orbie Ball model, in particular, enhances public health outcomes in low-income small-scale units by providing affordable, non-infrastructure-dependent treatment options. It encourages the local stakeholders, who are often marginalized from formal environmental governance, to reduce pollution at source, thus democratizing access to clean water and safer livelihoods for all social classes.

Economic Sustainability and Resource Circularity

From an economist's perspective, the model shows a significant cost–benefit. For instance, by using enzymes, the textile manufacturers can achieve operational savings of up to 20% relative to conventional treatment systems alone, primarily due to the lowered chemical input and energy demand, accompanied with the minimal maintenance requirements (Singh et al., 2020). The low cost of enzyme production (~ $0.18 per 1000 units) further enhances return on investment (ROI), especially for SMEs and decentralized operators challenged with the budgetary constrains.

Additionally, the use of an abundant, renewable, and otherwise problematic biomass—*P. juliflora*—introduces new revenue streams across the supply chain. Harvesting, preprocessing, and enzyme production generate employment in rural areas, integrating socio-economic upliftment with environmental action. This closed-loop model reduces dependency on synthetic or imported reagents and aligns with the national vision of self-reliance and green industrialization. Learning from the success and failure of other similar programs say various cookstove programs launched by different countries between 1980 and 2000, for reducing indoor pollution and improving the health of exposed population, Gyapa Stoves Project of Ghana has been immensely successful (Landrigan et al., 2018). The project was different from others as it aimed to create a sustainable business model that helped the local economy by creating jobs to manufacture stove. Our business model using Orbei balls and Gummy bears for treatment of affluents perfectly aligns with this framework that has been proven successful in reducing indoor air pollution and improving the health of woman and children in Africa besides improving the indoor air quality.

Considering this, cross-sectoral applications of MPx (e.g., in pulp and paper, leather, and food processing) could be visualised beyond the textile industry. It could open further avenues for diversification of product lines, and inclusion into Environmental, Social, and Governance (ESG) investment frameworks and enhancing long-term economic resilience besides promoting environmental justice.

3.4 Systemic Implications and Future Research Scope

The environmental and social gains of the MPx model are not expected to be merely additive, but they will be systemic, reinforcing multiple UN Sustainable Development Goals (SDGs), including SDG 6 (Clean Water), SDG 12 (Sustainable Consumption and Production), SDG 9 (Industry Innovation), and SDG 15 (Life on Land). This integrated framework would transform an environmental burden (invasive biomass) into a resource, while delivering economic and public health dividends, a rare combination of circular economy principles and practical industrial consideration.

However, some limitations need critical attention. The long-term stability and efficiency of enzyme formulations across different effluent compositions require thorough investigations. Further, ensuring uninterrupted supply chains in biomass collection and scalability under fluctuating climatic and storage conditions require further empirical analyses. Moreover, lifecycle assessment studies comparing MPx-based enzymatic treatment to the existing chemical alternatives will be essential to quantify the total environmental footprint and carbon offsets.

Future research should also focus on the existing policies of effluent discharge and ensuring their stringent implementation, besides exploring the digitization of treatment monitoring. Focus should also be on adopting decentralized enzyme production models using community-run cooperatives, and policy instruments (e.g., carbon credits, green tax exemptions) that can accelerate widespread adoption. Further, the interdisciplinary studies incorporating biotechnology, industrial ecology, and socioeconomic analyses will be helpful in refining the model's implementation and expanding its applicability across various sectors.

4 Conclusion

Employing mesquite peroxidase in textile industry for effluent treatment represents a scientifically grounded and economically viable application of circular economy principles. By transforming *P. juliflora*, an ecologically invasive species, into a value-added bioresource for effluent treatment, the business model demonstrates how environmental challenges can be reconceptualized as industrial opportunities. The dual-product approach—through Orbie Balls and Gummy Bears—addresses both informal and formal textile sectors, offering scalable solutions that are cost-effective and environmentally sound.

The model's strength lies in its adaptability to local resources and systemic integration that involves technological innovation and diverse production scales supported by stakeholder partnerships, policy instruments, and training programmes. Positioned within a growing global enzyme market and aligned with India's SDGs, this model favours pollution mitigation, supports green entrepreneurship and ensures ecosystem restoration.

Nevertheless, several challenges persist that warrant further research, for example, optimization of enzyme for stability, dose standardization for diverse effluent types, and decentralized enzyme manufacturing. Besides these, the socio-political factors, including regulatory enforcement and community engagement, will determine the model's adoption success and long-term impact. Longitudinal studies on environmental health outcomes, cost-efficiency, and industry uptake will further refine the implementation frameworks.

Finally, the mesquite peroxidase model serves as a replicable blueprint for sustainable industries and ensures sustainable production. Guided by the principles of biotechnology, ecology, and economics, it exemplifies the potential of circular innovation to align profitability with planetary health. Its successful adoption catalyzes a paradigm shift in wastewater management and positions the textile sector as a pioneer in green industrialization.

Acknowledgements Avi Sharma and Shivani Raheja thank Principal, Aryabhatta College University of Delhi South Campus, for his unwavering support. Vandana Mishara and Radhey Shyam Sharma thank the Institute of Eminence, University of Delhi, for Faculty Research Grant-Projects (Ref. NO./IoE/2023-24/12/FRP, 31.08.2023).

Authors Contributions **Avi Sharma**: Conceptualization, Methodology, Data curation, Formal analysis, Validation, Visualization, Writing—original draft, Writing—review and editing; **Shivani Raheja**: Conceptualization, Methodology, Supervision; Validation; Roles/Writing—review and editing; **Vandana Mishra**: Conceptualization; Funding acquisition; Methodology; Project administration; Resources; Supervision; Validation; Visualization; Writing – review & editing; **Radhey Shyam Sharma**: Conceptualization; Funding acquisition; Methodology; Project administration; Resources; Supervision; Validation; Visualization; Writing—review and editing.

Declaration of Competing Interests This chapter has been developed in line with Springer's ethical publishing standards. All listed authors have made substantial academic contributions to its conception and writing. The corresponding author collaborated with coauthors, including family members, based on shared subject expertise. Editorial independence and research integrity have been maintained throughout. The authors declare no conflicts of interest, financial or otherwise.

Consent to Publish Statement All authors have approved and consented to publish this paper. The data and materials are available upon request.

References

Aslam, S., Ali, A., Asgher, M., Farah, N., Iqbal, H. M., & Bilal, M. (2022). Fabrication and catalytic characterization of laccase-loaded calcium-alginate beads for enhanced degradation of dye-contaminated aqueous solutions. *Catalysis Letters, 152*(6), 1729–1741.

Bick, R., Halsey, E., & Ekenga, C. C. (2018). The global environmental injustice of fast fashion. *Environmental Health, 17*(1), 1–4. https://doi.org/10.1186/s12940-018-0433-7

Caniato, F., Caridi, M., Crippa, L., & Moretto, A. (2012). Environmental sustainability in fashion supply chains: An exploratory case-based research. *International Journal of Production Economics, 135*(2), 659–670. https://doi.org/10.1016/j.ijpe.2011.06.001

CPCB. (2017). *Charter for water recycling & pollution prevention in textile industriesin Ganga river basin*. https://cpcb.nic.in/ngrba/charter_textile.pdf

Darwesh, O. M., Matter, I. A., & Eida, M. F. (2019). Development of peroxidase enzyme immobilized magnetic nanoparticles for bioremediation of textile wastewater dye. *Journal of Environmental Chemical Engineering, 7*(1), 102805.

Department of Science and Technology [DST]. (2021). *Technology mission division (water & clean energy)*. Government of India. https://dst.gov.in/technology-mission-division

Ellsworth, S. W., Crandall, P. G., Lingbeck, J. M., & O'Bryan, C. A. (2018). Perspective on the control of invasive mesquite trees and possible alternative uses. *iForest-Biogeosciences and Forestry, 11*(5), 577.

Garg, S., Kumar, P., Singh, S., Yadav, A., Dumée, L. F., Sharma, R. S., & Mishra, V. (2020). *Prosopis juliflora* peroxidases for phenol remediation from industrial wastewater—An innovative practice for environmental sustainability. *Environmental Technology & Innovation, 19*, 100865.

Holkar, C. R., Jadhav, A. J., Pinjari, D. V., Mahamuni, N. M., & Pandit, A. B. (2016). A critical review on textile wastewater treatments: Possible approaches. *Journal of Environmental Management, 182*, 351–366. https://doi.org/10.1016/j.jenvman.2016.07.090

Joy, A., Sherry, J. F., Jr., Venkatesh, A., Wang, J., & Chan, R. (2012). Fast fashion, sustainability, and the ethical appeal of luxury brands. *Fashion Theory, 16*(3), 273–295.

Kabra, A. N., Khandare, R. V., & Govindwar, S. P. (2013). Development of a bioreactor for remediation of textile effluent and dye mixture: A plant–bacterial synergistic strategy. *Water Research, 47*, 1035–1048.

Karmakar, S., Mukherjee, P., Mishra, V., Gupta, R. K., Kumar, R., Srivastava, P., & Sharma, R. S. (2024). Microhabitat influences on phage-bacteria dynamics in an abandoned mine for ecorestoration. *Journal of Environmental Management, 370*, 122659.

Kaur, H., Rawat, D., Poria, P., Sharma, U., Gibert, Y., Ethayathulla, A. S., Dumée, L. F., Sharma, R. S., & Mishra, V. (2022). Ecotoxic effects of microplastics and contaminated microplastics—Emerging evidence and perspective. *Science of the Total Environment, 841*, 156593.

Kumar, P., Mishra, V., Yadav, S., Yadav, A., Garg, S., Poria, P., Farooqi, F., Dumée, L. F., & Sharma, R. S. (2022a). Heavy metal pollution and risks in a highly polluted and populated Indian river–city pair using the systems approach. *Environmental Science and Pollution Research, 29*(40), 60212–60231.

Kumar, P., Mishra, V., Bhardwaj, S., Garg, S., Dumée, L. F., & Sharma, R. S. (2022b). Five-wheel framework for system-based monitoring of heavy metal pollution in rivers. *Environmental Quality Management, 32*, 9–17.

Kundu, D., Thakur, M. S., & Patra, S. (2021). Textile fabric processing and their sustainable effluent treatment using enzymes—Insights and challenges. *Immobilization strategies: Biomedical, Bioengineering and Environmental Applications*, 645–666.

Landrigan, P. J., Fuller, R., Acosta, N. J., Adeyi, O., Arnold, R., Baldé, A. B., Bertollini, R., Bose-O'Reilly, S., Boufford, J. I., Breysse, P. N., & Chiles, T., (2018). The Lancet Commission on pollution and health. *The Lancet, 391*(10119), 462–512.

MacArthur, E. (2013). Towards the circular economy. *Journal of Industrial Ecology, 2*(1), 23–44.

MacArthur, E. (2017). *A new textiles economy: Redesigning fashion's future*. https://www.ellenmacarthurfoundation.org/a-new-textiles-economy

MarketsandMarkets. (2024). *Industrial enzymes market by type, application, source, and region—Global forecast to 2029*. https://www.marketsandmarkets.com/Market-Reports/enzyme-market-46202020.html

Ministry of Textiles. (2020). *Annual Report 2019–2020*. https://texmin.nic.in/sites/default/files/AR_MoT_2019-20_English.pdf

Mishra, V., Mudgal, N., Rawat, D., Poria, P., Mukherjee, P., Sharma, U., Kumria, P., Pani, B., Singh, M., Yadav, A., & Farooqi, F. (2023). Integrating microalgae into textile wastewater treatment processes: Advancements and opportunities. *Journal of Water Process Engineering, 55*, 104128.

Mishra, V., Sharma, U., Rawat, D., Benson, D., Singh, M., & Sharma, R. S. (2020). Fast-changing life-styles and ecotoxicity of hair dyes drive the emergence of hidden toxicants threatening environmental sustainability in Asia. *Environmental Research, 184*, 109253.

Mishra, V., Mukherjee, P., Bhattacharya, S., & Sharma, R. S. (2025). Innovative sustainable solutions for detoxifying textile industry effluents using advanced oxidation and biological methods. *Journal of Environmental Management, 380*, 124804.

Mukherjee, P., Sharma, R. S., Rawat, D., Sharma, U., Karmakar, S., Yadav, A., & Mishra, V. (2024a). Microbial communities drive flux of acid orange 7 and crystal violet dyes in water-sediment system. *Journal of Environmental Management, 351*, 119699.

Mukherjee, P., Sharma, R. S., & Mishra, V. (2024b). Deciphering the ecological impact of azo dye pollution through microbial community analysis in water–sediment microcosms. *Environmental Science and Pollution Research*, 1–19.

Narayanan, A., Sharma, R. S., & Mishra, V. (2025). Unmasking the masks of profits: Analyzing corporate social responsibility through an environmental justice framework. In *Integrating environmental sustainability and corporate social responsibility: A move beyond profit: corporate social responsibility for environmental sustainability* (pp. 65–94). Cham: Springer Nature Switzerland.

Niinimäki, K., Peters, G., Dahlbo, H., Perry, P., Rissanen, T., & Gwilt, A. (2020). The environmental price of fast fashion. *Nature Reviews Earth & Environment, 1*(4), 189–200. https://doi.org/10.1038/s43017-020-0039-9

Pichot, E., & Delarue, R. (2019). International labour organization (ILO). In Wessel, R.A., & Odermatt, J. (Eds.), *Research handbook on the European Union and international organizations* (pp. 101–119). Edward Elgar Publishing.

Porter, M. E., & Kramer, M. R. (2006). Strategy & society: The link between competitive advantage and corporate social responsibility. *Harvard Business Review, 84*(12), 78–92.

Rawat, D., Mishra, V., & Sharma, R. S. (2016). Detoxification of azo dyes in the context of environmental processes. *Chemosphere, 155*, 591–605.

Rawat, D., Sharma, R. S., Karmakar, S., Arora, L. S., & Mishra, V. (2018). Ecotoxic potential of a presumably non-toxic azo dye. *Ecotoxicology and Environmental Safety, 148*, 528–537.

Rawat, D., Sharma, U., Poria, P., Finlan, A., Parker, B., Sharma, R. S., & Mishra, V. (2022). Iron-dependent mutualism between *Chlorella sorokiniana* and *Ralstonia pickettii* forms the basis for a sustainable bioremediation system. *ISME Communications, 2*(1), 83.

Rawat, D., Sharma, U., Yadav, A., Poria, P., Farooqi, F., Pani, B., Mukherjee, P., Singh, M., Sharma, R. S., & Mishra, V. (2023). A bottom-up approach to select microbes from textile wastewater for detoxification of aromatic amine and azo dye in a single stage. *Journal of Water Process Engineering, 53*, 103668.

Rau, N., Mishra, V., Sharma, M., Das, M. K., Ahaluwalia, K., & Sharma, R. S. (2009). Evaluation of functional diversity in rhizobacterial taxa of a wild grass (*Saccharum ravennae*) colonizing abandoned fly ash dumps in Delhi urban ecosystem. *Soil Biology and Biochemistry, 41*(4), 813–821.

Sharma, R. S., Karmakar, S., Kumar, P., & Mishra, V. (2019a). Application of filamentous phages in environment: A tectonic shift in the science and practice of ecorestoration. *Ecology and Evolution, 9*, 2263–2304.

Sharma, R. S., Nayak, S., Malhotra, S., Karmakar, S., Sharma, M., Raiping, S., & Mishra, V. (2019b). Rhizosphere provides a new paradigm on the prevalence of lysogeny in the environment. *Soil and Tillage Research, 195*, 104368.

Shiferaw, W., Sebsebe, D., & Bekele, T. (2018). Invasive plant species in Ethiopia: Ecological impacts on biodiversity, a review paper. *International Journal of Molecular Biology, 3*, 169–176.

Singh, S., Mishra, R., Sharma, R. S., & Mishra, V. (2017). Phenol remediation by peroxidase from an invasive mesquite: Turning an environmental wound into wisdom. *Journal of Hazardous Materials, 334*, 201–211.

Singh, S., Malhotra, S., Mukherjee, P., Mishra, R., Farooqi, F., Sharma, R. S., & Mishra, V. (2020). Peroxidases from an invasive Mesquite species for management and restoration of fertility of phenolic-contaminated soil. *Journal of Environmental Management, 256*, 109908.

UN Environment Programme. (2019). *Sustainability and circularity in the textile value chain: Global stocktaking.*

Vijeyarasa, R., & Liu, M. (2022). Fast fashion for 2030: Using the pattern of the sustainable development goals (SDGs) to cut a more gender-just fashion sector. *Business and Human Rights Journal, 7*(1), 45–66.

World Bank Group. (2019). *How much do our wardrobes cost to the environment*? Retrieved from https://www.worldbank.org/en/news/feature/2019/09/23/costo-moda-medio-ambiente

Wolf, J., Johnston, R. B., Ambelu, A., Arnold, B. F., Bain, R., Brauer, M., Brown, J., Caruso, B. A., Clasen, T., Colford, J. M., & Mills, J. E. (2023). Burden of disease attributable to unsafe drinking water, sanitation, and hygiene in domestic settings: A global analysis for selected adverse health outcomes. *The Lancet, 401*(10393), 2060–2071.

World Economic Forum. (2020). *New nature economy report II: The future of nature and business.* World Economic Forum. http://www3.weforum.org/docs/WEF_The_Future_Of_Nature_And_Business_2020.pdf

Yadav, A., Kumar, P., Rawat, D., Garg, S., Mukherjee, P., Farooqi, F., Roy, A., Sundaram, S., Sharma, R. S., & Mishra, V. (2022). Microbial fuel cells for mineralization and decolorization of azo dyes: Recent advances in design and materials. *Science of the Total Environment, 826*, 154038.

Yaseen, D. A., & Scholz, M. (2019). Textile dye wastewater characteristics and constituents of synthetic effluents: A critical review. *International Journal of Environmental Science and Technology, 16*, 1193–1226.

Psychologically Informed CSR: A Conceptual Framework for Promoting Sustainable Consumer Behaviour in the Fashion Industry

Avi Sharma, Neera, Ankit Prakash, Vandana Mishra, and Radhey Shyam Sharma

Abstract The fashion industry significantly contributes to global environmental degradation driven by fast fashion, high-resource production methods, and consumption practices that can not be sustained. Despite growing awareness of these impacts, a persistent gap remains between consumers' pro-environmental attitudes and their actual behaviour. This conceptual paper asks: How can corporate social responsibility (CSR), when informed by psychological theories of consumer behaviour, be strategically integrated to promote responsible consumption in the fashion sector? The current study addresses this question by synthesizing insights from environmental psychology, specifically the theories of Value-Belief-Norm (VBN), Cognitive Dissonance, and Reasoned Action, and examines how cognitive, emotional, and socio-cultural factors shape consumer responses to CSR initiatives. It critically explores the psychological barriers that hinder sustainable fashion choices and identifies the mechanisms through which CSR can activate behavioural change. The paper further proposes a conceptual framework linking corporate responsibility with political consumerism, emotional engagement, and cultural resonance to drive circular economy transitions. By bridging CSR strategy with behavioural science, this study contributes a novel theoretical lens for understanding and influencing sustainable

---

A. Sharma · Neera · A. Prakash
Department of Psychology, Aryabhatta College, University of Delhi South Campus, New Delhi, India

V. Mishra · R. S. Sharma (✉)
Department of Environmental Studies, Bioresources and Environmental Biotechnology Laboratory, University of Delhi, Delhi, India
e-mail: rads26@hotmail.com

V. Mishra
Centre for Interdisciplinary Studies on Mountain & Hill Environment (CISMHE), University of Delhi, Delhi, India

DU-DDA Biodiversity Parks Programme, CEMDE, University of Delhi, Delhi, India

R. S. Sharma
Delhi School of Climate Change & Sustainability, Institute of Eminence, University of Delhi, Delhi, India

P. Singh et al. (eds.), *Circular Economy and Sustainable Innovation*,
https://doi.org/10.1007/978-3-032-00437-6_9

consumer behaviour in high-impact sectors. The findings have implications for corporate practice, environmental policy, and the evolving discourse in environmental psychology on fostering collective pro-environmental action.

Keywords Corporate social responsibility (CSR) · Sustainable consumer behavior · Environmental psychology · Circular economy · Fashion industry sustainability · Attitude–Behavior gap

1 Introduction

The fashion industry sits at the crossroads of culture and commerce, but its environmental and social costs are becoming impossible to ignore (Garg et al., 2020; Niinimäki et al., 2020; Rawat et al., 2016). Fast-fashion brands churn out cheap, quickly obsolete clothing, a model that strains ecosystems and labour systems alike (Mukherjee et al., 2024a, 2024b; Rawat et al., 2018). Textiles now rank among the leading sources of greenhouse-gas emissions, toxic chemicals, water misuse, and unfair working conditions (Kaur et al., 2022; Kumar et al., 2022; Singh et al., 2020). These problems have sparked calls for circular-economy practices that cut waste, curb pollution, and protect workers (Kumar et al., 2023; Mishra et al., 2023, 2025; Rawat et al., 2022, 2023; Yadav et al., 2022).

Corporate Social Responsibility (CSR) gives firms a way to account for the environmental and social costs that rarely show up on a balance sheet. In practice, companies pledge to use ethically sourced materials, publish detailed impact reports, and run take-back or repair schemes intended to close the loop on waste (Carroll, 1999; Narayanan et al., 2025; Porter & Kramer, 2006; Singh et al., 2025). Whether these efforts actually shift what people buy or deliver measurable environmental gains, however, is still uncertain (Desore & Narula, 2018; Dissanayake et al., 2021; Khagokpam et al., 2025). Outcomes hinge as much on how consumers interpret a company's actions as on the actions themselves. Although surveys show rising ecological concern, most shoppers still choose low-cost, resource-intensive fashion, revealing a stubborn gap between stated attitudes and real-world purchases (White et al., 2019; Kozar & Connell, 2013).

This attitude–behaviour gap points to a weakness in conventional CSR: campaigns often stop at information or moral exhortation and overlook the psychological forces that shape choice. Values, emotions and perceived social norms steer purchasing far more than facts alone (Mishra et al., 2020). When CSR fails to address those drivers, it risks becoming little more than a public-relations exercise, with limited power to change behaviour (Delmas & Burbano, 2011; Khagokpam et al., 2025). Closing the gap, therefore, requires CSR programmes designed with behavioural science in mind—initiatives that tap into personal values, leverage peer influence and offer easy, credible ways to act sustainably. In short, CSR must evolve from a reporting toolkit into a behaviour-change intervention. The framework proposed here, grounded in three well-established psychological theories, outlines how that shift can be achieved.

This paper argues that CSR should do more than report impacts or fund charities; it must actively steer consumer behaviour. To show how, we outline a framework that weaves together three well-established psychological lenses—Value-Belief-Norm (VBN) theory, the Theory of Reasoned Action (TRA), and Cognitive Dissonance Theory (CDT) (Fishbein & Ajzen, 1977; Morvan & O'Connor, 2017; Stern et al., 1999). Taken together, these perspectives explain how personal values, social expectations, and the need for internal consistency shape what people buy. By designing CSR initiatives that speak to those motives at once, fashion brands can guide consumers toward choices that are both environmentally responsible and psychologically satisfying. Illustrated with recent industry cases, the framework positions CSR as a practical tool for closing the attitude–behaviour gap and speeding the shift to circular, ethically grounded fashion systems.

2 Theoretical Foundations

To see how CSR can foster responsible consumption, we link three behavioural lenses—motivation (VBN), intention (TRA), and emotion (CDT).

2.1 Value-Belief-Norm (VBN) Theory

VBN theory starts with personal values. When people hold strong biospheric or altruistic values, they view environmental harm as morally unacceptable. That belief activates a sense of responsibility and, in turn, a felt obligation to act (Stern et al., 1999). In practice, brands can prime these values—through storytelling that highlights ecological or social stakes—to make sustainable choices feel not just sensible but morally necessary.

To internalize these moral norms, brands can authentically align their CSR initiatives with consumers' values and present clear links between consumption choices and ecological impact, similar to Patagonia's "Don't Buy This Jacket" campaign (Chouinard & Stanley, 2013). Such a campaign appeals to consumers to practice environmental ethics by reconsidering unnecessary purchases, disrupting the company's traditional marketing logic. The message also activated moral reflection and strengthened the alignment between values and behaviour of the consumers. By activating personal norms, such CSR initiatives help consumers to express their environmental values through deliberate choices for sustained pro-environmental behaviour (Fletcher, 2013; Kozar & Connell, 2013).

2.2 Theory of Reasoned Action

The TRA offers an additional perspective by emphasizing how our attitudes and perceived social pressures influence our intentions to act (Fishbein & Ajzen, 1977). According to TRA, our decision to behave a particular way is guided not only by our personal evaluation of that behaviour but also by our perception of how others view it. In the context of fashion, where clothing choices are inherently public and expressive of identity, perceived social approval or disapproval carries substantial weight (Beckmann, 2007).

When CSR programs use social influence tools like peer examples, endorsements from trusted influencers, or grassroots narratives, they're much more likely to embed sustainable habits into everyday life. For example, H&M's Conscious Collection employs sustainability-themed marketing alongside high-profile endorsements to signal that eco-friendly fashion is both stylish and socially valued. Such campaigns can shift subjective norms by redefining ethical consumption as aspirational and expected within one's social environment (Hur & Kim, 2017).

The effectiveness of these strategies, however, depends on the credibility of the messenger and the perceived sincerity of the brand's commitments. TRA highlights the importance of referent group alignment—consumers are more likely to adopt behaviours modelled by individuals or groups they identify with (Vallerand et al., 1992). Therefore, authenticity and cultural congruence are critical to shaping social norms that promote sustainable behaviour.

2.3 Cognitive Dissonance Theory

Cognitive Dissonance Theory (CDT) explains why a clash between beliefs and actions can be a powerful driver of behavioural change. When people who care about the environment keep buying fast-fashion items, the mismatch triggers psychological discomfort—dissonance—that they feel compelled to ease (Morvan & O'Connor, 2017).

Fashion brands can turn that discomfort into a lever for good. Effective CSR programmes do two things at once: (i) make the value–behaviour conflict salient and (ii) offer an easy, credible way to resolve it. Marks & Spencer's "Shwopping" scheme is a practical example: shoppers were reminded of textile waste and immediately given the option to drop off old clothes for reuse or recycling. The act of "shwopping" transforms unease into relief, realigning behaviour with pro-environmental values (Desore & Narula, 2018).

Such initiatives work best when they position dissonance not as guilt to be endured but as a chance to reaffirm a green self-image (Fletcher, 2013; Lichy et al., 2023). By lowering the psychological hurdles to change, they shift consumers from passive awareness to active participation. While VBN highlights the motivational role of internalized values, TRA emphasizes the influence of social norms, and CDT explains how emotional discomfort can prompt self-correction.

3 A Psychological Framework for CSR-Led Behaviour Change

Integrating VBN theory, the TRA, and CDT yields a tripartite framework that treats CSR as a behaviour-change tool rather than a public-relations add-on. The three components interact as follows: Activate personal moral norms (VBN)—it highlights how a purchase—or a take-back, repair, or rental option—serves biospheric and altruistic values, turning sustainability into a moral "ought." Shape perceived social expectations (TRA)—it uses transparent reporting, visible eco-labels, and community storytelling to signal that sustainable fashion is the emerging social norm, thereby strengthening behavioural intentions. Ease cognitive discomfort (CDT)- it confronts value–behaviour gaps openly and provides low-friction remedies—recycling bins in stores, clear care instructions, or subscription repair services—to convert dissonance into affirmative action. By moving simultaneously on these three fronts, CSR can shrink the well-documented attitude–behaviour gap and raise the bar for environmental and ethical performance across the fashion sector (Kozar & Connell, 2013).

Value-Belief-Norm (VBN) theory predicts that consumers are more likely to adopt sustainable practices when those practices reflect their core values and are framed as moral responsibilities. Effective CSR programmes therefore convert general ethical statements into transparent, routine actions that engage these internal norms. Toms' "One for One" model, which links each purchase to a verifiable donation, and Eileen Fisher's Renew initiative, which accepts used garments for resale or fibre-to-fibre recycling, demonstrate how companies can embed moral purpose in day-to-day transactions. Both examples encourage consumers to view purchasing as a participatory moral act rather than a purely economic decision, aligning product benefits with biospheric and altruistic values (Stern et al., 1999; Kingston & Guellil, 2016; Weber, 2018; Sunder et al., 2022).

The TRA complements this value activation by explaining how attitudes and perceived social norms translate moral intentions into observable behaviour. Because fashion is a prominent medium for social signalling, interventions that address norms can quickly influence consumer expectations. Patagonia's "Don't Buy This Jacket" campaign positions durability as a desirable attribute, while Allbirds' practice of printing the carbon footprint on each shoebox normalises carbon accounting and implies that peers now compare footwear by emissions as well as style. H&M's Conscious Collection and its visible in-store take-back bins similarly pair a favourable attitude toward sustainable design with clear cues that other consumers are already participating. These strategies present sustainable choices as both personally advantageous and socially endorsed, thereby strengthening behavioural intentions (Beckmann, 2007).

When the value activation outlined by VBN is combined with the normative influence described by TRA—and companies simultaneously offer low-effort actions that reduce the cognitive dissonance associated with unsustainable habits—the decision to choose the environmentally preferable option can shift from occasional to habitual. Under this integrated, theory-based approach, CSR functions not merely as a reputational strategy but as a structured behaviour-change mechanism capable of accelerating the transition from fast fashion to genuinely circular and ethically responsible industry practice.

Examples such as Everlane's "Radical Transparency" initiative illustrate how CSR can reshape perceived norms. By openly disclosing cost structures, sourcing details, and environmental impacts, Everlane has positioned ethical consumption as both fashionable and normative within a broader movement of conscious consumers (Richards, 2021). Likewise, Mud Jeans' leasing model redefines ownership and consumption, positioning shared responsibility as a mark of innovation and ethical leadership. By showcasing sustainability as desirable and mainstream, these companies leverage the attitudinal and normative pathways emphasized in TRA (Thatta & Polisetty, 2022).

The third component, drawn from CDT, focuses on the psychological tension experienced when individuals' behaviours contradict their environmental values. Many consumers who express concern for sustainability still purchase from fast fashion brands, resulting in dissonance that, if unresolved, may lead to rationalization or disengagement. CSR initiatives that make these inconsistencies salient, while simultaneously offering accessible alternatives, can transform this discomfort into a force for behavioural change (Delmas & Burbano, 2011; Morvan et al., 2017).

For example, ThredUp's *Fashion Footprint Calculator* quantifies the environmental impact of an individual's wardrobe and immediately suggests corrective actions such as second-hand shopping or clothing donation (Table 1). This tool transforms guilt into guided action by framing dissonance not as a moral failure, but as a solvable problem (ThredUp, 2022). Similarly, Tentree's *Impact Wallet* provides personalized environmental feedback—such as the number of trees planted from past purchases—which helps reinforce a pro-environmental identity and reduce the psychological discomfort associated with prior unsustainable choices (Karamally & Robertson, 2023).

Together, these three mechanisms—moral norm activation, social norm construction, and dissonance resolution—form a dynamic psychological feedback loop. For example, value activation among consumers increases their receptivity to social cues, whereas social validation helps them become aware of their behavioural inconsistencies. On the other hand, opportunities for dissonance resolution reinforce both their moral identity and sustainable habits. How CSR can simultaneously address moral, social, and cognitive dimensions of behaviour is nicely exemplified by Brands such as Outland Denim, which integrated ethical labour practices and environmentally conscious production (Xu & Bolton, 2024).

Table 1 CSR initiatives in the fashion industry categorized by psychological theory (VBN, TRA, CDT), indicating the mechanisms of influence and anticipated consumer responses

Theory	CSR initiative	Fashion industry example	Psychological mechanism engaged	Expected consumer impact	References
VBN	Leasing and take-back	MUD Jeans Lease-a-Jeans	Activates personal moral norms by linking reuse to environmental responsibility	Higher lease renewals; reduced purchase guilt; stronger brand engagement	Thatta and Polisetty (2022)
	One-for-One giving	TOMS "One for One"	Engages moral obligation by tying each purchase to a social good	Stronger emotional connection; increased purchase intention	Marquis and Park (2014)
	Social-enterprise employment	Outland denim social enterprise model	Reinforces pro-social values by highlighting empowerment of marginalized workers	Increased willingness to pay premium; stronger advocacy	Barrett and Moores (2022)
TRA	Circular take-back and recycling rewards	Marks and Spencer "Shwopping" campaign	Shapes subjective norms by making recycling socially visible and rewarded	Increased recycling participation; positive brand association	Morgan (2022)
	Radical transparency	Everlane "Radical Transparency"	Builds favourable attitudes and normative beliefs through open cost breakdowns	Higher trust; greater intention to choose Everlane over less transparent brands	Richards
CDT	Take-back and resale	Eileen Fisher Renew programme	Provides dissonance-reducing alternatives so values ("buy less") align with behaviour	Greater participation in reuse; reduced guilt; enhanced brand loyalty	Weber (2018)
	In-store drop-off rewards	H&M Garment Collecting	Offers clear pathways to resolve dissonance between eco-values and purchase habits	Higher collection rates; repeat store visits	Seo and Jin (2024)

(continued)

Table 1 (continued)

Theory	CSR initiative	Fashion industry example	Psychological mechanism engaged	Expected consumer impact	References
	Online impact calculator	ThredUP Fashion Footprint Calculator	Exposes behaviour–value gaps and offers actionable steps to reduce dissonance	Shift toward lower-impact purchases; improved consumer self-efficacy	ThredUP (2022)
CDT	Rental platform (sharing economy)	Flyrobe fashion rental (India)	Gives values-congruent alternative. To ownership, reducing purchase–value conflict	Increased rental adoption; fosters circular mindset	Gupta et al.
VBN/TRA	Community-based storytelling	Tentree impact wallet	Reinforces both personal values and perceived norms through impact feedback	Stronger brand loyalty; repeat purchases motivated by visible impact	Karamally and Robertson (2023)

These psychological processes, although distinct, operate recursively, reinforcing one another throughout the consumer decision-making process. This integrated approach enables CSR to function not merely as a tool of corporate communication but as a catalyst for systemic behavioural transformation. Therefore, effective CSR must be designed to engage with moral values, influence social attitudes, and provide mechanisms for behavioural alignment.

In the fast fashion industry, overproduction, trend obsolescence, and disposability are structurally embedded. Adopting the present psychologically-informed CSR framework can normalize ethical consumption and facilitate consumers' transitions away from unsustainable behaviours. By resonating with their personal values, fashion brands can meaningfully contribute to achieving the intended targets of the circular economy. It bridges the gap between sustainability commitments and the psychological gaps in consumer behaviour by designing psychologically informed and behaviourally effective CSR strategies.

4 Operationalizing the Framework: CSR Domains and Psychological Mechanisms

This section outlines the psychological mechanisms that are activated by the strategic CSR initiatives applied in the fashion industry (Fig. 1). Rather than revisiting the conceptual framework, the focus here is on translating psychological insights into actionable CSR strategies. By mapping each psychological pathway to a

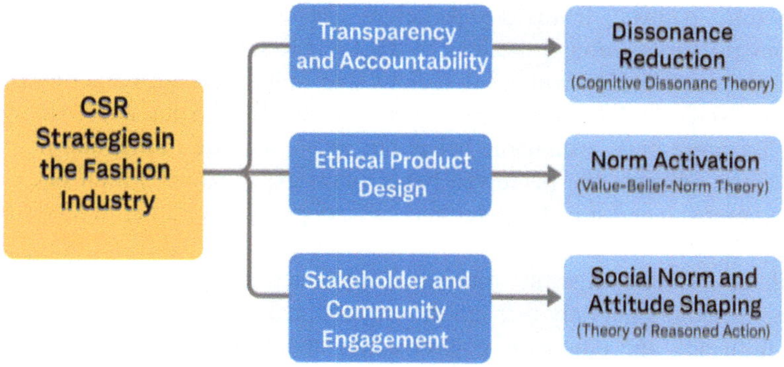

Fig. 1 Flow chart showing how and why CSR strategies in the fashion industry should involve ensuring transparency to reduce cognitive dissonance, designing ethically to activate norms, and engaging communities to shape social attitudes

corresponding domain of CSR practice, the framework demonstrates how brands can design initiatives that not only communicate ethical intent but also influence consumer behaviour in measurable and lasting ways.

The framework posits that CSR efforts are most impactful when they operate simultaneously at three interrelated psychological levels: personal norms and values, social attitudes and expectations, and cognitive dissonance resolution. At its core lies consumer engagement, conceptualized as a dynamic progression from awareness to evaluation, intention, and ultimately to sustainable action. The goal is to reduce the well-documented attitude–behaviour gap by aligning CSR practices with the cognitive and emotional drivers that shape consumer decisions (Connell, 2019; White et al., 2019).

The first pathway emphasizes transparency and accountability, involving the disclosure of environmental performance, labour practices, and supply chain operations (Fig. 1). These practices, when candid and substantive, can elicit cognitive dissonance among environmentally conscious consumers by drawing attention to inconsistencies between their values and past purchasing behaviours (Delmas & Burbano, 2011). If companies pair transparency with accessible and ethical alternatives—such as sustainable collections or take-back programmes—they enable consumers to resolve this dissonance, thus encouraging value-aligned behavioural change.

The second pathway focuses on ethical product design, directly engaging the mechanisms outlined in VBN theory. CSR initiatives that prioritize eco-friendly materials, fair labour, and product longevity resonate with consumers who hold strong biospheric or altruistic values (Niinimäki et al., 2020). When such initiatives are communicated effectively—framing sustainability as both urgent and actionable—they can activate personal moral norms and foster enduring commitment to responsible consumption (Fletcher, 2013). Examples such as Eileen Fisher's *Renew* program illustrate how design-centred CSR can trigger long-term value alignment (Sunder et al., 2022; Weber, 2018).

The third pathway involves stakeholder and community engagement, which functions primarily through the lens of TRA. CSR strategies such as educational campaigns, partnerships with civil society, and influencer-led marketing serve to shift perceived social norms and consumer attitudes (Beckmann, 2007; Hur & Kim, 2017). By presenting sustainable behaviour as popular, desirable, and normative, these efforts enhance the perceived behavioural expectations within one's social group, thereby increasing the likelihood of adoption (Everlane, 2023; Thatta & Polisetty, 2022).

Collectively, this tripartite framework underscores that effective CSR must engage consumers not only through information and values but also by leveraging social influence and emotional-cognitive alignment. When executed cohesively, CSR becomes a mechanism for building trust, activating intrinsic motivations, and fostering behavioural consistency, thereby contributing to systemic shifts in fashion consumption (Carroll, 1999; Porter & Kramer, 2006).

4.1 Core Components of the Framework

To operationalize three core components of the proposed framework, three primary domains of CSR messaging and practice are focused and designed to activate specific psychological mechanisms, influencing consumer behaviour. For example, the first domain, *Transparency and Accountability*, involves practices such as sustainability reporting, supply chain disclosures, and third-party ethical certifications (GRI, 2021; Dissanayake et al., 2021). These efforts aim to build trust among consumers and promote awareness of existing cognitive dissonance. By making ethical performance visible, consumers are encouraged to reconcile their values with their purchasing decisions (Delmas & Burbano, 2011). Ethical Product Design, the second focus area, emphasizes choosing eco-friendly materials, ensuring fair labor practices, and building products to last longer. These practices provide clear pathways for value-consistent consumption (Fletcher, 2012; Niinimäki, 2010) because they closely align with consumers' moral and environmental values, activating personal norms. Finally, *Stakeholder and Community Engagement* represents the third domain that includes partnerships with advocacy groups, integration with educational initiatives, and campaigns by social actors. These strategies will help shape social norms and consumer attitudes, aligning with TRA (Fishbein & Ajzen, 1975). Consequently, sustainable behavior is presented as normative, desirable, and socially endorsed (Beckmann, 2007; Hur & Kim, 2017). In this way, these three domains explain how psychologically informed CSR practices can drive pro-environmental behavior synergistically and strengthen consumers' sustainable engagement in the fashion industry.

5 Discussion and Implications for Theory and Practice

The proposed psychologically-informed framework is to be integrated within CSR for influencing the behaviour of the consumer by adopting the principles of the circular economy. By integrating three distinct yet connected psychological theories, i.e., VBN Theory, the Theory of Reasoned Action, and Cognitive Dissonance Theory, the framework aims to address the attitude–behaviour gap. Consequently, the consumers' pro-environmental concerns and their actual purchasing patterns remain aligned (White et al., 2019; Kozar & Connell, 2013) by engaging values, social norms, and internal cognitive tensions during CSR activities to support long-term behavioural change.

The reconceptualization of CSR presented here represents how multi-level psychological processes can be mobilized through CSR activities to make them effective by using psychological processes as a critical complement to normative and instrumental motives (Agudelo et al., 2019; Meseguer-Sánchez et al., 2021). Therefore, the current framework will help contribute to understanding how consumer behaviour can be shaped through ethically aligned business strategy.

5.1 Theoretical Implications

The present analysis fosters interdisciplinary integration between environmental psychology and corporate sustainability. While psychologists traditionally focused on understanding individual behaviour, and business analysts focused on linking CSR with firm accountability and performance (Carroll, 1999; Eccles et al., 2014), this framework bridges these domains by conceptualizing consumer behaviour as shaped by strategic CSR interventions. It provides a roadmap for using the environmental psychological theories in different sectors. In the fashion industry, unique behavioural challenges like rapid change in trends, seasonal collections (summer/winter/spring/autumn), identity signalling, and high consumptions- pose extreme sress on the environment (Joung, 2014). By adopting this psychological framework, theories originally developed for general pro-environmental behavior can be tailored to fashion-specific interventions, contributing to environmental protection. Finally, the model repositions CSR as a psychologically active system, not merely a reputational or compliance tool. This shift enables new directions for theory-building that explore consumer–firm co-evolution in sustainability transitions, emphasizing reciprocal influence and shared responsibility for behavioral outcomes (Beckmann, 2007; Freeman, 2010; Kramer and Porter, 2011).

5.2 Practical Implications

For corporate practitioners, the framework offers actionable insights. CSR initiatives must be designed not only to meet ethical and environmental standards but also to resonate with the psychological drivers of consumer behaviour (Table 1). Effective CSR engages emotional, moral, and social dimensions through transparent communication, ethical practices, and participatory platforms (Fletcher, 2012; Hur & Kim, 2017). Brands such as Levi Strauss & Co., which integrate product lifecycle data with repair programs, or Suay Sew Shop, which anchors sustainability in community-based upcycling, exemplify how CSR can shape identity and behaviour simultaneously (Brown & Vacca, 2022a).

These insights also extend to policymakers and regulators. Behaviorally informed sustainability policies can enhance CSR's effectiveness by requiring firms to assess not just environmental outcomes but also psychological impact—how messages are perceived, interpreted, and acted upon by consumers (Dissanayake et al., 2021; Fonseca et al., 2023). Education campaigns, certification schemes, and public–private partnerships can be designed to reinforce the social desirability of sustainable behaviours, particularly among young and values-driven demographics (Connell, 2019).

5.3 Research Gaps and Future Research Directions

As a conceptual contribution, this study does not include empirical validation of the proposed framework. Future research could employ qualitative case studies, field experiments, or longitudinal methods to test how different psychological mechanisms operate across contexts and consumer segments (Kozar & Connell, 2013). Investigating the relative impact of norm activation, social influence, and dissonance reduction among different demographic cohorts (e.g., Gen Z vs. older consumers) would offer practical insights into segmentation strategies (Chakraborty & Sadachar, 2023).

However, the cross-cultural applicability of the framework needs to be verified and strengthened by empirical studies. Studies have shown that consumer values, norms, and dissonance thresholds differ across cultures; therefore, CSR strategies successful in Western markets may not be equally effective elsewhere (Halder et al., 2020). Therefore, modifying the framework considering the drivers and factors dominant in non-Western, emerging, or collectivist markets would enhance its global relevance.

6 Strategic Implications and Directions for Future Research

This final section analyzes how the psychological framework can lead to practical and research-oriented pathways, supporting the circularity- and sustainability-driven CSR in the fashion industry. Integrating environmental psychology into CSR strategy design highlights the need to take into account the cognitive, emotional, and social drivers of consumer behaviour while formulating corporate actions to address sustainability challenges (Hur & Kim, 2017; Stern et al., 1999). Finally, this section offers directions for future empirical and conceptual research to develop strategic corporate practices, aligning the fashion industry with circular economy principles.

6.1 Strategic Implications for CSR Practice

To implement a psychologically informed CSR strategy, fashion companies must not only meet environmental goals but also influence consumer perceptions, values, and behaviours. Transparency and data disclosure are essential. However, beyond compliance with frameworks like the Global Reporting Initiative (GRI, 2021), firms should communicate their environmental performance in accessible, emotionally resonant, and value-aligned ways. For example, framing disclosures around impacts on local communities or future generations can activate moral norms and enhance engagement with sustainable practices.

CSR strategies should aim to activate and reinforce social norms through tools like educational campaigns, sustainability labels, and social actors' narratives, which shape consumer attitudes and frame ethical fashion as socially desirable (Beckmann, 2007; Thatta & Polisetty, 2022). Aligned with the TRA, such strategies influence both attitudes and perceived norms, strengthening behavioural intentions (Fishbein & Ajzen, 1977).

Equally important is creating opportunities for dissonance resolution. Visible alternatives, such as take-back schemes, repair services, and resale programmes, help consumers reconcile their values with their purchasing behaviour, promoting change without inducing moral crisis or defensiveness (Fletcher, 2013; ThredUp, 2022). Finally, combining blockchain with IoT makes sustainability statements more transparent and dependable. These tools support accountability while empowering consumers with greater agency and control (Saberi et al., 2019).

6.2 Future Research Directions

Although the present study is conceptual, it lays a strong foundation for empirical research exploring how CSR initiatives shape consumer behaviour. Whether survey-based, experimental, or using behavioral data, one needs to examine how different CSR messages based on the VBN theory, TRA, and CDT influence consumer decisions.

Another important area is consumer segmentation. Given the variation in consumer values, attitudes, and practices, future studies should investigate how demographic factors such as age, cultural background, or socioeconomic status shape psychological responses to CSR messaging (Chakraborty & Sadachar, 2023; Halder et al., 2020). Such insights would be invaluable for designing more targeted and effective CSR strategies. Further, longitudinal research to track consumer responses to CSR strategies aligned with personal values can reveal whether initial shifts in attitude translate into sustained behaviour change and brand loyalty (White et al., 2019). Cross-cultural validation represents another necessary step. Cultural norms strongly influence perceptions of corporate responsibility, so comparative studies across different regions can reveal where and how CSR strategies must be adapted (Halder et al., 2020; Hur & Kim, 2017). Without this validation, any conceptual model remains limited in scope.

Finally, future work should explore how CSR interacts with evolving regulatory provisions. As governments implement policies like extended producer responsibility or eco-labelling mandates, it becomes important to understand how these frameworks influence consumer psychology and corporate behaviour (Fonseca et al., 2023). It will reveal whether regulation enhances or reduces the psychological effects of CSR activities, offering insights relevant to both corporate strategy and public policy. Psychological framework-based CSR strategies opens promising pathways for advancing both research and real-world impact, driving the circularity and sustainability in the fashion industry by building lasting consumer trust, encouraging sustainable behaviour, and contributing meaningfully to environmental goals.

7 Conclusion

The fashion industry faces a critical moment where environmental perspective intersects with the need for shifts in consumer behaviour. Despite rising awareness, a clear gap persists between pro-environmental attitudes and actual purchasing choices. This paper addresses the gap by presenting a conceptual framework integrating CSR with key environmental psychology theories: the Value-Belief-Norm Theory, the Theory of Reasoned Action, and Cognitive Dissonance Theory.

The central argument is that the fashion industry should not only focus on transparency or branding goals in CSR but also aims to drive behavioural change by gaining psychological insight into consumers. By activating moral norms, shaping social expectations, and resolving value-behaviour conflicts, CSR can engage consumers across cognitive, emotional, and social dimensions. This makes CSR a more active force in sustainability and circularity transitions.

Positioning consumer psychology at the heart of CSR strategy, the framework offers a novel way to align corporate action with personal motivations for sustainable consumption. Although conceptual, it lays the groundwork for empirical testing and cross-cultural validation, while also pointing to the role of policy in reinforcing corporate initiatives. Ultimately, the model redefines CSR as a strategic lever for advancing sustainable fashion and circular economy objectives through deeper psychological engagement.

Acknowledgements Avi Sharma, Neera and Ankit Prakash thank the Principal, Aryabhatta College, University of Delhi South Campus, for all the support. Vandana Mishra and Radhey Shyam Sharma thank the Institute of Eminence, University of Delhi, for Faculty Research Grant-Projects (Ref. No./IoE/2024-25/12/FRP dated 06.09.2024 and Ref. NO./IoE/2023-24/12/FRP, 31.08.2023).

AuthorContribution **Avi Sharma**: Conceptualization, Methodology, Data curation, Formal analysis, Validation, Visualization, Writing—original draft, Writing—review and editing; **Neera**: Conceptualization, Methodology, Supervision; Validation; Writing—review and editing; **Ankit Prakash**: Conceptualization, Methodology, Supervision; Validation; Writing—review and editing; **Vandana Mishra**: Conceptualization, Methodology, Supervision; Validation; Writing—review and editing; **Radhey Shyam Sharma**: Conceptualization; Funding acquisition; Methodology; Project administration; Resources; Supervision; Validation; Visualization; Writing—review and editing.

Competing Interests
This chapter complies with Springer's ethical publishing standards. All authors contributed substantially and collaborated based on subject expertise, including familial co-authors. Editorial independence and research integrity were fully maintained. The authors declare no conflicts of interest, financial or otherwise.

References

Agudelo, L. M. A., Jóhannsdóttir, L., & Davídsdóttir, B. (2019). A literature review of the history and evolution of corporate social responsibility. *International Journal of Corporate Social Responsibility, 4*, 1–23.

Barrett, M., & Moores, K. (2022). Australia: Family Business and Sustainability: The case of outland denim. In *Attaining the 2030 sustainable development goal of decent work and economic growth* (pp. 53–68). Emerald Publishing Limited.

Beckmann, S. C. (2007). Consumers and corporate social responsibility: Matching the unmatchable? *Australasian Marketing Journal., 15*, 27–36.

Brown, S., & Vacca, F. (2022). Cultural sustainability in fashion: Reflections on craft and sustainable development models. *Sustainability: Science, Practice and Policy, 18*, 590–600.

Carroll, A. B. (1999). Corporate social responsibility: Evolution of a definitional construct. *Business & Society, 38*, 268–295.

Chakraborty, S., & Sadachar, A. (2023). Can a connection with the indigenous cultural values encourage sustainable apparel consumption? *Journal of Fashion Marketing and Management: An International Journal, 27*, 80–99.

Chouinard, Y., & Stanley, V. (2013). *The Responsible Company: What we've learned from Patagonia's first 40 years.* Ventura (CA): Patagonia.

Connell, K. Y. H. (2019). Utilizing political consumerism to challenge the 21st-century fast fashion industry. In M. Micheletti & A. Boström (Eds.), *The Oxford handbook of political consumerism* (pp. 293–308). Oxford University Press.

Delmas, M. A., & Burbano, V. C. (2011). The drivers of greenwashing. *California Management Review, 54*, 64–87.

Desore, A., & Narula, S. A. (2018). An overview on corporate response towards sustainability issues in textile industry. *Environment, Development and Sustainability, 20*, 1439–1459.

Dissanayake, D., Tilt, C. A., & Qian, W. (2021). How do public companies respond to national challenges through sustainability reporting?–The case of Sri Lanka. *Qualitative Research in Accounting & Management, 18*, 455-483.

Eccles, R. G., Ioannou, I., & Serafeim, G. (2014). The impact of corporate sustainability on organizational processes and performance. *Management Science, 60*, 2835–2855.

Fishbein, M., & Ajzen, I. (1977). *Belief, attitude, intention, and behavior: An introduction to theory and research.* Addison-Wesley.

Fletcher, K. (2012). Durability, fashion, sustainability: The processes and practices of use. *Fashion Practice, 4*, 221–238.

Fletcher, K. (2013). *Sustainable fashion and textiles: Design journeys.* Routledge.

Fonseca, L., Carvalho, F., & Santos, G. (2023). Strategic CSR: Framework for sustainability through management systems standards—Implementing and disclosing sustainable development goals and results. *Sustainability, 15*, 11904.

Freeman, R. E. (2010). *Strategic management: A stakeholder approach.* Cambridge University Press.

Garg, S., Kumar, P., Singh, S., Yadav, A., Dumée, L. F., Sharma, R. S., & Mishra, V. (2020). *Prosopis juliflora* peroxidases for phenol remediation from industrial wastewater—An innovative practice for environmental sustainability. *Environmental Technology & Innovation, 19*, 100865.

GRI. (2021). *Consolidated Set of GRI Standards 2021.* Global Reporting Initiative.

Gupta, R., Sinha, P. K., Sahu, A., & Sood, V. (2021). Success drivers of urban haats selling craft products. *International Journal of Rural Management, 17*, 120–139.

Halder, P., Hansen, E. N., Kangas, J., & Laukkanen, T. (2020). How national culture and ethics matter in consumers' green consumption values. *Journal of Cleaner Production, 265*, 121754.

Hur, W. M., & Kim, Y. (2017). How does culture improve consumer engagement in CSR initiatives? The mediating role of motivational attributions. *Corporate Social Responsibility and Environmental Management, 24*, 620–633.

Joung, H. M. (2014). Fast fashion consumers' post-purchase behaviors. *International Journal of Retail & Distribution Management, 42*, 688–697.

Karamally, T., & Robertson, J. L. (2023). Green Leadership. In MacKie, D. (Eds.), *The handbook of climate change leadership in organisations* (pp. 179–202). Routledge.

Kaur, H., Rawat, D., Poria, P., Sharma, U., Gibert, Y., Ethayathulla, A. S., Dumée, L. F., Sharma, R. S., & Mishra, V. (2022). Ecotoxic effects of microplastics and contaminated microplastics—Emerging evidence and perspective. *Science of the Total Environment, 841*, 156593.

Khagokpam, D., Sharma, A., & Sharma, R. S. (2025). Corporate social responsibility: A pathway to sustainability. In *Circular economy and environmental resilience: Solutions for a sustainable tomorrow* (p. 343). Cham: Springer Nature, Switzerland

Kingston, L. N., & Guellil, J. (2016). TOMS and the citizen-consumer: Assessing the impacts of socially-minded consumption. *Journal of Human Rights Practice, 8*, 284–297.

Kozar, J. M., & Hiller Connell, K. Y. (2013). Socially and environmentally responsible apparel consumption: Knowledge, attitudes, and behaviors. *Social Responsibility Journal, 9*, 315–334.

Kramer, M. R., & Porter, M. E. (2011). Creating shared value. *Harvard Business Review, 89*, 62–77.

Kumar, P., Mishra, V., Bhardwaj, S., Garg, S., Dumée, L. F., & Sharma, R. S. (2023). Five-wheel framework for system-based monitoring of heavy metal pollution in rivers. *Environmental Quality Management, 32*, 9–17.

Kumar, P., Mishra, V., Yadav, S., Yadav, A., Garg, S., Poria, P., Farooqi, F., Dumée, L. F., & Sharma, R. S. (2022). Heavy metal pollution and risks in a highly polluted and populated Indian river–city pair using the systems approach. *Environmental Science and Pollution Research, 29*, 60212–60231.

Lichy, J., Ryding, D., Rudawska, E., & Vignali, G. (2023). Resale as sustainable social innovation: understanding shifts in consumer decision-making and shopping orientations for high-end secondhand clothing. *Social Enterprise Journal.*

Marquis, C. & Park, A. (2014). Inside the Buy-One Give-One Model. *Stanford Social Innovation Review.*

Meseguer-Sánchez, V., Gálvez-Sánchez, F. J., López-Martínez, G., & Molina-Moreno, V. (2021). Corporate social responsibility and sustainability: A bibliometric analysis of their interrelations. *Sustainability, 13*, 1636.

Mishra, V., Mudgal, N., Rawat, D., Poria, P., Mukherjee, P., Sharma, U., Kumria, P., Pani, B., Singh, M., Yadav, A., Farooqi, F. & Sharma, R. S. (2023). Integrating microalgae into textile wastewater treatment processes: Advancements and opportunities. *Journal of Water Process Engineering, 55*, 104128.

Mishra, V., Mukherjee, P., Bhattacharya, S., & Sharma, R. S. (2025). Innovative sustainable solutions for detoxifying textile industry effluents using advanced oxidation and biological methods. *Journal of Environmental Management, 380*, 124804.

Mishra, V., Sharma, U., Rawat, D., Benson, D., Singh, M., & Sharma, R. S. (2020). Fast-changing life-styles and ecotoxicity of hair dyes drive the emergence of hidden toxicants threatening environmental sustainability in Asia. *Environmental Research, 184*, 109253.

Morgan, E. (2022). 'Plan A': Analysing business model innovation for sustainable consumption in mass-market clothes retailing. In *New business models for sustainable fashion* (pp. 73–98). Routledge.

Morvan, C., & O'Connor, A. (2017). *An analysis of Leon Festinger's A Theory of Cognitive Dissonance.* Macat Library.

Mukherjee, P., Sharma, R. S., Mishra, V. (2024a). Deciphering the ecological impact of azo dye pollution through microbial community analysis in water–sediment microcosms. *Environmental Science and Pollution Research,* 1–19.

Mukherjee, P., Sharma, R. S., Rawat, D., Sharma, U., Karmakar, S., Yadav, A., & Mishra, V. (2024b). Microbial communities drive flux of acid orange 7 and crystal violet dyes in water-sediment system. *Journal of Environmental Management, 351*, 119699.

Narayanan, A., Sharma, R. S., & Mishra, V. (2025). Unmasking the masks of profits: Analyzing corporate social responsibility through an environmental justice framework. In: *Integrating environmental sustainability and corporate social responsibility: A move beyond profit: corporate social responsibility for environmental sustainability* (pp. 65–94). Cham: Springer Nature Switzerland

Niinimäki, K., Peters, G., Dahlbo, H., Perry, P., Rissanen, T., & Gwilt, A. (2020). The environmental price of fast fashion. *Nature Reviews Earth & Environment, 1*, 189–200.

Niinimäki, K. (2010). Eco-clothing, consumer identity, and ideology. *Sustainable Development, 18*, 150–162.

Porter, M. E., & Kramer, M. R. (2006). The link between competitive advantage and corporate social responsibility. *Harvard Business Review, 84*, 78–92.

Rawat, D., Mishra, V., & Sharma, R. S. (2016). Detoxification of azo dyes in the context of environmental processes. *Chemosphere, 155*, 591–601.

Rawat, D., Sharma, R. S., Karmakar, S., Arora, L. S., & Mishra, V. (2018). Ecotoxic potential of a presumably non-toxic azo dye. *Ecotoxicology and Environmental Safety, 148*, 528–537.

Rawat, D., Sharma, U., Poria, P., Finlan, A., Parker, B., Sharma, R. S., & Mishra, V. (2022). Iron-dependent mutualism between *Chlorella sorokiniana* and *Ralstonia pickettii* forms the basis for a sustainable bioremediation system. *ISME Communications, 2*, 83.

Rawat, D., Sharma, U., Yadav, A., Poria, P., Farooqi, F., Pani, B., Mukherjee, P., Singh, M., Sharma, R. S., & Mishra, V. (2023). A bottom-up approach to select microbes from textile wastewater for detoxification of aromatic amine and azo dye in a single stage. *Journal of Water Process Engineering, 53*, 103668.

Richards, H. (2021). Rethinking value: 'Radical transparency' in fashion. *Continuum, 35*, 914–929.

Saberi, S., Kouhizadeh, M., & Sarkis, J. (2019). Blockchains and the supply chain: Findings from a broad study of practitioners. *IEEE Engineering Management Review, 47*(3), 95–103.

Seo, H., & Jin, B. E. (2024). Engaging in fashion take-back programs: the role of loyalty and perceived benefits from a social exchange perspective. *Sustainability, 16*(22), 10031.

Singh, P., Daga, S., & Yadav, K., (Eds.). (2025). *Integrating environmental sustainability and corporate social responsibility: A move beyond profit, corporate social responsibility for environmental sustainability* 2nd ed. Springer Nature, Switzerland.

Singh, S., Malhotra, S., Mukherjee, P., Mishra, R., Farooqi, F., Sharma, R. S., & Mishra, V. (2020). Peroxidases from an invasive Mesquite species for management and restoration of fertility of phenolic-contaminated soil. *Journal of Environmental Management,* 256, 109908.

Stern, P. C., Dietz, T., Abel, T., Guagnano, G. A., & Kalof, L. (1999). A value-belief-norm theory of support for social movements: The case of environmentalism. *Human Ecology Review, 6*, 81–97.

Sundar, A., Pothier, W. G., Pfeiffer, B., & Lei, L. (2022). The renewal workshop: Consumer considerations in a global supply chain. *SAGE Business Cases Originals.*

Thatta, S., & Polisetty, A. (2022). The future is circular: A case study on MUD jeans. *FIIB Business Review., 11*, 137–146.

ThredUp. (2022). Fashion footprint calculator [Internet]. [cited 2025 Jun 27]. Available from: https://www.thredup.com/fashionfootprint

Vallerand, R. J., Deshaies, P., Cuerrier, J. P., Pelletier, L. G., & Mongeau, C. (1992). Ajzen and Fishbein's theory of reasoned action as applied to moral behavior: A confirmatory analysis. *Journal of Personality and Social Psychology, 62*, 98.

Weber, S. (2018). A circular economy approach in the luxury fashion industry: a case study of Eileen Fisher. In: *Sustainable luxury: Cases on circular economy and entrepreneurship*(pp. 127–60). Singapore: Springer.

White, K., Hardisty, D. J., & Habib, R. (2019). The elusive green consumer. *Harvard Business Review, 11*, 124–135.

Xu, H., & Bolton, L. E. (2024). How do consumers react to production waste? *Journal of Consumer Research, 51*, 212–237.

Yadav, A., Kumar, P., Rawat, D., Garg, S., Mukherjee, P., Farooqi, F., Roy, A., Sundaram, S., Sharma, R. S., & Mishra, V. (2022). Microbial fuel cells for mineralization and decolorization of azo dyes: Recent advances in design and materials. *Science of the Total Environment., 826*, 154038.

Larkai Healthcare: Enabling AI in Diagnostics and Making Healthcare Accessible and Affordable with the Circular Economy and Disruptive Technology

Mark Esposito, Jordi Diaz, Tatiana Harkiolakis, and Daphne Halkias

Abstract This case study delves into the entrepreneurial careers of Pritam Dhalla and Abhilash Chakraverty, the founder and co-founder of Larkai Healthcare. Larkai was founded in 2020 due to Pritam's exploration into the realm of entrepreneurship and his idea for easily accessible healthcare. The company's flagship product, WREN, is a smart multiparameter portable device that seamlessly integrates measures of respiratory rate, body temperature, NIBP (Non-invasive blood pressure), and ECG (electrocardiogram) to revolutionize patient monitoring. The benefits of WREN are seen in better patient care, more accessibility to healthcare, and more efficiency. The case explores the origins of Larkai, Pritam's product developmental history, Abhilash's business and financial heft, and the difficulties encountered in early business when artificial intelligence was still in the realm of theory. It illustrates the company's ability to secure funds through strategic initiatives and collaborations. Due to its global ambitions, Larkai faces a unique challenge: how to balance quick

M. Esposito (✉)
HULT International Business School, Berkman Klein Center for Internet & Society, Harvard University, Cambridge, MA, USA
e-mail: markesposito@fas.harvard.edu

J. Diaz
EADA Business School, Barcelona, Spain
e-mail: jdiaz@eada.edu

T. Harkiolakis · D. Halkias
Center for Policy and Competitiveness, Ecole Des Ponts ParisTech Business School, Paris, France
e-mail: tatiana.harkiolaki@iscparis.com

D. Halkias
e-mail: daphne.halkias@faculty.eim.education

T. Harkiolakis
ISC Paris Business School, Paris, France

D. Halkias
EIM—European Institute of Management, Valletta, Malta

P. Singh et al. (eds.), *Circular Economy and Sustainable Innovation*,
https://doi.org/10.1007/978-3-032-00437-6_10

innovation with a variety of legal environments. The case showcases Larkai's dedication to the three Ps: Purpose, Profit, and Planet. Using a circular business model, revenue targets, and an emphasis on making a difference in millions of lives, Larkai became a disruptive force in the healthcare industry. Larkai's case sets an example for the industry by demonstrating how a purpose-driven approach may result in financial success while having a beneficial global impact.

Keywords Larkai Healthcare · Healthcare · India · Artificial intelligence · Circular economy · Diagnostics · Disruptive technology

1 Introduction

Being explorative and adventurous from his young days, being inspired by his surroundings and naturally inclined towards healthcare, and being motivated by an able mentor, guide, and support, Pritam, age 28 years, took the leap of faith and entered the world of entrepreneurship. Believing in "The how is easy if the why is powerful", he had a clear idea of his end goal, besides being absolutely determined, motivated, and optimally confident of his abilities to give shape and life to his idea. Buckling up for the undulating terrain ahead, with a rigorous market survey of the gaps in healthcare and realizing the potential of ever-expanding scope in cardiac care, he focused his product on being diagnostic and preventive rather than a cure and in making healthcare affordable, accessible, and non-invasive, but not compromising on the factors of conclusively and comprehensively of the detection and diagnosis. What inspired him exponentially in taking this giant leap was the logarithmic potential of Artificial Intelligence in healthcare, and thereby, he began his journey of sowing, nurturing, and giving birth to Larkai.

2 Theoretical Foundation

2.1 Disruptive Innovation Theory

The concept of disruptive innovation is not novel, having been popularized by Professor Clayton Christensen in his seminal work, *The Innovator's Dilemma* (1997). Christensen introduced a new paradigm for innovation, which focused on employing sophisticated technologies to optimize processes and create a competitive advantage. He emphasized the importance of integrating disruption into a company's daily operations, processes, stakeholder relationships, and communication channels. At that time, it was clear that organizations needed to take the lead in breaking away from traditional innovation models. Although the idea of disruption was introduced two decades ago, it remains highly relevant today, continuing to distinguish one organization from another and serving as a lasting framework in business literature. Scholars such as Markides (2006) and King and Baatartogtokh (2015) have further developed Christensen's original notion of "the innovator's dilemma," refining it into a theory of disruptive innovation (Christensen et al., 2015). They argue that disruptive innovations are not limited to technological advancements but also encompass 'new-to-the-world products' and 'business model innovations.'

To understand organizational progress, three types of innovation must be considered: sustaining innovation (which is not disruptive), low-end disruptive innovation, and new-market disruptive innovation (Christensen et al., 2015). First, sustaining innovation refers to improvements designed for the most demanding customers—those already willing and able to purchase the product or service and pay for its enhancements. Second, low-end disruptive innovation targets customers' basic needs, deliberately avoiding additional features that would increase both the value and the cost of the offering. Finally, new-market disruptive innovation addresses non-customers—individuals or groups whose needs have not yet been met by existing companies. Sustaining innovation generally benefits industry incumbents and is characterized by three key attributes: (i) it enhances an already successful product, (ii) it targets the most profitable customer segments, and (iii) it improves profit margins (Christensen et al., 2015).

Technology adoption has enabled both low-end and new-market disruptive innovations to thrive across various industries (Christensen et al., 2015). The first path of innovation emerges when new entrants introduce products or services to the lower end of the market, prompting incumbents to retreat. This form of innovation is characterized by three key elements: (i) it offers products that are "good enough" to meet basic customer needs, (ii) it targets "over-served" customers or those at the bottom of the market who require fewer product features, and (iii) it operates through a low-cost business model (Diaz & Halkias, 2021). Guided by disruptive innovation theory and accelerated by rapid technological advancements, this value proposition—centered on affordability and simplicity at scale—has successfully captured the low-end market segment (Christensen et al., 2015).

3 Disruptive Innovation in Healthcare

Disruptive innovation theory has been applied to explain recent technology-driven innovations in healthcare (Pezzuto, 2019). Technological innovations powered by breakthroughs in hardware, software, services, and emerging technologies are increasingly giving birth to disruptive innovations that are agile, convenient, lean, and a fraction of the price compared to traditional solutions (Ganguly & Kumar, 2022). Business models enabled by these innovations disrupt industry incumbents, spurring industry convergence and establishing new dimensions of consumer experience (Pattnaik et al., 2024).

Technological advances in healthcare have adversely resulted in rising healthcare costs, increasing problems of healthcare accessibility for citizens in the Global

South (Pezzuto, 2019). The healthcare industry worldwide has focused on sustaining innovation and developing more expensive and sophisticated procedures and devices that wealthier citizens can only afford (Bellucci, 2022). Innovative tech companies, particularly start-ups, have risen to disrupt these trends with medical devices that are more affordable and efficient, initially targeting a niche consumer base that cannot afford more expensive options (Pattnaik et al., 2024).

Scholars have utilized disruptive innovation theory to explore disruptive technologies in healthcare. Some previous works that have explored disruptive innovations in the Global South include Pezzuto's (2019) exploration of disruptive innovation in the healthcare sector in Nigeria, Iizuka and Hane's (2021) studies of disruptive start-ups in Asia, and Sounderajah et al.'s (2021) bibliometric literature review on disruptive innovations in the healthcare literature. Nevertheless, a gap remains in the literature regarding the application of disruptive innovation theory to circular digital health technologies in the Global South (Pezzuto, 2019).

4 Literature Review

4.1 Digital Health Technologies in Developing Economies

An increase in the volume and variety of medical services available, tightening national and international legislation in the industry, and growing consumer demand are driving forward notable changes in the healthcare industry (Kulkova et al., 2023). An increasing number of patients causes medical personnel to endure a growing and unpredictable workload despite the need to provide high-quality services to individual patients (Mathur & Sutton, 2017). While nations continually increase funding for national health systems and competition grows in the healthcare industry, companies producing medical solutions and products attempt to bring novelties to the market quickly (Tarkkala et al., 2019).

The healthcare industry can respond to these pressing challenges now and in the future if it embraces technological change (Flessa & Huebner, 2021; Tortorella et al., 2022). Technology entrepreneurs can respond faster to new challenges with increased scalability and flexibility in delivering services through their in-depth understanding of the healthcare industry, their technological tools, and the policy environment in which they will operate (Lee & Yoon, 2021). Understanding national and international healthcare industry regulations is especially crucial for digital healthcare start-ups, as regulations can enable or constrain the degree and type of innovation possible in healthcare technology (Cozzolino & Geiger, 2024).

Developing economies present unique opportunities and challenges in the nexus of entrepreneurship and digital health technologies. The diffusion of digital health technologies in developing economies can enable better access to healthcare products and services, increased quality of care, and lower costs (Preko et al., 2020). Nevertheless, enabling policy environments and entrepreneurial initiatives notably affect how effectively these technologies can be integrated into pre-existing healthcare systems (Jiang et al., 2022; Ravindranath, 2024).

Expanding markets and technology readiness play crucial roles in digitalization's impact on healthcare technology entrepreneurship (Jafari-Sadeghi et al., 2021). Developing economies are compelled to foster environments where digital innovation can thrive in order to take advantage of the benefits of healthcare technology entrepreneurship and technological market expansion (Preko et al., 2020). Telepresence robots in healthcare were one of the first applications of healthcare technology in developing countries, which provided crucial in addressing challenges posed by the COVID-19 pandemic (Wang et al., 2021). Such a use case proved the practical application of healthcare technology entrepreneurship in serving the particular healthcare needs of patients in low-resource settings (Wang et al., 2021).

Yeganegi et al. (2021) analyzed the role of information availability in technology entrepreneurship, noting how innovation in the healthcare sector of developing economies requires high-quality, accessible data. The implications of Big Data in healthcare digitization are far-reaching, with Healthcare 4.0 being an emerging topic in the scholarly research literature (Juneja et al., 2020; Kulkov, 2023). Developing economies can use Big Data to vastly improve healthcare outcomes and service delivery (Sharma & Kshetri, 2020). In the long run, healthcare outcomes in developing economies can be improved through research and development, increased funding, renewed policies, and public and private support for healthcare innovations and healthcare technology entrepreneurship (Tarkkala et al., 2019).

Digitalizing healthcare technologies has also led to the rise of low-cost medical devices and diagnostic tools in developing economies (Olatunji et al., 2024). Embracing a strategy of disruptive innovation, the start-ups behind these tools utilize emerging technologies and groundbreaking product design to create portable, cost-effective devices that are easy to use (Santra et al., 2019). These devices significantly make healthcare accessible to patients living in underserved or rural areas (Pezzuto, 2019).

The availability of accurate diagnostics tools is a particular challenge in settings with limited resources. Portable, battery-powered diagnostics tools that non-specialized healthcare workers can handle are crucial in providing accurate diagnostics to underserved populations (Oppong, 2020). These devices also allow such populations to be treated in their homes or immediate communities without being referred to distant hospitals. Many of these devices utilize digital health technologies, such as artificial intelligence, IoT, and cloud computing, to store, process, and analyze diagnostic data (Olatunji et al., 2024). For example, handheld ultrasound devices allow for pregnancy complications, cardiac issues, and liver disease to be diagnosed without expensive imaging equipment (Wilkinson & Saxhaug, 2021). These devices are often combined with smartphone technology and can provide the same diagnostic capabilities at a fraction of the cost of traditional equipment (Najjar, 2024).

5 The Circular Economy and Digital Health Technologies in the Global South

Sustainable development remains a persistent challenge in the Global South, particularly in healthcare and resource management (Uwaoma et al., 2023). Circular economy (CE) business models, which involve optimizing resources and reducing waste, offer a way for Global South economies to achieve environmental sustainability (Simamindra & Rajaonarivo, 2024). CE models require organizations and entrepreneurs to recycle and reuse their resources, which has proved crucial for the survival of ventures in resource-constrained regions (Emilsson et al., 2020; Nikolova & Ness, 2022).

At the same time, digital health technology is emerging in developing economies to enhance the efficiency and delivery of healthcare outcomes (Hendl & Shukla, 2024). Digital health technologies improve the sustainability of the healthcare sector by increasing the efficiency of resource use and reducing waste, such as by digitalizing previously paper-based processes (Kong et al., 2023). One recent example of an innovative, sustainable solution in healthcare is using biochar applications, which promote carbon neutrality and circularity in healthcare waste disposal (Kurniawan et al., 2023).

In response to the COVID-19 pandemic, digital health technologies in the Global South have rapidly advanced, lowering the carbon footprint of healthcare services and streamlining healthcare processes (Petrakaki et al., 2023). Artificial intelligence, the Internet of Things (IoT), electronic health records, and telemedicine are the most promising technologies in this area. In addition to advancing sustainability, these technologies also make healthcare more accessible and affordable to citizens in low-resource settings or who face various barriers to receiving care, such as those who live in rural areas (Zwack et al., 2023). These technologies have been proven to facilitate improved diagnosis, tracking, and treatment, allowing for tailored healthcare solutions (Masanneck et al., 2023). The efficiency-improving aspects of digital health technologies work in tandem with CE goals. For example, AI can reduce food waste while simultaneously improving nutritional outcomes and reducing malnutrition among vulnerable populations (Onyeaka et al., 2023).

6 Challenges of Implementing Circular Digital Health Initiatives in the Global South

The Global South faces numerous challenges in implementing digital health technologies, including regulatory challenges, lack of digital literacy, and inefficient infrastructure. Specific policy frameworks are needed to ensure these technologies do not jeopardize patient privacy or data security (Borges do Nascimento et al., 2023). A lack of robust infrastructure is a persistent challenge in the Global South that hinders the uptake of sustainable digital health technologies and the creation of a sustainable digital health ecosystem (Hendl & Shukla, 2024).

Efficient waste management systems, for example, are crucial for repurposing digital health devices and recycling their physical components (Nikolova & Ness, 2022). Many economies in the Global South, classified as low- and middle-income, struggle with such systems, coupled with a shortage of government regulations regarding circular production models and the sustainable processing of e-waste. Immediate health concerns are often prioritized in these regions at the expense of long-term sustainability plans, especially given the scarcity of technical and financial resources (Jakovljevic et al., 2021). The rapid pace of technological obsolescence and advancement increases the amounts of e-waste, often sent from around the world, to be dumped in developing economies, exacerbating these economies' existing waste management problems (Esposito et al., 2023).

The funding and political interests behind digital health initiatives in the Global South can pose further barriers to implementing circular economy business models in this sector (Godinho et al., 2022). International organizations and donor funding behind such initiatives tend not to prioritize circular economy principles. A disconnect is often found between the goals of long-term sustainability and more immediate short-term goals for healthcare delivery, which are prioritized by the entities providing the much-needed external funding for such initiatives (De Carvalho et al., 2021). A lack of widespread education and awareness about circular economy goals further detracts from implementing sustainable digital health projects (Kong et al., 2023).

7 Implementing Circular Digital Health Initiatives in India

Like other economies in the Global South, India faces several challenges to implementing circular digital health initiatives. This emerging economy faces the dual challenges of inadequate healthcare access in rural areas combined with rapid urbanization that puts pressure on the national healthcare infrastructure (Ghosh et al., 2021). Managing medical waste remains a pressing challenge as rural areas lack the necessary infrastructure, technology, and facilities and a constrained budget that limits the available resources to invest in advanced recycling and recovery technologies (Chauhan et al., 2021). Urban medical facilities are beginning to embrace such initiatives, but an inadequate regulatory framework provides no financial incentives for hospitals to implement circular economy initiatives (Siddiqui & Pandit, 2021). While private hospitals may have more financial resources, circular economy technologies present the problem of having a long-term return on investment and staggering upfront costs (Ghosh et al., 2021).

Many urban and rural medical institutions in India rely on disposable medical equipment due to its low price compared to reusable equipment; however, disposable equipment results in increasing amounts of waste that is not recycled (Hossain et al., 2022). There are also cultural barriers to adopting reusable medical equipment in India, with patients and staff tending to perceive reused or recycled equipment as lacking adequate hygiene or safety standards (Rohit et al., 2021). This resistance presents another barrier that is difficult to overcome without adequate public awareness or staff training programs on the benefits of circular equipment (Hossain et al., 2022).

While digital health initiatives may hold promise for ensuring greater sustainability in healthcare in India without requiring expensive equipment, medical devices are often not designed with the principles of circularity in mind (Mehra & Sharma, 2021). Most medical devices used in India are imported and made from single-use plastics, contributing to a vicious cycle of consumption and waste (Syama & Prakash, 2022). Insufficient collaboration among healthcare stakeholders and fragmented supply chains magnify the challenges of implementing the circular economy in India's digital health initiatives and devices (Mehra & Sharma, 2021).

8 Biography and Developmental Background of the Protagonists

Pritam Dhalla, a passionate leader and innovator in the healthcare sector, currently steers Larkai Healthcare as its founder. His journey began in 2013, fueled by a burning desire to make healthcare sustainable and accessible. While pursuing his Bachelor's in Technology and Master's in Biotechnology, which he completed in 2018, this vision grew stronger. He was joined by Abhilash Chakraverty, who had leadership roles in Philips Eindhoven Nederlands and General Electric in Milwaukee, USA, and has worked globally. He infused Larkai Healthcare with innovative and diverse healthcare solutions, solidifying its unique position in the industry. His global outlook significantly shaped his approach, seamlessly blending a deep understanding of real-world healthcare needs with scientific rigor.

9 Background Information on the Product

Born into a family of doctors, Pritam naturally gravitated towards a healthcare career. Partnering with his pulmonologist father, he brought to life the company's first device, the CardioMate. This innovative technology could detect heart and lung sounds, producing a printed ECG report. However, a crucial question arose: who would be their target customers? Cardiologists, pulmonologists, or general physicians? The market research revealed an essential insight: while high-grade doctors preferred specialized devices, a significant segment craved multi-functionality. They desired ECG alongside capabilities like blood pressure or SpO_2 (oxygen saturation) monitoring. This realization sowed the seeds for the WREN, a device designed to cater to this broader market demand.

Keenly observant of market demands and the usual problems ordinary people face, Pritam envisioned a solution that was both advanced and accessible to all. WREN, the world's smartest multiparameter portable device, was born from this vision. It seamlessly measured five vital signs: ECG, SpO_2, body temperature, NIBP, and respiratory rate. WREN transcended a simple device; it is a complete ecosystem, seamlessly connecting with an app and cloud system for comprehensive health monitoring. While CardioMate held a special place in Pritam's heart, it was commercially a failure. In early 2022, Abhilash, who had two decades of experience in global healthcare giants like Philips and General Electric, joined him as co-founder and pivoted the company product and market access strategy. Whereas CardioMate was meant for rural low-resource and primary care, WREN resonated with the market's needs. Widely embraced and cherished, WREN has experienced an impressive global expansion, reaching new corners of the world.

The goal of Larkai Healthcare is to use artificial intelligence (AI) in healthcare so that people can receive medical diagnosis and treatment from any location at any time while incorporating the circular economy into every stage of the product lifecycle (Larkai Healthcare, 2024). With its cutting-edge AI technology, the company aims to bridge the gap between treatment and diagnosis by providing immediate and accurate diagnosis. This goal and their dedication to sustainability align with a larger industry trend that uses AI to improve the effectiveness and accessibility of medical services while minimizing environmental waste.

10 About WREN: Bridging the Gap Between Diagnosis and Treatment

Larkai Healthcare's WREN is not aiming to replace traditional hospital ECG monitors. Instead, it fills a crucial gap in the healthcare landscape by offering real-time patient monitoring through an advanced, affordable, and accessible solution. Bulky ECG machines, while powerful diagnostic tools, lack portability and are often absent from smaller healthcare facilities like nursing homes and clinics.

WREN, in contrast, is a smart, portable multiparameter monitor equipped with a touchscreen, a dedicated app, and a cloud system, creating a complete patient monitoring ecosystem (Larkai Healthcare, 2024). This compact device offers a significant advantage over traditional ECGs:

- Portability: WREN's small size (380 g) and wireless connectivity make it ideal for use in ambulances, nursing homes, and even patients' homes. This versatility allows for continuous monitoring during transport or in settings where bulky equipment is impractical.
- Cost-effectiveness: Compared to traditional ECG machines, WREN is significantly more affordable ($550 versus $2200). A wider range of healthcare providers and patients can thus access this device.

- Real-time monitoring: WREN's cloud-based system provides doctors with immediate access to patient vitals, allowing them to change treatment plans and make informed decisions in real-time, even remotely.

These features combined made WREN a game-changer in the healthcare market in 2022. Its impact can be seen in:

- Improved patient care: WREN's real-time monitoring capabilities allow for earlier detection of potential complications and quicker response times, which leads to improved patient outcomes.
- Increased access to care: WREN's affordability and portability make it accessible to underserved communities and patients who might not have access to traditional ECG monitoring.
- Enhanced healthcare efficiency: WREN's ability to streamline patient monitoring processes can save healthcare providers valuable time and resources.

Overall, WREN represented a significant step forward in patient monitoring technology. It could revolutionize healthcare delivery and improve patient care by bridging the gap between diagnosis and treatment.

11 Larkai: Company History and Profile

The spark of inspiration struck Pritam back in 2014, his university days fueled by the "fail fast, grow early" mantra and continuous inspiration and enthusiasm instilled by one of his KIIT (Kalinga Institute of Industrial Technology) professors. Embracing the philosophy of "read, learn, launch", Pritam dove headfirst into the world of code, learning to interpret the intricate language of ECG graphs and their peaks. By the end of his fourth year, in 2017, Pritam's brainchild bloomed into existence—his first product, capable of detecting and analyzing heart sounds. This innovation marked a triumphant year, culminating in his first pitch and the subsequent funding for his invention. However, fate threw a curveball. Another company scooped up the idea, forcing Pritam to pivot his design. Instead of viewing this as a setback, Pritam saw an opportunity to innovate. He channeled his creative energy into crafting something even better, something more universally needed—CardioMate. His dedication paid off again in 2019, with CardioMate securing its second round of funding.

Fast forward to 2020, when Pritam felt the time was ripe to formalize his endeavors. Thus, the seeds of Larkai were sown. On January 1, 2020, Larkai Healthcare officially sprouted, ready to flourish and bear fruit, nurtured by Pritam's unwavering passion and the lessons learned along the way. When the company was initially starting and had no capital, Larkai Healthcare was able to develop its first product and introduce it to the market with the help of Start-up Odisha and the Odisha government (Odisha is a state in Eastern India). This assistance helped the company get traction and caught the attention of investors. Start-up Odisha arranged the first pitching, which began late because of some mismanaged planning. Abhilash, the co-founder and director of the company, seized the chance and took the initiative to make the product pitch. He performed ECG on an Investor, Amit Singhal, and other people as soon as he could go center stage. He presented his idea to five investors, one of whom was Mr. Amit Singhal from Realtime Angel Funds, who was enthralled with the technology and concept of WREN, and this marked the first funding for WREN.

Thus, the company has raised finances from venture capital firms, angel funds, and international market funds in that order. The start-up began with a modest proposal to angel investors, and now it has a healthy cash chest of millions of dollars. FAAD Network, a significant backer of Larkai Healthcare's seed round, focuses on early-stage investments in various industries and regions. Their calculated investment in Larkai Healthcare demonstrates their belief in the company's ability to transform the medical technology industry. More government funds supported Larkai Healthcare's seed funding round in addition to private investments. This cooperative strategy showed a more thorough understanding of the company's possible influence on the healthcare environment.

12 The Protagonist's Dilemma—Balancing Local and Global Aspirations

Larkai Healthcare leverages cutting-edge solutions to bridge the gap between diagnosis and treatment. However, Larkai's start-up status, global aspirations, and circular economy orientation present a unique challenge: keeping pace with global trends while adapting to the Indian market's specific needs. Is it possible for Larkai to balance its goal of helping underprivileged areas in India with its quick worldwide expansion, all while guaranteeing a sustainable product? How can they enter profitable markets while ensuring their solutions are still available?

In the fast-paced world of innovation, ideas are fluid, and the first to market often secures the strongest foothold. To achieve global prominence, Larkai must work smart and hard, pioneering sustainable solutions for India and the wider world. How can Larkai strike a balance between innovation, navigability, and sustainability? High-end solutions offer a tempting springboard into lucrative developed markets, accelerating Larkai's global expansion and brand recognition.

Nevertheless, venturing across borders is not without hurdles. Each country's regulatory landscape features distinct approval processes involving lengthy clinical trials, data analysis, and safety assessments. These processes can be expensive, time-consuming, and prone to unforeseen complications. Moreover, regulatory requirements vary significantly, forcing Larkai to adapt their solutions, evidence, and sustainability characteristics to each specific standard—a resource-intensive and potentially frustrating endeavor. Furthermore, even with meticulous preparation, there is no guarantee of approval, leading to delays, additional costs, and potentially scrapped projects.

Being the first mover with a groundbreaking and sustainable solution can solidify Larkai's position as a med-tech leader and offer previously unmet healthcare solutions, improving patient outcomes and generating significant demand. Is it crucial for Larkai to be the first in every market with new solutions, or could learning from competitors' successes and failures inform a more strategic approach? Larkai's journey demands a delicate balancing act: innovating rapidly and sustainably while navigating complex regulations. Through careful planning, strategic partnerships, and a data-driven approach, they can bridge the gap between their local and global aspirations, securing their place at the forefront of healthcare innovation and the circular economy.

13 Conclusion: Larkai's Healthcare Innovation with Social and Environmental Impact

Larkai Healthcare Private Limited has established challenging yet purpose-driven objectives to make a long-lasting impression on the connection between technology and healthcare. The short-term goal of building a strong IT team complements the bigger vision of making healthcare accessible globally. A comprehensive approach to business, where success can be measured in societal and environmental benefits and financial gain, is underscored by Larkai's dedication to the 3 Ps: Purpose, Profit, and Planet.

Larkai's purpose in bridging the gap between technology and healthcare is driven by the company's ambitious goal of impacting 10 million lives by the end of 2024 and 100 million lives in the next two years (Crunchbase, 2024). This commitment positions Larkai as a transformative force in the global healthcare landscape, striving for positive change. On the profit front, the established revenue goals of $10 million by the end of 2024, followed by a 10× increase to $100 million over the next three years, show a strategic business plan.

Larkai secured enough funding to invest in scaling their operations, securing an investment of 500 k USD for the seed funding round from FAAD Network, QI Ventures, RTAF, and additional government grants (Vardhan, 2024). Larkai is also recognized and lauded in the business, healthcare, and technology communities, receiving many product innovation awards for WREN. The company was recognized as one of the top 3 finalists in the category of Innovative Diagnostic Solutions for WREN for the Aegis Graham Bell Awards.

Larkai seeks to demonstrate that social responsibility and financial success can be strengthened when combined in the same venture. Larkai is dedicated to the circular economy through reusing, replenishing, refurbishing, and recycling, which promotes responsible consumption and aligns with environmental aims (Larkai Healthcare, 2024). Allowing returns of the product after five years is an inventive way to demonstrate environmentally conscious thinking. Moreover, Larkai incorporates sustainability into all business operations, from reducing paper usage to using digital substitutes like electronic cards and brochures. Their commitment to environmental preservation is further demonstrated by the recycled and non-hazardous materials used in creating their products. In 2023, Larkai Healthcare was awarded the GCPIT Global Sustainability Award 2023 for designing and building sustainable, green medical products that limit hazardous substances, minimize wastage, and champion the circular economy.

Lastly, Larkai's emphasis on diversity and inclusion, as indicated by achieving an equal male-female ratio and receiving workplace excellence awards, demonstrates its dedication to building an inclusive and innovative workplace. Larkai Healthcare Private Limited emerges as a pathfinder in a new era of business, where purpose, profit, and the planet combine to create a model of success that is not only financially rewarding but also socially responsible and environmentally sustainable. Their journey sets a notable example for the industry, showcasing that a purpose-driven approach can lead to both business success and positive global impact.

Larkai Healthcare's disruptive innovation with WREN and Bluetail significantly advances patient monitoring and diagnostics, yet challenges persist in making these technologies accessible to the broader population. As a B2B company, Larkai primarily interfaces with healthcare providers and institutions, which poses a barrier to direct access for the general population, particularly in areas where poverty prevails. Despite the strides made, there is a recognition of the need to address this accessibility gap. Larkai is committed to overcoming this disparity by exploring the development of a more accessible lite version of WREN, aligning with their dedication to inclusivity and the circular economy and their mission to bridge the healthcare divide and promote sustainable digital healthcare in economically challenged regions.

Acknowledgements While preparing this work, the authors used Grammarly to edit the text to fit academic writing style and English-language spelling and grammar. After using this tool/service, the authors reviewed and edited the content as needed and take full responsibility for the publication's content.

References

Bellucci, N. (2022). Disruptive innovation and technological influences on healthcare. *Journal of Radiology Nursing, 41*(2), 98–101. https://doi.org/10.1016/j.jradnu.2022.02.008

Borges do Nascimento, I. J., Abdulazeem, H., Vasanthan, L. T., Martinez, E. Z., Zucoloto, M. L., Østengaard, L., Azzopardi-Muscat, N., Zapata, T., & Novillo-Ortiz, D. (2023). Barriers and facilitators to utilizing digital health technologies by healthcare professionals. *NPJ Digital Medicine, 6*(1), 161. https://doi.org/10.1038/s41746-023-00899-4

Chauhan, A., Jakhar, S. K., & Chauhan, C. (2021). The interplay of circular economy with Industry 4.0 enabled smart city drivers of healthcare waste disposal. *Journal of Cleaner Production, 279*, 123854. https://doi.org/10.1016/j.jclepro.2020.123854

Christensen, C. M. (1997). *The innovator's dilemma: When new technologies cause great firms to fail.* Harvard Business Review Press.

Christensen, C. M., Raynor, M., & McDonald, R. (2015). What is disruptive innovation? *Harvard Business Review*. https://hbr.org/2015/12/what-is-disruptive-innovation

Cozzolino, A., & Geiger, S. (2024). Ecosystem disruption and regulatory positioning: Entry strategies of digital health start-up orchestrators and complementors. *Research Policy, 53*(2), Article 104913. https://doi.org/10.1016/j.respol.2023.104913

Crunchbase. (2024). *Larkai Healthcare*. Retrieved September 1, 2024, from https://www.crunchbase.com/organization/larkai-healthcare

De Carvalho, G., Schmid, A., & Fischer, J. (2021). Classifications of health care systems: Do existing typologies reflect the particularities of the Global South? *Global Social Policy, 21*(2), 278–300. https://doi.org/10.1177/1468018120969315

Diaz, J., & Halkias, D. (2021). Reskilling and upskilling 4IR leaders in business schools through an innovative executive education ecosystem: An integrative literature review. *SSRN Electronic Journal*. https://doi.org/10.2139/ssrn.3897059

Emilsson, M., Ernstson, U., Gustavsson, L., & Svensson, A. (2020). Sustainable innovations in small enterprises for the transformation of the primary healthcare sector. *Sustainability, 12*(16), 6391. https://doi.org/10.3390/su12166391

Esposito, M., Halkias, D., Tse, T., & Harkiolakis, T. (2023). Environmental and climate impacts of the Metaverse. *SSRN Electronic Journal*. https://doi.org/10.2139/ssrn.4616695

Flessa, S., & Huebner, C. (2021). Innovations in health care—A conceptual framework. *International Journal of Environmental Research and Public Health, 18*(19), 10026. https://doi.org/10.3390/ijerph181910026

Ganguly, A., & Kumar, C. (2022). Disruptive innovation in healthcare: A review and future research agendas. *International Journal of Innovation Management, 26*(01), 2250002. https://doi.org/10.1142/S1363919622500025

Ghosh, S. K., Ghosh, S. K., & Baidya, R. (2021). Circular economy in India: Reduce, reuse, and recycle through policy framework. In S. K. Ghosh & S. K. Ghosh (Eds.), *Circular economy: Recent trends in global perspective* (pp. 183–217). Springer Singapore. https://doi.org/10.1007/978-981-16-0913-8_6

Godinho, M. A., Martins, H., Al-Shorbaji, N., Quintana, Y., & Liaw, S. T. (2022). "Digital health diplomacy" in global digital health? A call for critique and discourse. *Journal of the American Medical Informatics Association, 29*(5), 1019–1024. https://doi.org/10.1093/jamia/ocab282

Hendl, T., & Shukla, A. (2024). Can digital health democratize health care? *Bioethics, 38*(6), 491–502. https://doi.org/10.1111/bioe.13266

Hossain, R., Islam, M. T., Shanker, R., Khan, D., Locock, K. E. S., Ghose, A., Schandl, H., Dhodapkar, R., & Sahajwalla, V. (2022). Plastic waste management in India: Challenges, opportunities, and roadmap for circular economy. *Sustainability, 14*(8), 4425. https://doi.org/10.3390/su14084425

Iizuka, M., & Hane, G. (2021). Towards attaining the SDGs: Cases of disruptive and inclusive innovations. *Innovation and Development, 11*(2–3), 343–364. https://doi.org/10.1080/2157930X.2021.1954751

Jafari-Sadeghi, V., Garcia-Perez, A., Candelo, E., & Couturier, J. (2021). Exploring the impact of digital transformation on technology entrepreneurship and technological market expansion: The role of technology readiness, exploration and exploitation. *Journal of Business Research, 124*, 100–111. https://doi.org/10.1016/j.jbusres.2020.11.020

Jakovljevic, M., Liu, Y., Cerda, A., Simonyan, M., Correia, T., Mariita, R. M., Krstic, K., Osabohien, R., Toan, T. K., Adhikari, C., Chuc, N. T. K., Khatri, R. B., Chattu, V. K., Wang, L., Wijeratne, T., Kouassi, E., Khan, H. N., & Varjacic, M. (2021). The Global South political economy of health financing and spending landscape–history and presence. *Journal of Medical Economics, 24*(sup1), 25–33. https://doi.org/10.1080/13696998.2021.2007691

Jiang, C., Chang, H., & Shahzad, I. (2022). Digital economy and health: Does green technology matter in BRICS economies? *Frontiers in Public Health, 9*, Article 827915. https://doi.org/10.3389/fpubh.2021.827915

Juneja, S., Juneja, A., & Anand, R. (2020). Healthcare 4.0-Digitizing healthcare using big data for performance improvisation. *Journal of Computational and Theoretical Nanoscience, 17*(9–10), 4408–4410. https://doi.org/10.1166/jctn.2020.9087

King, A. A., & Baatartogtokh, B. (2015). How useful is the theory of disruptive innovation? *MIT Sloan Management Review, 57*(1), 77–90. https://sloanreview.mit.edu/article/how-useful-is-the-theory-of-disruptive-innovation/

Kong, J. D., Akpudo, U. E., Effoduh, J. O., & Bragazzi, N. L. (2023). Leveraging responsible, explainable, and local artificial intelligence solutions for clinical public health in the Global South. *Healthcare, 11*(4), 457. https://doi.org/10.3390/healthcare11040457

Kulkov, I. (2023). Next-generation business models for artificial intelligence start-ups in the healthcare industry. *International Journal of Entrepreneurial Behavior & Research, 29*(4), 860–885. https://doi.org/10.1108/IJEBR-04-2021-0304

Kulkova, J., Kulkov, I., Rohrbeck, R., Lu, S., Khwaja, A., Karjaluoto, H., & Mero, J. (2023). Medicine of the future: How and who is going to treat us? *Futures, 146*, Article 103097. https://doi.org/10.1016/j.futures.2023.103097

Kurniawan, T. A., Othman, M. H. D., Liang, X., Goh, H. H., Gikas, P., Chong, K. K., & Chew, K. W. (2023). Challenges and opportunities for biochar to promote circular economy and carbon neutrality. *Journal of Environmental Management, 332*, Article 117429. https://doi.org/10.1016/j.jenvman.2023.117429

Larkai Healthcare. (2024). *Larkai Healthcare pitch deck* (confidential).

Lee, D., & Yoon, S. N. (2021). Application of artificial intelligence-based technologies in the healthcare industry: Opportunities and challenges. *International Journal of Environmental Research and Public Health, 18*(1), 271. https://doi.org/10.3390/ijerph18010271

Markides, C. (2006). Disruptive innovation: In need of better theory. *Journal of Product Innovation Management, 23*(1), 19–25.

Masanneck, L., Gieseler, P., Gordon, W. J., Meuth, S. G., & Stern, A. D. (2023). Evidence from ClinicalTrials.gov on the growth of digital health technologies in neurology trials. *NPJ Digital Medicine, 6*(1), 23. https://doi.org/10.1038/s41746-023-00767-1

Mathur, S., & Sutton, J. (2017). Personalized medicine could transform healthcare. *Biomedical Reports, 7*(1), 3–5. https://doi.org/10.3892/br.2017.922

Mehra, R., & Sharma, M. K. (2021). Measures of sustainability in healthcare. *Sustainability Analytics and Modeling, 1*, Article 100001. https://doi.org/10.1016/j.samod.2021.100001

Najjar, R. (2024). Digital frontiers in healthcare: Integrating mHealth, AI, and radiology for future medical diagnostics. In T. F. Heston & C. E Doarn (Eds.), *A comprehensive overview of telemedicine* (pp. 1–33). InTech Open. https://doi.org/10.5772/intechopen.114142

Nikolova, A. S., & Ness, D. (2022). 'The impossible dream': Can the circular economy alone solve waste management complexities of the Global South? In H. Lehmann, C. Hinske, V. de Margerie, & A. S. Nikolova (Eds.), *The impossibilities of the circular economy: Separating aspirations from reality* (pp. 23–36). Routledge.

Olatunji, A. O., Olaboye, J. A., Maha, C. C., Kolawole, T. O., & Abdul, S. (2024). Revolutionizing infectious disease management in low-resource settings: The impact of rapid diagnostic technologies and portable devices. *International Journal of Applied Research in Social Sciences, 6*(7), 1417–1432. https://www.fepbl.com/index.php/ijarss/issue/view/226

Onyeaka, H., Tamasiga, P., Nwauzoma, U. M., Miri, T., Juliet, U. C., Nwaiwu, O., & Akinsemolu, A. A. (2023). Using artificial intelligence to tackle food waste and enhance the circular economy: Maximising resource efficiency and minimising environmental impact: A review. *Sustainability, 15*(13), 10482. https://doi.org/10.3390/su151310482

Oppong, J. (2020). Disruptive technologies and the African healthcare crisis: A path to sustainability. In P. Arthur, K. Hanson, & K. Puplampu (Eds.), *Disruptive technologies, innovation and development in Africa* (pp. 117–145). Palgrave Macmillan. https://doi.org/10.1007/978-3-030-40647-9_6

Pattnaik, P., Patnaik, R., & Mohanty, P. K. (2024). The digital disruption model: Redefining the healthcare sector with innovation. In S. K. Baral, R. Goel, T. Singh, E. Kaynak, & V. Jain (Eds.),

Changing competitive business dynamics through sustainable big data analysis (pp. 70–88). Bentham Books.

Petrakaki, D., Chamakiotis, P., & Curto-Millet, D. (2023). From 'making up' professionals to epistemic colonialism: Digital health platforms in the Global South. *Social Science & Medicine, 321*, Article 115787. https://doi.org/10.1016/j.socscimed.2023.115787

Pezzuto, I. (2019). Making healthcare systems more efficient and sustainable in emerging and developing economies through disruptive innovation: The case of Nigeria. *Journal of Management and Sustainability, 9*(2), 1–25. https://heinonline.org/HOL/P?h=hein.journals/jms9&i=140

Preko, M., Osei-Boateng, R., & Durosinmi, A. E. (2020). Digitalising healthcare in developing economies: Challenges and mitigating strategies. In R. Boateng (Ed.), *Handbook of research on managing information systems in developing economies* (pp. 332–353). IGI Global.

Ravindranath, M. (2024, February 20). *As regulators play catch up with AI, tech start-ups say uncertainty is slowing them down.* STAT. https://www.statnews.com/2024/02/20/health-ai-regulation-tech-startups-compliance/

Rohit, A., Rajasekaran, S., Shenoy, S., Rai, S., Iddya, K., & Dorairajan, S. K. (2021). Reprocessing of N95 masks: Experience from a resource-limited setting in India. *International Journal of Infectious Diseases, 104*, 41–44. https://doi.org/10.1016/j.ijid.2020.12.070

Santra, S., Mandal, T. S., & Das, P. (2019). *Leveraging disruptive technology innovations for healthcare delivery in Sub-Saharan Africa* [Issue Brief No. 298]. Observer Research Foundation. https://www.orfonline.org/public/uploads/posts/pdf/20240724104454.pdf

Sharma, R., & Kshetri, N. (2020). Digital healthcare: Historical development, applications, and future research directions. *International Journal of Information Management, 53*, Article 102105. https://doi.org/10.1016/j.ijinfomgt.2020.102105

Siddiqui, A., & Pandit, R. K. (2021). Smart cities in India: Linkages with circular economy. In U. Iyer-Raniga (Ed.), *Sustainability in the built environment in the 21st century: Lessons learned from India and the region* (pp. 185–200). Springer. https://doi.org/10.1007/978-3-030-61891-9_12

Simamindra, R. S., & Rajaonarivo, L. (2024). A global analysis of circular economy initiatives: Weak or strong sustainability? *Journal of Cleaner Production, 467*, Article 142830. https://doi.org/10.1016/j.jclepro.2024.142830

Sounderajah, V., Patel, V., Varatharajan, L., Harling, L., Normahani, P., Symons, J., Barlow, J., Darzi, A., & Ashrafian, H. (2021). Are disruptive innovations recognised in the healthcare literature? A systematic review. *BMJ Innovations, 7*(1), 208–216. https://doi.org/10.1136/bmjinnov-2020-000424

Syama, U. S., & Prakash, D. (2022). Scope of circular economy in the health sector of India. In *7th International conference on "Economic growth and sustainable development: Emerging trends"* (pp. 1–10). Shri Dharmasthala Manjunatheshwara Institute for Management Development.

Tarkkala, H., Helén, I., & Snell, K. (2019). From health to wealth: The future of personalized medicine in the making. *Futures, 109*, 142–152. https://doi.org/10.1016/j.futures.2018.06.004

Tortorella, G. L., Fogliatto, F. S., Espôsto, K. F., Mac Cawley, A. F., Vassolo, R., Tlapa, D., & Narayanamurthy, G. (2022). Healthcare costs' reduction through the integration of healthcare 4.0 technologies in developing economies. *Total Quality Management & Business Excellence, 33*(3–4), 467–487. https://doi.org/10.1080/14783363.2020.1861934

Uwaoma, P. U., Eboigbe, E. O., Kaggwa, S., Akinwolemiwa, D. I., & Eloghosa, S. O. (2023). Ecological economics in the age of 4IR: Spotlight on sustainability initiatives in the Global South. *International Journal of Advanced Economics, 5*(9), 271–284. https://doi.org/10.51594/ijae.v5i9.647

Vardhan, G. (2024, January 3). *Larkai Healthcare raises $500k in seed round.* EnTrackr. https://entrackr.com/2024/01/larkai-healthcare-raises-500k-in-seed-round/

Wang, M., Pan, C., & Ray, P. K. (2021). Technology entrepreneurship in developing countries: Role of telepresence robots in healthcare. *IEEE Engineering Management Review, 49*(1), 20–26. https://doi.org/10.1109/EMR.2021.3053258

Wilkinson, J. N., & Saxhaug, L. M. (2021). Handheld ultrasound in training—The future is getting smaller! *Journal of the Intensive Care Society, 22*(3), 220–229. https://doi.org/10.1177/1751143720914216

Yeganegi, S., Laplume, A. O., & Dass, P. (2021). The role of information availability: A longitudinal analysis of technology entrepreneurship. *Technological Forecasting and Social Change, 170*, Article 120910. https://doi.org/10.1016/j.techfore.2021.120910

Zwack, C. C., Haghani, M., Hollings, M., Zhang, L., Gauci, S., Gallagher, R., & Redfern, J. (2023). The evolution of digital health technologies in cardiovascular disease research. *NPJ Digital Medicine, 6*(1), 1. https://doi.org/10.1038/s41746-022-00734-2

Integrating Education, Technology, and Sustainable Living: Advancing the Circular Economy for Future Generations

Shalini Singh, Satyam Anand, Anandi, and Shubhanker Yadav

Abstract Sustainable living is crucial in reducing society's reliance on natural resources, a goal intimately connected to the principles of the circular economy. This conceptual framework highlights the interplay between education, technology, and the standard of living as fundamental aspects driving sustainable practices. Education empowers individuals with the knowledge and skills necessary to raise awareness and encourage sustainable behavior, shaping beliefs and intentions towards resource efficiency. However, recent studies suggests, increased access to information can paradoxically lead to greater complexities and dilemmas, creating challenges in enacting sustainable consumption and inadvertently slowing sustainable development efforts. Furthermore, the role of technology in the circular economy is indispensable. Advanced technological solutions facilitate the efficient use of resources, offering innovative approaches to mitigate environmental impact while enhancing the quality of life. Earlier literatures underscores the importance of education, funding, and innovation in green technology to achieve Sustainable Development Goals (SDGs) by 2030, highlighting the current state and research hotspots in this domain. A comprehensive literature review utilizing the Web of Science database reveals global contributions to SDG research, emphasizing the need for a balanced approach that integrates education, technology, and a sustainable standard of living. Together, these elements contribute to the development of a society that fosters a culture of sustainability. By promoting a circular economy, where resources are reused, recycled, and repurposed, we can ensure that the needs of the present are met without compromising the ability of future generations to thrive. This approach not only preserves environmental health but also cultivates resilient communities capable of navigating the complexities of sustainability in a rapidly changing world.

S. Singh (✉) · S. Anand · Anandi
School of Commerce, Finance and Accountancy, CHRIST (Deemed to be University), Bengaluru, India
e-mail: shalini.singh@christuniversity.in

S. Yadav
Jaipuria Institute of Management, Lucknow, India

P. Singh et al. (eds.), *Circular Economy and Sustainable Innovation*,
https://doi.org/10.1007/978-3-032-00437-6_11

Keywords Circular economy (CE) · Sustainable living · Green technology · Education for sustainability · Green finance · Resource efficiency · Renewable energy · Innovation in sustainability · Environmental awareness · Digital transformation · Eco-centric business models

1 Introduction

1.1 Definition and Importance of Sustainable Living

Sustainable living refers to a lifestyle and a policy framework that minimizes the depletion of Earth's natural resources while fostering long-term ecological balance and economic viability (Bănescu & Ţiţan, 2023; Valadez-Martínez, et al., 2017). It is rooted in the principles of resource conservation, responsible consumption, and environmental stewardship, all aimed at securing a livable planet for future generations (Behera, 2023; Lodh, 2024).

Beyond individual behavioral changes such as reducing waste or saving energy, sustainable living in today's context necessitates structural transformation across social, economic, and technological systems (Kaur et al., 2023; Wahyono, 2022). It involves the widespread adoption of sustainable innovation and circular economy practices that promote regenerative design, efficient resource utilization, and waste minimization. Green technologies, digital tools, and eco-centric business models are no longer peripheral—they are central to advancing a sustainable society.

However, achieving sustainability at scale is contingent upon enabling frameworks such as education, technological access, infrastructure, and most critically, green finance (Dauda et al., 2024; Odeyemi et al., 2023). Supportive renewable energy uptake policies, carbon-neutrality, and circular innovation must be backed by strong financing mechanisms—like green bonds, climate funds, and sustainability-linked loans—which can trigger investment into sustainable transformations. Without this support infrastructure, sustainable living is mere theory, not practical reality.

Thus, coupling circular economy principles with sustainable innovation and targeted financial instruments offers an integrated approach toward achieving ecological resilience, social well-being, and economic change for India and other nations (Aiwei Wang, 2023; Kholijah, 2024; Sahar Afshan, n.d.).

The circular economy (CE) is a new model that re-thinks conventional production and consumption based on the concept of "closing the loop" of product life cycles. Rather than the conventional take-make-dispose, CE focuses on reduction, reuse, recycling, and regeneration. CE seeks to maximize material flows, minimize ecological pressure, and generate socio-economic value through extended product life, energy efficiency, and lower reliance on virgin materials (Aluchna Maria, 2019).

CE is becoming a real potential way toward sustainable development, especially for developing economies such as India, where population increase, urbanization, and ecological deterioration coincide. CE presents a template to separate economic

Table 1 Green finance investment trends in India (2020–2024)

Year	Green bonds issued (INR Cr)	% Growth YoY	Notable milestone
2020	6000	–	Launch of sovereign green bond framework
2021	7400	23	SEBI mandates ESG disclosures for top firms
2022	9800	32	India joins international platform on SF
2023	12,300	26	Climate finance added in Union Budget
2024	14,200 (est.)	15	Second sovereign green bond tranche issued

Source RBI Bulletins, SEBI Reports, Union Budget 2023–24

growth from ecological damage and stimulate circular business models, resource productivity, and eco-innovation. In spite of its strategic potential, CE suffers from difficulties in practice, such as diversified stakeholder participation, the absence of uniform frameworks, and insufficient access to funding for environmentally friendly initiatives (Chigbu, 2024).

This is where green finance comes into play. Green bonds and ESG-based investments, which are focused on environmental considerations, play a crucial role in scaling circular practices across industries. As business houses become more open to CE models, financial mechanism innovations can drive their uptake by reducing risk, enhancing investor confidence, and promoting long-term sustainability (Chirea Ionela Mirela, 2022).

In short, the circular economy, harmonized with sustainable living and paired with green finance, offers a strong pillar of constructing inclusive, climate-resilient economies. It promotes not just environmental renewal but also economic competitiveness and social justice—fundamental principles of the circular transformation necessary for sustainable development (Table 1).

1.2 Purpose and Objectives of the Chapter

This chapter seeks to examine the pivotal nexus of technology, education, and standard of living in providing impetus to India's move away from fossil fuels and towards a green economy. It highlights how these pillars, underpinned by financial instruments like green bonds and sustainability-linked loans, assist in circular economy practices and sustainable innovation.

The specific goals are:

1. To examine how education promotes sustainable practices and makes communities ready for engagement in circular economy activities.

2. To assess the impact of technological advancements on environmental conservation and green resource utilization.
3. To evaluate how living standards, particularly through access to infrastructure, finance, and digital tools, influence sustainable choices.
4. To examine India's policy and financial environment in supporting renewable energy adoption through green finance.

Through these objectives, the chapter contributes to understanding the systemic enablers of circular economy transitions within India's sustainable development agenda.

1.3 Structure of the Chapter

- **Brief overview of the chapter's structure and key topics**

The analysed articles cumulatively justify the multidimensional paradigm for developing circular economy values and sustainable innovation. An integral component of this change is encouraging ethical consumption and promoting energy efficiency, both of which are imperative to reduce environmental footprint and enhance social value (Akhtar, 2023). For instance, Internet of Things (IoT)-based technologies facilitate intelligent management of resources, optimizing environmental observation and sustainability methods. With IoT and cloud computing complemented with artificial intelligence (AI), it enhances the potential to predict to initiate preventive action of environmental conservation measures. Additionally, policy interactions facilitate key enablers driving circular economy opportunities in green buildings, for example, development (Solomon et al., 2024b).

These policies guarantee that sustainable construction integrates technological innovations and legal systems, minimizing environmental impacts while conserving natural resources (Bukhari, 2024). Agroforestry approaches are also making their contributions towards soil conservation and sustainable crop yield, illustrating how the applications of the circular economy are not limited to industrial approaches. This series of studies emphasizes the need for governmental intervention and support through regulations in propelling sustainable innovations in various industries. As highlighted by (Solomon et al., 2024a), green buildings are the embodiment of the synergy of technology, policy, and resource preservation, providing a potential path toward a greener and more resilient future.

2 Conceptual Framework

2.1 The Role of Education in Sustainability

- **Importance of education in raising awareness and shaping sustainable behavior**

Education is key in the development of consciousness and attitudes that promote the circular economy and sustainable innovation. Studies identify that education raises public consciousness regarding environmental and social matters, enhances sustainable practices, and stimulates active engagement in sustainability activities (Emon & Hassan, 2023). Education institutions also contribute significantly by integrating sustainability into school curricula and promoting Sustainable Development Goals (SDGs) in study and practice (Suklun & Bengü, 2024; Veckalne & Tambovceva, 2022). The Turkish curriculum, for example, targets developing awareness towards sustainability, showing the ability of education systems to promote positive change. Specific awareness-raising campaigns also have an influence on consumers' attitudes and consumption habits, enhancing the utilization of green products and circular economy solutions (Uniya, 2024) Challenges such as resource limitation, financial constraints, and resistance to change, however, hinder the adoption of sustainability education, particularly in nations such as Bangladesh (Emon & Hassan, 2023). Despite these challenges, education is a significant driver in embedding circular economy concepts in various industries.

Initiatives like sustainability-oriented curricula at Saudi Arabian schools illustrate how learning can prepare individuals to adopt sustainable practices and set the pace of innovation in minimization of wastes, optimization of resources, and regenerative designs (Altassan, 2023). The combined effect of these researches emphasizes that the incorporation of sustainability in learning alters attitudes and behaviors, paving the way toward a more resource-efficient and circular economy (Sami, 2024; Salonen & Hakari, 2018). However, to make the most difference, programs must be able to break down obstacles such as accessibility, funding, and public participation. By integrating sustainability education at all levels, societies are empowered to create a highly qualified workforce and an enlightened citizenry, both of which are required to advance the transition to a more sustainable, cleaner tomorrow (Emon & Hassan, 2023).

- **The paradox of increased information leading to complexity**

The greater availability of information on sustainable innovation and the circular economy is creating a paradox—though it enhances learning, it also creates complexity that may be inimical to effective learning. The broad range of sustainability-related themes, ranging from resource efficiency to regenerative design, makes it hard for learners and teachers to make sense of the material (Hsu & Chen, 2021; Anastasiadis et al., 2020). Teachers usually find it challenging to break down complex ideas without sacrificing their fundamental nature, as conventional curricula are more inclined towards linear problem-solving strategies than the systems thinking needed for circular economy learning (Austin & Brown, 1999; Marques et al., 2019). Moreover, ethical challenges in sustainability decision-making bring an additional layer of complexity, necessitating students to acquire knowledge and critical thinking skills at the same time (Rahmawati et al., 2022).

Information overload also makes learning more challenging, as it does in other disciplines where too much data impacts behaviour and decision-making. Too much information can cause cognitive overload in circular economy learning, rendering students unable to effectively absorb essential principles (Kunrath & Beliatis, 2022). The difficulty is in finding the balance—sufficient depth of knowledge without information overload. Teachers need to develop new pedagogical strategies that enhance active learning and systems thinking and tackle the holistic nature of circular economy strategies (Wang et al., 2021). By making educational processes more relevant and streamlined to highlight key, actionable information, institutions can better prepare students with the skills necessary for inducing sustainable innovation without flooding them with too much complexity.

- **Case studies or examples of successful educational initiatives**

Case studies and best practice examples are necessary to comprehend the real-world application and effect of circular economy concepts and sustainable innovation (Aulia et al., 2023; Oliveira, 2024). The report of (Levett-Jones et al., 2018) is a comprehensive review of seven case studies from five nations demonstrating novel strategies for circular economy practice implemented successfully and assessed. These case studies identify strategies to break down barriers to sustainability and enhance resource efficiency and waste minimization (Levett-Jones et al., 2018). Likewise (Díez-Palomar et al., Pathways to Equitable and Sustainable Education through the Inclusion of Roma Students in Learning Mathematics, 2018), analysed six instances of sustainable urban planning projects, determining the most important features that contributed to success and how the principles of circular economy contribute to long-term environmental and social gains. Under the focus of sustainable business models, Oliveira highlights that it was seen that in Brazil there were identified good strategies to attract industries towards practices of the circular economy, wherein case studies attributed the favourable relation between financial outcomes and corporate sustainability initiatives. There have been devised case studies even under the scope of green technology to bring down sustainability effects while using creative design for products as well as including eco-friendly elements (Prakash & Kumar, 2024). Moreover, a case study and musing about planning and executing community-led sustainability initiatives identified how local knowledge and conventional ecological practices were integrated in the approach to circular economies (Jessy, 2023). Moreover (Meylani, 2024), contains several case studies that examine both successful and unsuccessful adoptions of sustainable production and consumption models critically, such as factors making circular economy use effective (Meylani, 2024). Together, these studies highlight the value of case studies to establish the real-world application of circular economy innovations in different industries. They offer useful lessons on how new and innovative methods, community participation, and inclusive sustainability measures can result in effective environmental and economic performances. These results are a reference for policymakers, companies, and stakeholders when planning and putting into practice efficient circular economy measures towards a more sustainable future (Díez-Palomar et al., 2018; Jessy, 2023; Levett-Jones et al., 2018; Meylani, 2024; Prakash & Kumar, 2024).

2.2 Technology as a Catalyst for Sustainable Living

- **Overview of green technologies and their impact on resource efficiency**

Circular economy values and sustainable innovation are crucial in reducing environmental harms and improving resource efficiency. Such strategies involve extensive goals of reducing waste, maximizing resource utilization, and promoting long-term environmental sustainability. The effect of circular economy practices on resource efficiency is two-pronged, as it encourages reuse, recycling, and regeneration of resources, hence minimizing natural resource pressure. The literature emphasizes varied evidence on the role of sustainable innovation in facilitating circular economy transitions. For example, the use of digital finance has been found to enhance the development and uptake of circular economy solutions, resulting in enhanced resource efficiency and waste minimization (Cai, 2023). As a result, so too has increased material efficiency in the manufacturing sector and optimized supply chain management brought about by application of big data in manufacturing as well as in mining helped limit resource depletion (Wang et al., 2023). Still, technological strides in digital means have further made industrial structures overhaul through advancing more sustainable forms of production as well as innovation circles in the economies (Shi et al., 2024). In addition, policy-based programs, like China's industrial agglomeration pilot policy, have illustrated how purposeful interventions could increase carbon efficiency and decrease pollution through circular economy patterns (Li et al., n.d.). Such findings emphasize the capabilities of circular economy systems and sustainable innovations in facilitating resource efficiency and environmental sustainability across industries. In summary, sustainable innovation and circular economy are central forces of a greener future since they allow industries and enterprises to shift towards regenerative and low-impact operations. The empirical evidence indicates that the strategies contribute to notable resource efficiency improvements and waste reduction, hence furthering global sustainability agendas. In order to reap maximum benefits, it is important to promote ongoing innovation, enact right policies, and drive large-scale adoption of circular economy values across sectors (Behera, 2023; Cai, 2023; Shi et al., 2024; Wang et al., 2023) (Table 2).

- **Discussion on innovation and its role in achieving SDGs**

Table 2 Growth in India's renewable energy capacity (in GW)

Year	Solar	Wind	Biomass and others	Total RE capacity	Share of total energy (%)
2020	37	38	10	85	21
2021	44	40	11	95	23
2022	52	42	12	106	24.5
2023	61	44	13	118	26
2024	70 (est.)	46	14	130	28 (est.)

Source Ministry of New and Renewable Energy (MNRE), CEA Reports

Innovation has a vital role in developing the foundations of the circular economy and inducing sustainable innovation (Jack Williams, 2024; Said, et al., 2024), both of which are necessary for realizing global sustainability aspirations. Literature suggests that proper use of science, technology, and innovation (STI) can contribute to meeting major challenges like depletion of resources, management of waste, climate change, and environmental degradation, facilitating progress toward a sustainable future (Adenle et al., 2023). There needs to be sustainable innovation in creating innovative and pragmatic solutions that are congruent with the circular economy principles, reducing waste and increasing resource efficiency. Green technology adoption and the development of infrastructure have been found to contribute positively to sustainability, affecting economic, environmental, and social factors. Yet, the interaction between innovation and the implementation of circular economy is not straightforward. Policies in the economic sector, like taxation, can facilitate or deter innovation by influencing the financial strength of companies to invest in sustainable alternatives (Godinho et al., 2024). Furthermore, digital resources and intelligent technologies are important for facilitating the shift towards circular economic patterns through maximizing production processes, improving supply chain efficiency, and aiding waste minimization. But the effective application of these implies huge investment in digital competencies and technological developments (Aparicio-Gómez et al., 2024). On balance, circular economy practices and green innovation are essential measures toward a greener world that provide realistic solutions to global challenges of sustainability while promoting economic resilience and environmental care.

- **Examples of technological advancements supporting the circular economy**

Technology development is one of the major propellants in circular economy growth, offering new solutions for enhancing the use of resources and facilitating the perpetual cycling of materials. Carbon capture, utilization, and storage technologies are especially important to help reach net-zero emissions while supporting circular economy principles by mitigating the environmental footprint (Munir et al., 2024). New digital technologies such as the Internet of Things (IoT), blockchain, and artificial intelligence (AI) play a crucial role in resource management through closed-loop systems that maximize efficiency and minimize waste (Williams et al., Enhancing Circular Economy with Digital Technologies: A PLS-SEM Approach,

2024). Furthermore, sophisticated sorting, composting, and waste-to-energy technologies have significantly enhanced urban sustainability for cities like San Francisco, showing actual use of circular economy practices. While these new technologies are available, such developments are not without challenges, with issues of data protection, the digital divide, and disparities in access to technology potentially slowing adoption and advantage (Williams et al., Enhancing Circular Economy with Digital Technologies: A PLS-SEM Approach, 2024). Financial and infrastructure limitations are also hindrances to implementing these innovations. In spite of such challenges, coupling Industry 4.0 with circular economy will propel sustainable economic development by facilitating smarter, more resource-efficient production systems (Zhou et al., Driving force for China's economic development under Industry 4.0 and circular economy: Technological innovation or structural change?, 2020). One prominent case of the adoption of circular economy is electric vehicle (EV) battery recycling, which enhances resource efficiency and facilitates the achievement of several Sustainable Development Goals (Chigbu, 2024). These technological innovations and cutting-edge solutions underscore the promise of circular economy models for leading the way toward a more sustainable and resource-efficient future.

2.3 Sustainable Standard of Living

2.3.1 Defining a Sustainable Standard of Living

Sustainable living is a way of life that seeks to minimize the use of Earth's natural resources while maintaining long-term environmental health and economic sustainability. It focuses on equilibrating resource utilization and the capacity of the planet to replenish itself, thereby reducing the ecological footprint and ensuring a sustainable future for generations to come (Behera, 2023; Lodh, 2024).

In the realm of sustainable innovation and the circular economy, sustainable living goes beyond consumer-level practices such as energy saving and waste reduction. It is deeply intertwined with broader economic, technological, and financial systems that drive systemic transformation. Innovations in sustainable business models, green technologies, and renewable infrastructure are fundamental in facilitating large-scale transitions to sustainable living. Ideas like regenerative design, effective use of resources, and circular production patterns are key to creating a low-impact and resilient society.

Moreover, policy and governance play a fundamental role in setting environmental policies, promoting renewable energy, and encouraging green business practices. Green finance—via the use of green bonds, ESG-linked loans, and climate funds—becomes an enabling force in this regard, connecting the investments and pushing the funding for circular and sustainable practices. Anything less than concerted policy effort, finance effort, and innovation effort leaves the movement toward a low-carbon, inclusive, and sustainable future unfinished. Hence, synthesizing circular economy practices with sustainable innovation and financial mechanisms presents a holistic approach towards ensuring long-term ecological equilibrium and resource productivity (Aiwei Wang, 2023; Behera, 2023; Kholijah, 2024; Sahar Afshan, n.d.).

The circular economy is an innovative approach to incorporating sustainability into industrial and economic systems through "closing the loop" of product life cycles. The system offers numerous environmental, economic, and social advantages like reduced consumption of resources, reduced ecological footprint, cost reductions, and creation of jobs (Aluchna Maria, 2019). In contrast to the traditional linear economy—which exhibits a take-make-dispose philosophy—the circular economy is centered around interdependence among industries, maximization of the efficiency of resources, and stimulating sustainable innovation.

As a sustainability strategy, the circular economy also remains to be implemented, particularly with regard to stakeholder coordination, technological uptake, and availability of financial support (Calzolari et al., 2023). Though the concept has attracted much interest in policy and academic discourse, actual implementation still lags behind due to disconnected efforts and weak integration across sectors (Aluchna Maria, 2019).

However, companies are increasingly seeing the long-term benefits of circular approaches. Besides increasing operational effectiveness, integrating sustainable innovation into circular structures enables companies to achieve changing environmental requirements and consumer demands. Financial instruments—most notably green finance instruments—are crucial in facilitating this transition by providing investment in renewable energy, sustainable infrastructure, and resource-efficient technologies.

As industries continue to explore opportunities for circularity, green innovation remains a decisive factor in overcoming existing barriers and realizing the full potential of the circular economy. With the right combination of policy alignment, financial innovation, and stakeholder collaboration, the circular economy can serve as a pathway to a greener and more sustainable future (Chirea Ionela Mirela, 2022).

3 Circular Economy and Its Intersection with Sustainable Living

3.1 Understanding the Circular Economy

- **Definition and principles of the circular economy**

The circular economy (CE) is an innovative model that encourages sustainable development by avoiding waste and maximizing the use of resources. In contrast to the traditional linear economy with its "take-make-dispose" strategy, CE focuses on redesigning products and processes to maximize longevity, efficiency, and environmental regeneration. By means of maintenance, repair, reuse, remanufacturing, refurbishment, and recycling strategies, CE avoids resource input while diminishing waste, emissions, and energy leakage. This model not only enhances environmental sustainability but also inspires innovative solutions across industries, leading the way to a greener tomorrow (Mandych et al., 2023; Solomon et al., 2024a).

- **How the circular economy aligns with sustainable living**

The circular economy (CE) and green living are complementary by nature since the principles of CE are geared towards minimizing waste, avoiding emissions, and limiting the exploitation of resources in order to ensure environmental sustainability (Mandych et al., 2023). Along with environmental sustainability, CE also provides economic and social advantages—allowing companies to reduce costs, drive innovation, and access new markets while driving social responsibility and job creation (Mandych Oleksandra, 2023). Transition to a circular economy is achieved in all stages of production, transformation, and consumption, which addresses the three pillars of sustainability—environmental, economic, and social (Stankevičienė & Nikanorova, 2020). While others view CE from a systemic or regenerative perspective, others see it from a business perspective, and hence the variation in how CE reacts to sustainability (Salminen et al., 2023).

In general, the circular economy is a revolutionary way of rethinking resource use in the service of sustainable living. By embedding economic activities into environmental and social factors, CE offers a robust platform for sustainable development. Its influence is manifested in various industries like manufacturing and fashion, where it promotes innovation for sustainability (Brydges, 2021; Nwankwo Constance Obiuto, 2024). However, actualizing the full potential of CE calls for sustained discourse, policy, and collective efforts to bolster its implementation (Solomon et al., 2024a; Toni, 2023).

3.2 Integration of Education and Technology in the Circular Economy

- **Role of education in promoting circular economy practices**

Education is fundamental in driving engagement with circular economy (CE) activities through promoting information exchange, facilitating multidisciplinary collaboration, and developing creative thinking—integral factors in managing sustainable resources. Through providing knowledge and skills required by individuals and organizations, education facilitates the efficient application of CE principles (Ritter et al., 2024).

Although literature often emphasizes the significance of education in CE, most of the attention is often given to organizational and technical learning dimensions. For example (Ritter et al., 2024), highlight the significance of bringing together various disciplinary approaches to deal with the larger challenges involved in implementing CE. Likewise (Mensah et al., 2021), appreciate that practices of CE should advance continuously, highlighting the need for intra- and inter-organizational learning as key to sustainability and identity formation. Implications of these results are that education goes beyond formal education—it involves lifelong learning, continuous adjustment, and knowledge sharing within and between organizations, all important in driving CE and sustainable innovation.

- **Technological innovations enabling the circular economy**

The principles of the circular economy (CE) can be seamlessly integrated into innovative, resource-efficient, and sustainable models rather than relying on conventional linear systems, particularly with advancements in technology. For instance, blockchain technology enhances market positioning and transparency, reinforcing CE principles (Souza et al., 2024). Similarly, despite challenges such as resource limitations, the digital divide, and the circularity gap, adopting Industry 4.0 digital technologies—especially among small and medium-sized enterprises (SMEs)—is essential for advancing CE practices (Neri et al., 2023). However, while technology plays a pivotal role in facilitating this transition, addressing supply chain integration, cultural attitudes, and regulatory frameworks is equally crucial for building a functional circular economy.

Cultural factors significantly influence the effectiveness of CE initiatives, as human interaction and communication are vital to their success (Naveed et al., 2022). Additionally, the degree of supply chain integration and institutional barriers can either support or hinder CE implementation. This suggests that while technology is an important enabler, it alone is not sufficient to ensure CE's successful adoption—broader systemic changes in policies, corporate culture, and supply chain structures are necessary for full implementation (Calzolari et al., 2023).

3.3 Challenges and Opportunities

- **Discuss challenges such as the information paradox**

A number of barriers impede the technological and educational development of the circular economy (CE). Education for Sustainable Development (ESD) is necessary to empower the workforce with the skills needed to facilitate CE practices. Nonetheless, even more substantial challenges remain, including a shortage of specialization in key areas—seen in the intricacies of recycling lithium-ion batteries (LIBs) for electric vehicles (EVs)—and in incorporating CE principles within current education structures (Chigbu, 2024). Technologically, the demand for improved waste treatment solutions is still a challenge, as evidenced by China's experiment with creating its CE model (Jinhui Li, 2011) and the current technological bottlenecks in recycling processes (Chigbu, 2024).

Although technology enhances resource productivity and promotes sustainable economic development, it also brings along risks, such as data protection issues, the digital divide, and disproportionate access to advanced technologies, which might create disparities in the CE benefit distribution (Williams et al., Enhancing Circular

Economy with Digital Technologies: A PLS-SEM Approach, 2024). In order to successfully combine Industry 4.0 with CE, it is important to evaluate the stakeholder engagement and regional initiative readiness to ensure a balanced and equitable transition towards a sustainable economy (Zhou et al., Driving force for China's economic development under Industry 4.0 and circular economy: Technological innovation or structural change?, 2020).

- **Opportunities for advancing the circular economy through education and technology**

From every study, there appears to be much that can be learned from there to be able to spread this circular economy with technology and education. From (Williams et al., Enhancing Circular Economy with Digital Technologies: A PLS-SEM Approach, 2024), the two principles of the circular economy, that of resource efficiency and sustainable economic development, can be further enhanced in addition to the incorporation of technologies like blockchain, artificial intelligence, and Internet of Things. This would also involve strategic deployment of the concepts of the digital economy across the product and service life cycle in a bid to enhance the longevity of the products and minimize wastage (Wilson et al., 2024). The new methods and approaches that come with the shift can drive the concepts of a circular economy forward to facilitate for the expediency of EdTech in the education sector since it is bound to revolutionize existing theories regarding education (Ranbir, 2024; Shustova et al., 2023). Even though technology and education are well conceded to advance a circular economy, problems and discrepancies enter the scene (Wilson & Garcia, 2011; Yurt & Deniz, 2021).

For instance, the EdTech itself does not necessarily have to demonstrate to bring about sustainability by virtue of the concept itself in its construction, and therefore the excessive use of resources and energy can be compounded so as to add to the climate disaster (Gomma & Gauthier, 2024). There has to be further research to carry on verifying this as additional evidence of EdTech's role towards education quality within various fields, for example, medical education (Hong & Kim, 2021). There is a sincere prospect for the circular economy through the opportunity it offers for technology and education to cross paths. Care must be taken to the concerns of EdTech sustainable design and evidence of the effects that technology is having on education across a range of disciplines, but in order to truly fulfil this promise. This would have corporations, governments, and educators bring with them a role to enable the circular economy objectives to be advanced instead of going against the use of technology in learning (Gomma & Gauthier, 2024; Hong & Kim, 2021; Williams et al., Enhancing Circular Economy with Digital Technologies: A PLS-SEM Approach, 2024; Wilson et al., 2024).

4 Case Studies and Practical Applications

4.1 Global Case Studies

Many countries and regions have targeted integration of technology, education, and standard of living to improve economic development along with raising the standards of living. Two international case studies that demonstrate good integration are the study on how technology affects living standards in industrialized nations as well as the Education 4.0 project.

Case Study 1—World Economic Forum's Education 4.0

Technology and education are critical drivers of enhanced living standards and sustainable innovation. The World Economic Forum's Education 4.0 program is a prime example of the revolutionizing power of innovative education reform to push these objectives. In accordance with the United Nations Sustainable Development Goals (SDGs), Education 4.0 encourages the development of innovative, technological, and critical thinking skills through problem-solving and learner-focused learning approaches (Joshi et al., 2024; Karampelas, 2022).

This model is charged with preparing students to develop the skills of an increasingly complex and rapidly changing global economy. Via innovation and global citizenship, Education 4.0 allows students to tackle real-world issues head-on, such as sustainability and the circular economy.

An example of this paradigm in real life is the Living City Project, which employs STEAM education to facilitate design-based learning. With this, creativity, collaboration, and problem-solving—the most essential skills required to design sustainable cities and systems—are enhanced. Education 4.0 therefore not only enhances the quality of education but also long-term social and environmental sustainability by empowering future generations with the mindset and ability to design a greener and more inclusive future.

Case Study 2—Technology and Living Standard

Recent research has established that technological advancement is proportional to the standards of living in affluent nations. Various research on the Human Development Index, i.e., the overall general level of development of a nation, demonstrated how innovation and ICT improved the quality of living conditions for the citizens of affluent nations immensely (Zelenkov & Lashkevich, 2020). Promoting knowledge, education, as well as infrastructure and technology within the already existing political and social systems is seen to enhance the standards of living among individuals (Zelenkov & Lashkevich, 2020; Martínez-Navarro & Oliver-Márquez, 2021).

4.2 Lessons Learned

Key takeaways from the case studies.

Case 1:

- Position: global citizenship, creativity, and technology proficiencies; embrace student-led, problem-orienting learning.
- By including design-centered learning, STEM education promotes student autonomy and creativity.
- Compliant with the Sustainable Development Goals as set by the UN.
- Practical activities like the Living City program allow for real-world applications of STEAM principles to high school problems.
- Living standards are raised through integrating technology with education.

Case 2:

- Technology is positively related to living standards, especially in industrialized countries.
- This is a measure to estimate well-being called the Human Development Index.
- High-standard living requirements are improved in developing countries with appropriate political and social structures, thanks to innovation and ICT.
- The quality of life is improved by developments in the knowledge, education, and technological infrastructure sectors.
- When bolstered by robust institutional frameworks, innovation propels advancements in living standards.

5 Conclusion

This chapter has examined the transformative role of education, technology, and standards of living in facilitating India's shift from fossil fuels to a green economy, framed within the principles of circular economy and sustainable innovation. Education contributes by nurturing environmental awareness and critical thinking, equipping individuals with the capacity to adopt and advocate for sustainable practices. Technology, particularly digital innovations such as IoT, AI, and blockchain, enhances resource efficiency, supports clean production systems, and enables data-driven sustainability solutions (Kim et al., 2010; She, 2017). Meanwhile, improvements in living standards—reflected in access to infrastructure, finance, and digital tools—create the socio-economic foundation for the adoption of green lifestyles and eco-conscious choices.

Green finance is the economic force behind this change. Through avenues such as green bonds, ESG-tied investments, and policy-based incentives to sustainability, the financial flows are being redirected to invest in renewable energy, sustainable infrastructure, and circular business models. These financing tools not only enable massive investment to be made, but they also bring accountability and long-term value creation together with the sustainability agenda.

Even with these advances, policy fragmentation, financial exclusion, and information overload continue to be major impediments. We require a synthesizing approach—an approach that ties together education reform, technological diffusion, inclusive finance, and adaptive policy frameworks—so that the full potential of a circular and sustainable economy may be realized.

Finally, India's journey to a greener tomorrow needs systemic alignment among the actors—industry, government, academia, and civil society. Through coordination of finance, education, and innovation with the circular economy agenda, the nation can establish a robust, inclusive, and environmentally sustainable growth model. Subsequent research needs to further analyze the synergies between these domains and provide lessons for scalable and equitable uptake of circular economy principles.

6 Policy Recommendation

For the advancement of India's circular economy and facilitating sustainable innovation, certain targeted policy interventions are recommended. Circular economy thinking and sustainability literacy must be encouraged by formal and informal education of all types. Incorporating the subjects in school curricula, vocational education, and higher education can facilitate the culture of environmental stewardship and systems thinking at the early stage.

Second, the government should increase financial incentives and technical assistance to green technology and circular business model sectors. This involves offering tax rebates, grants, and concessionary finance to firms with measurable environmental footprints, thereby encouraging more green transformation participation.

Third policies have been needed to ensure access to digital infrastructure and digital skills training is made available, particularly to low-income communities. Closing this digital divide is critical to empower all communities to benefit from this technology leapfrogging and fully engage in circular economy opportunities.

Fourth, alignment by sectors is necessary. Circular economy objectives must be aligned with national development planning to ensure consistent governance and avoid overlaps in regulation. Synergies may be unleashed and systems change triggered by coordinated action among industry stakeholders, local governments, and ministries.

Last but not least, decentralizing green finance through grassroots efforts and local efforts toward sustainability will promote bottom-up participation. Promoting cooperative models, microfinancing initiatives, and place-specific investment can make the dividends of green finance inclusive and equitable, promoting an equitable transition toward sustainability.

References

Adenle, A. A., De Steur, H., Mwongera, C., Rola-Rubzen, F., de Barcellos, M. D., Vivanco, D. F., Timilsina, G. R., Possas, C., Alders, R., Chertow, M., & Poons, S. (2023). Global UN 2030 agenda: How can Science, Technology and Innovation accelerate the achievement of sustainable development goals for all? *PLOS Sustainability and Trasformation*. https://doi.org/10.1371/journal.pstr.0000085

Aiwei Wang, N. I. (2023). Does the individual effect of resource rents imperative in the attainment of environmental sustainability? Evidence of Southeast Asian economies. *Enviroment Science and Pollution Research*. https://doi.org/10.1007/s11356-023-29605-3

Akhtar, S. (2023). Sustainable lifestyle: A path towards environmental sustainability and sustainable development. *Managemant Journal for Advanced Research*. https://doi.org/10.54741/mjar.3.4.8

Altassan, A. (2023). Sustainable integration of solar energy, behavior change, and recycling practices in educational institutions: A holistic framework for environmental conservation and quality educatio. *Sustainability*. https://doi.org/10.3390/su152015157

Aluchna Maria, R. B. (2019). Closing the loop: Circular economy through sustainable innovation lens. https://doi.org/10.1007/978-3-030-11217-2_2

Anastasiadis, S., Perkiss, S., Dean, B. A., Bayerlein, L., Gonzalez-Perez, M. A., Wersun, A., Acosta, P., Jun, H., & Gibbons, B. (2020). Teaching sustainability: Complexity and compromises. *Journal of Applied Research in Higher Education*. https://doi.org/10.1108/jarhe-02-2020-0029

Aparicio-Gómez, O.-Y., Ortiz, O. L., & Feigenblatt, O. F. (2024). Building a sustainable future: The role of digital resources in achieving the sustainable development goals (SDGs). https://doi.org/10.24140/issn.1645-7250.rle61.08

Aulia, R. N., Jasin, F. M., Priyanto, S., Effendi, M. R., & Afifah, I. (2023). Empowering communities for Islamic eco-friendly house development: a case study in a tourist village. *Indionesian Journal of Cultural and Community Development*. https://doi.org/10.21070/ijccd.v14i2.972

Austin, M., & Brown, L. D. (1999). Internet Plagiarism: Developing strategies to curb student academic dishonesty. *The Internet and Higher Education*. https://doi.org/10.1016/s1096-7516(99)00004-4

Bănescu, C. E., & Ţiţan, E. (2023). The impact of the standard of living on digital performance in the European Union. *Journal of Social Economics Statistics*. https://doi.org/10.2478/jses-2023-0004

Behera, D. K. (2023). Promoting sustainable development through environmental policy, green technologies, and effective waste management: A comprehensive review. *Journal of Multidisplenary Sciences*. https://doi.org/10.58578/mikailalsys.v1i2.1675

Brydges, T. (2021). Closing the loop on take, make, waste: Investigating circular economy practices in the Swedish fashion industry. *Journal of Cleaner Production*. https://doi.org/10.1016/j.jclepro.2021.126245

Bukhari, S. D. (2024). Circular economy through smart and sustainable supply chain management. https://doi.org/10.4018/979-8-3693-0669-7.ch002

Cai, Y. (2023). Green innovations and environmentally friendly technologies: Examining the role of digital finance on green technology innovation. *Environmental Science and Pollution Research International*. https://doi.org/10.1007/s11356-023-31094-3

Calzolari, T., Pinis, M. B., Genovese, A., & Brint, A. (2023). Understanding the relationship between institutional pressures, supply chain integration and the adoption of circular economy practices. *Journal of Cleaner Production*. https://doi.org/10.1016/j.jclepro.2023.139686

Chigbu, B. I. (2024). Advancing sustainable development through circular economy and skill development in EV lithium-ion battery recycling: a comprehensive review. *Frontiers in Sustainability*. https://doi.org/10.3389/frsus.2024.1409498

Chirea Ionela Mirela, D. P. (2022). A critical examination of the importance of sustainability in the circular economy: UK perspective. *International Journal of Business & Economic Development*. https://doi.org/10.24052/ijbed/v010n01/art-02

Dauda, H., Bala, B., & Danladi, S. (2024). Financial inclusion and technological influence on SDG 4—Quality education attainment in Africa: A case study of Gombe State. *Journal of Economocs, Management and Trade.* https://doi.org/10.9734/jemt/2024/v30i51205

Díez-Palomar, J., Sanmamed, A. F., García-Carrión, R., & Molina-Roldán, S. (2018). Pathways to equitable and sustainable education through the inclusion of Roma students in learning mathematics. *Sustainability.* https://doi.org/10.3390/su10072191

Echanique, B. B. (2020). Students' knowledge of Plagiarism: Basis for SPUP honor code framework. *Universal Journal of Educational Research.* https://doi.org/10.13189/ujer.2020.080931

Emon, M. M., & Hassan, F. (2023). The role of education in promoting sustainable behavior change in Bangladesh: An exploratory study. *Education, Sustainability & Society.* https://doi.org/10.26480/ess.01.2023.41.43

Ercegovac, Z., & Richardson, J. V. (2004). Academic dishonesty, Plagiarism included, in the digital age: A literature review. *College & Research Libraries.* https://doi.org/10.5860/crl.65.4.301

Godinho, V., Monteiro, C., & Azevedo, G. M. (2024). Effects of taxation on innovation and implications for the sustainable development goals. https://doi.org/10.4018/979-8-3693-0847-9.ch014

Gomma, O., & Gauthier, A. (2024). Sustainability in education technology design. https://doi.org/10.4018/979-8-3693-2113-3.ch004

Hong, H., & Kim, Y. (2021). Applications and effects of EdTech in medical education. *Korean Medical Education Review.* https://doi.org/10.17496/kmer.2021.23.3.160

Hsu, C.-L., & Chen, M.-C. (2021). Advocating recycling and encouraging environmentally friendly habits through gamification: An empirical investigation. *Technology in Society.* https://doi.org/10.1016/j.techsoc.2021.101621

Jack Williams, M. R. (2024). Enhancing circular economy with digital technologies: A PLS-SEM approach. *International Transaction on Education Technology (ITEE).* https://doi.org/10.33050/itee.v2i2.590

Jessy, T.M.A.V. (2023). Enhancing tribal education through community-based learning initiatives. *International Journal of Advance Research.* https://doi.org/10.21474/ijar01/17613

Jinhui Li, K. Y. (2011). A study on legislative and policy tools for promoting the circular economic model for waste management in China. *Journal of Material Cycles and Waste Management.* https://doi.org/10.1007/s10163-011-0010-4

Joshi, B. M., Pokhrel, R. K., & Khatiwada, S. P. (2024). Influence of socioeconomic factors on access to digital resources for education. *Rupantaran: A Multidisciplinary journal.* https://doi.org/10.3126/rupantaran.v8i01.65197

Karampelas, A. (2022). Living City project: Integrating education 4.0 content and experiences into the STEAM classroom. *Hellenic Journal of STEM Education.* https://doi.org/10.51724/hjstemed.v2i1.16

Kaur, A., Paruchuri, R. G., Nayak, P., Devi, K. B., Upadhyay, L., Kumar, A., Pancholi, R., & Yousuf, M. (2023). The role of agroforestry in soil conservation and sustainable crop production: A comprehensive review. *International Journal of Environment and Climate Change.* https://doi.org/10.9734/ijecc/2023/v13i113478

Kholijah, S. K. (2024). Analysis of economic and environmental benefits of green business practices in the hospitality and tourism sector. *Involvement International Journal of Business.* https://doi.org/10.62569/iijb.v1i1.7

Kim, T., Hong, H., & Magerko, B. (2010). Design requirements for ambient display that supports sustainable lifestyle. https://doi.org/10.1145/1858171.1858192

Kunrath, K., & Beliatis, M. (2022). Design for impact (D4i): A framework for teaching sustainability in engineering design. https://doi.org/10.5821/conference-9788412322262.1424

Levett-Jones, T., Burdett, T., Chow, Y. L., Jönsson, L., Lasater, K., Mathews, L. R., McAllister, M., Pooler, A., Tee, S., & Wihlborg, J. (2018). Case studies of interprofessional education initiatives from five countries. *Journal of Nurshing Scholarship: An Official Publication of Sigma Theta Tau Internation Honor Society of Nursing.* https://doi.org/10.1111/jnu.12384

Li, C., Li, Q., Wang, H., & Liu, Q. (n.d.). Does innovative industrial agglomeration promote environmentally-friendly development? Evidence from Chinese prefecture-level cities. *Sustainability*. https://doi.org/10.3390/su142013571

Lodh, D. D. (2024). Sustainable waste management practices and their profound impact on environmental conservation. https://doi.org/10.58532/v3bcag21p1ch2

Mahmood, S., El-kenawy, E.-S. M., Khafaga, D. S., Alharbi, A. H., Iqba, A., & Sun, H. (n.d.). The role of green finance in promoting sustainable development goals (SDGs) through the mediation effect of green technology innovation. https://doi.org/10.21203/rs.3.rs-4194345/v1

Mandych Oleksandra, G. O. (2023). Circular economy model: Generating benefits for sustainable development. *Economic Analysis*. https://doi.org/10.35774/econa2023.01.252

Mandych, O., Girzheva, O., Taran, O., Babko, N., & Pasemko, G. (2023). Circular economy model: Generating benefits for sustainable development. *Economic Analysis*. https://doi.org/10.35774/econa2023.01.252

Marques, T., Gomes, J., & Reis, N. (2019). A bibliometric study on academic dishonesty research. *Journal of Academic Ethics*. https://doi.org/10.1007/s10805-019-09328-2

Martínez-Navarro, D., & Oliver-Márquez, F. J. (2021). 13—Economic indicators of sustainable resource management. *Sustainable Resource Management*. https://doi.org/10.1016/b978-0-12-824342-8.00002-x

Mensah, Y. A., Baah, C., Tang, L., Dacosta, E., & Afum, E. (2021). Organisational identity and circular economy: Are inter and intra organisational learning, lean management and zero waste practices worth pursuing? *Sustainable Production and Consumption*. https://doi.org/10.1016/j.spc.2021.06.018

Meylani, R. (2024). An in-depth literature review of remote, online, and hybrid learning with case studies of successful and failed attempts. *Internation Journal of Studies in Education and Science*. https://doi.org/10.46328/ijses.87

Munir, M. T., Naqvi, U., & Li, B. (2024). A converging path: A decade's reflection on net zero emissions and the circular economy. *Frontiers in Energy Research*. https://doi.org/10.3389/fenrg.2024.1332174

Naveed, W., Naveed, N., Juan, A. A., & Ammouriova, M. (2022). Circular economy and information technologies: Identifying and ranking the factors of successful practices. *Sustainability*. https://doi.org/10.3390/su142315587

Neri, A., Kumar, V., Franzò, S., Negri, M., Cagno, E., Bassani, C. A., & Lampertico, T. (2023). The role of digital technologies in supporting the implementation of circular economy practices by industrial small and medium enterprises. *Business Strategy and the Environment*. https://doi.org/10.1002/bse.3388

Nwankwo Constance Obiuto, N. N.-E.-L. (2024). Implementing circular economy principles to enhance safety and environmental sustainability in manufacturing. *International Journal of Advanced Multidisciplinary Research and Studies*. https://doi.org/10.62225/2583049x.2024.4.2.2432

Odeyemi, O., Mhlongo, N. Z., Usman, F. O., Elufioye, O. A., & Ike, C. U. (2023). Sustainable entrepreneurship: A review of green business practices and environmental impact. *World Journal of Advanced Research and Reviews*. https://doi.org/10.30574/wjarr.2024.21.2.0461

Oliveira, T. (2024). Exploring family engagement and learning outcomes in early childhood education programs in Brazil. *Research and Advances in Education*. https://doi.org/10.56397/rae.2024.02.05

Prakash, R., & Kumar, D. S. (2024). Enhancing learning outcomes through effective E-learning strategies. https://doi.org/10.55524/csistw.2024.12.1.24

Rahmawati, Y., Taylor, P. C., Ridwan, A., Mardiah, A., & Taylor, E. (2022). Students' engagement in education as sustainability: Implementing an ethical dilemma-STEAM teaching model in chemistry learning. *Sustainability*. https://doi.org/10.3390/su14063554

Ranbir, D. (2024). Educational technology integration: Challenges and opportunities. *Innovative Research Thoughts*. https://doi.org/10.36676/irt.v10.i2.1406

Rezanejad, A., & Rezaei, S. (2013). Academic dishonesty at universities: The case of plagiarism among Iranian language students. *Journal of Academic Ethics*. https://doi.org/10.1007/s10805-013-9193-8

Ritter, M., Schilling, H., Brüggemann, H., Fröhlich, T., Goldmann, D., Henze, R., Kuhlmann, M., Mennenga, M., Mrotzek-Blöß, A., Niemeyer, J. F., & Schmidt, K. (2024). Towards achieving the sustainable development goals: A collaborative action plan leveraging the circular economy potentials. *GIO*. https://doi.org/10.1007/s11612-024-00733-9

Sahar Afshan, I. O. (n.d.). Facilitating renewable energy transition, ecological innovations and stringent environmental policies to improve ecological sustainability: Evidence from MM-QR method. *Renewable Energy*. https://doi.org/10.1016/j.renene.2022.06.125

Said, M., Hassan, F. U., & Isack, A. D. (2024). Biotechnological advancements for environmental conservation in East Africa. *Journal of Chemical, Enviromental and Biological Engineering*. https://doi.org/10.11648/j.jcebe.20240801.13

Salminen, H., Kujala, J., & Heikkinen, A. (2023). Connecting the circular economy and sustainability: Finnish stakeholder perceptions. https://doi.org/10.1007/978-3-031-31937-2_13

Salonen, A. O., & Hakari, S. (2018). Early childhood educators and sustainability—Sustainable living and its materialising in everyday life. *Utbilding & Demokrati*. https://doi.org/10.48059/uod.v27i2.1103

Sami, A. M. (2024). Perspective chapter: The impact of digital education on modern educational institutions. https://doi.org/10.5772/intechopen.108058

She, C. (2017). The role of multimedia technology in the construction of digital community education resources. *DEStech Trasactions on Social Science, Education and Human Science*. https://doi.org/10.12783/dtssehs/icsste2017/9374

Shi, G., Li, Q., & Yu Wei, M. A. (2024). Does digital technology advancement promote natural resource utilization efficiency? *Resource Policy*. https://doi.org/10.1016/j.resourpol.2024.105088

Shustova, I., Polevaya, E., Ivanov, A., Pukhkal, V., & Uvarova, S. (2023). The impact of Edtech: Strengths and opportunities. *E3S Web of Conferences*. https://doi.org/10.1051/e3sconf/202345806012

Sindhu, C. M., & Binoy, K. (2017). Do digital natives aware on digital technology. *Scholarly Research Journal for Humanity Science & English Language*. https://doi.org/10.21922/srjhsel.v4i24.10324

Singh, M., & Vallarasu, K. (2023). Environmental conservation and sustainability: Strategies for a greener future. *International Journal for Multidimentional Research Perspectives*. https://doi.org/10.61877/ijmrp.v1i3.97

Solomon, N. O., Adenekan, O. A., Simpa, P., & Obasi, S. (2024a). Circular economy principles and their integration into global supply chain strategies. *Finance & Accounting Research Journal*. https://doi.org/10.51594/farj.v6i5.1133

Solomon, O. N., Adenekan, A. O., Obasi, C. S., & Simpa, P. (2024b). Sustainable nanomaterials' role in green supply chains and environmental sustainability. *Engineeering Science & Technology Jounal*. https://doi.org/10.51594/estj.v5i5.1136

Souza, E. B., Scur, G., Carlos, R. L., & Mattos, C. D. (2024). The role of blockchain platform in enabling circular economy practices. *Corporate Social Responsibility and Environment Management*. https://doi.org/10.1002/csr.2885

Stankevičienė, J., & Nikanorova, M. (2020). Eco-innovation as a pillar for sustainable development of circular economy. *Business Theory and Practice*. https://doi.org/10.3846/btp.2020.12963

Suklun, H., & Bengü, E. (2024). Raising awareness of sustainable development goals in higher education institution. *Research in Education Administration and Leadership*. https://doi.org/10.30828/real.1357661

Toni, M. (2023). Conceptualization of circular economy and sustainability at the business level. Circular economy and sustainable development. *International Journal of Empirical Research Methods*. https://doi.org/10.59762/ijerm205275791220231205140635

Uniya, D. J. (2024). Exploring the role of education and awareness campaigns in shaping consumer behavior towards green products. *International Journal for Multidisciplinary Research*. https://doi.org/10.36948/ijfmr.2024.v06i02.15324

Valadez-Martínez, L., Padley, M., & Penagos, M. F. (2017). A dignified standard of living in Mexico: results of a pilot study of the minimum income standard approach. *Social Indicators Research*. https://doi.org/10.1007/s11205-017-1780-4

Veckalne, R., & Tambovceva, T. (2022). The role of digital transformation in education in promoting sustainable development. *Virtual Economics*. https://doi.org/10.34021/ve.2022.05.04(4)

Vernyuy, A. (2024). Impact of technological advancements on human existence. *International Journal of Philosophy*. https://doi.org/10.47941/ijp.1874

Wahyono, T. (2022). Management of quality human resources to improve the community of life in Klambir Lima Kebun Village. *International Journal of Research and Review*. https://doi.org/10.52403/ijrr.20220515

Wang, J., Yu, H., Ren, D., & Zhang, J. (2023). Promoting mineral resources consumption efficiency: Evidence from technology of big data. *Resource Policy*. https://doi.org/10.1016/j.resourpol.2023.104210

Wang, Y., Xu, X., Lu, S., Zhang, X., Liu, X., & Zhou, Q. (2021). Information complexity and behavior intention to prescribe antibiotics based on the antimicrobial susceptibility testing report: The mediating roles of information overload and attitude. *Frontiers in Pharmacology*. https://doi.org/10.3389/fphar.2021.778664

Williams, J., Rodriguez, M., Prawiyogi, A. G., & Kovac, I. (2024). Enhancing circular economy with digital technologies: A PLS-SEM approach. *International Transactions on Education Technology (ITEe)*. https://doi.org/10.33050/itee.v2i2.590

Wilson, A., Ming, L. W., & Kask, R. (2024). Exploring circular digital economy strategies for sustainable environmental, economic, and educational technology. *International Transaction on Education Technology*. https://doi.org/10.33050/itee.v2i2.579

Wilson, E. D., & Garcia, A. C. (2011). 'Going Green' in food services: Can health care adopt environmentally friendly practices? *Canadian Journal of Dietetic Practice and Research*. https://doi.org/10.3148/72.1.2011.43

Yurt, C., & Deniz, D. (2021). Product-based and knowledge-based sustainable living practices: The case of IKEA. *Sustinere: Journal of Environment and Sustainability*.

Zelenkov, Y., & Lashkevich, E. (2020). Fuzzy regression model of the impact of technology on living standards. *Business Informatics*. https://doi.org/10.17323/2587-814x.2020.3.67.81

Zhou, X., Song, M., & Cui, L. (2020). Driving force for China's economic development under Industry 4.0 and circular economy: Technological innovation or structural change? *Journal of Cleansen Production*. https://doi.org/10.1016/j.jclepro.2020.122680

E-Waste Management in the Circular Economy: A Transition Towards Sustainable Future

Adori Khatun, Mamun Mandal, Anamika Roy, and Abhijit Sarkar

Abstract Electronic waste (e-waste) has become a fast-growing type of problematic solid waste worldwide. The improper use of e-waste disposal practices, as well as their increased potential to create environmental contamination and human health dangers, has resulted in a global problem. Eliminating the present e-waste management issue is feasible through "sustainable production and consumption," which requires countries to undergo significant transformations in their societies' ability to manufacture and consume commodities. This is done via the collective efforts of governments, international organisations, commercial firms, and people to transform outdated production and consumption habits. This study will cover e-waste issues and discuss the dangerous elements contained inside them. The problem of electronic and electrical waste management in developed and emerging nations (namely, India, Japan, and the United States) and appropriate routes, including recyclable components and expanded producer responsibility, will be investigated, culminating in establishing a circular economy. The current state of the e-waste sector will be evaluated in order to examine the issues involved with e-waste recycling and circular economy, such as illegal e-waste trade-in from developed nations and careless discarding, both of which cause significant environmental impact.

Keyword Electronic waste · Management · Recycle · Circular economy

A. Khatun · M. Mandal · A. Roy · A. Sarkar (✉)
Laboratory of Applied Stress Biology, Department of Botany, University of Gour Banga, Malda, West Bengal, India
e-mail: abhijitbhu@gmail.com; abbhijitbot@ugb.ac.in

M. Mandal
e-mail: mamun.mandal9933@gmail.com

P. Singh et al. (eds.), *Circular Economy and Sustainable Innovation*,
https://doi.org/10.1007/978-3-032-00437-6_12

1 Introduction

Waste is a disposable or unusable element located in our surroundings. Waste is mainly considered solid waste, including waste from our houses, schools, offices, and municipal areas, as well as from industries and factories. In the same way, e-waste is also seen in our locality, such as electrical or electronic equipment that has been discarded. This includes working and non-working items, like broken items that are thrown in the garbage or donated to charity resellers (He et al., 2024). Waste and e-waste are particularly different. Waste cannot be recycled or reused; E-waste can be recycled not only for its metal content but also for its valuable items. E-waste production in certain developing countries is a similar reason for relation at this moment since the smaller number and longer half-life of electronic equipment in particular countries remain economically restricted on both local community and national scales (Dixit et al., 2023).

The annual rate of generation of E-waste is 74.7 million tons (mt) globally. Three major continents that produce e-waste are Asia, Europe, and America. In India, about 800,000 tons of E-waste were generated, which was trouble for developing countries (Heeks et al., 2015). E-waste recycling is the management of e-products, and it is a good time for developing countries to remove from their glass, copper, nickel, chromium, silver, and many more. To manage the production of e-waste recycling, the best process is to minimize it. The recycling process is eco-friendly and protects our environment. It reduces garbage, E-waste, and energy consumption. If we do not recycle E-waste, recycling does not work (Ilankoonet al., 2018). E-waste is a significant problem in the world. It impacts the environment and human health. E-waste is the reverse construction process by re-engineering and renovating the waste material to new, polished products for future use (Kumar et al., 2019; Mandal et al., 2025). This renovation process helps to control the ecosystem. Conservative concerning the recycling process of e-waste is the greatest successful management process that decreases the negative effect of e-waste on the environment (Kiddee et al., 2013). E-waste contains many precious metals like gold, silver, platinum, copper, and so on, which are extracted by the recycling process. If the recycling process is managed by experts, we can make a profitable business (Chatterjee, 2012). In developing countries, e-waste, where the discarding cost is high and the environmental management system regulation is high for recycling, such countries are China, India, and other countries (Liu et al., 2023). Recycling is the best processing in many developed countries such as China, where the recycling rate is 16%; Germany, where the recycling rate is 52%; France, where the recycling rate is 56%; and France, where the recycling rate is 56% (Patel et al., 2023).

According to Sengupta et al. (2022), the lower recycling rates of e-waste lead to the disposal of vital essential components like rare earth elements and valuable, recoverable materials like important and base metals, which are valued at approximately 57 billion USA dollar, being dumped in landfills, burning, or using rudimentary recycling methods. If extracted with the right scientific methods, these metals can be a significant source of income (Islam et al., 2020). Furthermore, there is a high demand for natural minerals in the production of new electronic gadgets, which can be detrimental to both human health and the environment (Wang et al., 2022). The challenges posed by traditional e-waste disposal underscore the pressing requirement for a strong infrastructure to minimize, recycle, and dispose of end-of-life (EoL) e-waste sustainably. One paradigm that might help with these issues is the circular economy. Transforming the linear "take-make-use-dispose" product flow paradigm into a circular pattern with the goal of reducing waste by extending the useful life of materials is the foundation of the circular economy (Cheah et al., 2022; Chatterjee et al., 2024).

This study examines the challenges posed by e-waste while investigating sustainable solutions, by employing the circular economy to manage it effectively. Studying the detrimental effects of improper e-waste disposal methods on the environment and human health, assessing the efficacy of global recycling programs and regulations, and identifying technological advancements that enhance e-waste recycling procedures are the objectives of the research. While offering suggestions to improve the recycling system in poor countries where inappropriate waste treatment is still prevalent, this study highlights the significance of (Extended producer responsibility) EPR for sustainable e-waste management.

2 Sustainable Development

E-wastes are very harmful to the environment and causes environmental damage due to the use of toxic substances in manufacturing toxic material (Ashfaq & Khatoon, 2014). E-waste contains more than 1000 types of toxic materials, which can cause many problems, diseases, etc. Some materials contain very toxic substances that can contaminate the environment and human health if not properly managed. In recycling, many toxic materials are produced into the atmosphere, which may affect human health (Kidee et al., 2013). The most common substances which are associated with e-waste include toxic metals such as barium (Ba), beryllium (Be), cadmium (Cd), cobalt (Co), chromium (Cr), copper (Cu), iron (Fe), lead (Pb), lithium (Li), lanthanum (La), mercury (Hg), manganese (Mn), molybdenum (Mo), nickel (Ni), silver (Ag), hexavalent chromium (Cr), and persistent organic pollutant (POPs) such as dioxide, brominated flame retardants (BFRs), polycyclic aromatic hydrocarbons (PAHs), polychlorinated biphenyls (PCBs), polychlorinated dibenzo-p-dioxins and dibenzofurans (PCDD/FS), and polyvinyl chloride (PVC) (Kidee et al., 2013). Bioaccumulation of several chemical elements, such as PBCs and BFRs in the food chains, affects human health, especially for pregnant ladies and nurses. Lead and mercury contain highly potent neurotoxins, particularly among children, who can be affected by this and suffer IQ deficiency and developmental abnormalities (Baldé et al., 2017). Dioxins affect the methylation status of DNA. More changes in the serum levels of mothers and newborns are a potential hazard to maternal health and child development and produce hormone effects (Hossain et al., 2015; Mandal et al., 2022a). In landfilled areas, much of E-waste is present, such as computers, mobiles, and others, producing contaminated leachates that are harmful to the environment's groundwater. When computers melt into the ground, many acids and sludge, if disposed of on the ground, may cause soil acidification (Kumar et al., 2019). In areas where e-waste is recycled, many toxic substances are produced, and residents and workers are affected, causing the incidence of skin damage, headaches, vertigo, nausea, chronic gastritis, gastric and duodenal ulcers. A large number of workers and children have been found to disassembling e-waste activities (Hossain et al., 2015; Sharma et al., 2023a, 2023b). In the future, electric and electronic equipment technology will be very advanced. Around 50 million metric tons of e-waste are generated globally per year, with an average of more than 6 kg per person. In developed countries, more e-waste products are produced than this. For instance, Norway produces 28.5 kg per person per year; compared to the African countries, it is less than 2 kg. Furthermore, often referred to as the fastest-growing solid waste stream, the growth of e-waste is not surprising, as the amount of e-waste products is increasing. It is also expanding to developing countries, which a huge producers of waste in this industry and the huge amount of equipment sales (Kumar et al., 2019).

3 Production of E-Waste

E-waste production is mainly in developed countries and industrial areas. This waste stream can be characterized as one of the problematic wastes that cause environmental pollution according to various global, regional, and national reports (Song & Li, 2014; Baldé et al., 2017; Quinto et al., 2025). Hossain et al. (2015) referred to according to the USEPA (United States Environmental Protection Agency), the US is the world's largest e-waste producer, generating 3.16 mt in 2008. This e-waste stream volume is increased by different causes, such as short life spans or replacement intervals. For example, mobile phones and personal computers are the most used electronic equipment nowadays. Electronic computers, refrigerators, and electric cook stoves have an average life cycle of a few years (He et al., 2024; Mandal et al., 2024a). The European Union's e-waste generated 8.9 mt in the year 2010, and this rate is increasing by 3–5% yearly and is projected to be 12 mt in the future (Hossain et al., 2015). This rapidly growing trend of e-waste production is increasing in the biggest and fastest-growing economies, such as China and India. After the USA, China was the second-ranked to generate 2.3 mt of e-waste in 2010, and India generated 400,000 tons in 2011. E-waste generation increased from 2014 to 2019 globally (Dixit et al., 2023). After that, many countries produced a high amount of e-waste, where the USA generated 7 mt, China generated 10 mt, and India generated 3 mt in 2019. Global E-waste volume is projected to grow to 74.7 mt by 2030, meaning it will have almost doubled in only 2016 years. Three major continents are used for generating e-waste: Asia (24.9 mt), Europe (12 mt), and America (13.1 mt), when measured by total weight (Table 1; Forti et al., 2020).

Table 1 E-waste generation in different countries

Country	E-waste generation (mt)	Year
USA	6.3	2016
	6.9	2019
China	7.2	2016
	10.1	2019
France	1.4	2016
	1.6	2019
Japan	2.1	2016
	2.5	2019
India	2.0	2016
	3.2	2019
Brazil	1.5	2016
	2.1	2019
Germany	1.9	2016
	1.6	2019
Indonesia	1.3	2016
	1.6	2019

Source Author's own creation

4 Management of E-Waste

Managing e-waste to reduce its volume is the best way to deal with increasing e-waste. Recycling and reuse are the best processes that reduce the volume of e-waste (Yonget al., 2019). From recycling to recovery, the metals from computers, TVs, refrigerators, washing machines, and so on are selected for export in developed countries (Oteng-Ababio, 2010). Some appliance components, such as plastics, metals, and condensers, are used directly and reused (Oteng-Ababio, 2010). These methods can potentially conserve energy and protect and clean from the toxic elements in the environment (Yong et al., 2019). Various techniques are developed for e-waste management and also applied, including LCA (Life Cycle Assessment), MFA (Material flow analysis), MCA (Multi-criteria analysis), and EPR. All these methods are environmentally friendly to keep the environment safe. EPR programs are at the national scale level for e-waste management applied in developed countries such as Japan, the European Union, etc. The Organisation for Economic Co-operation and Development (OECD) has also supported EPR as environmentally friendly and published a guidance manual for the government (Kiddee et al., 2013; OECD, 2001). The current management of e-waste is described.

4.1 Recycling

Recycling is the process of reusing e-waste, which contains usable parts that are useful for other equipment used for work (Forti et al., 2020). Computers, laptops, TVs, monitors and CRTs, keyboard, fridge, mobiles, washing machine, telephones, hard drive, fax machines, printers, compact disks, CPUs, memory chips, and connecting wires can all be recycled. In the recycling process, remove different parts of e-waste containing destructive materials such as PCB, Hg, and other detachments of plastic, move the CRT, and segregate ferrous and nonferrous metals. Strong acids are used to remove precious metals such as gold, silver, platinum, copper, and palladium (Kumar et al., 2019). Cyanide, thiourea, and thiosulfate acids are used for gold and silver extraction.

$$4AU + 8CN^- + O_2 + 2H_2O\ 4AU(CN)_2^- + 4OH^-$$

For palladium and platinum extraction, a hydrochloric and nitric acid mixture (aqua regia) is used (Mandal et al., 2024b; Quinto et al., 2025; Yong et al., 2019).

$$Pd + 2/3HNO_3 + 4Cl^- + 2H^+ \rightarrow PdCl_4^{2-} + 2/3NO + 4/5H_2O.$$

E-waste recycling has many steps, as shown in Fig. 1. We don't find e-waste in general waste; when discarding e-waste, we must dispose of it in designated collection places, usually a particular recycling bin or a verified collection site. We should follow the legislation of a valid country (Sharma et al., 2023a, 2023b). In the e-waste recycling process, collection is the first step, and then the mix of e-waste products is dispatched to the expert electronics recyclers. In this step, we take care of mixed products that may cause damage, such as batteries that are harmful to other products. After that, store them in various categories. Then, mixed e-products categorization, such as batteries and bulbs (Zeng et al., 2017). After that, some e-waste products are dismantled by hand, and precious materials and some materials for reuse. In the segregation process, plastic, metals, cables and wires, glass, and PCB are separated (Mandal et al., 2025). After that, e-waste is shredded into small pieces. This is the vital stage of the process in that the e-products contain many precious or toxic materials that can dissociate mechanically when they break into centimeter-sized pieces (Kang et al., 2020; Mandal et al., 2022b). Mechanical separation involves two steps: magnetic separation and water separation. After shredding e-waste in the magnetic stage to pill out the ferrous metals such as iron and steel, the other nonferrous metals are differentiated using eddy currents. After that, this metal is dispatched to an expert recycling solution for blending (Suja et al., 2014). Extra materials like circuit boards, metal sets, and plastic are separated throughout this stage. Later, through magnetic separation, the other solid waste contains plastic and glass. After the separation step, water is used to refine the waste and differentiate the various plastics. When differentiated, all materials are prepared for reuse and sale. Some materials, such as plastic and steel, are managed by other recycling streams.

Then, the different materials can be manufactured onsite and dispatched directly, and the usable parts can be extracted at an earlier stage (He et al., 2024).

Recycling methods have advantages and disadvantages. The recycling process is eco-friendly, which protects the environment. It reduces the amount of e-waste and decreases the amount of global warming which protects us from increasing warming. Landfills are protected from e-waste by recycling and conserving natural resources (Gaidajis et al., 2010). In the recycling method, many strong acids are used, which are unsafe for the environment. Some materials are costly to recycle (Table 2). When landfills are filled with e-waste, many chemical reactions are produced, which cause

Fig. 1 Steps of E-waste recycling method

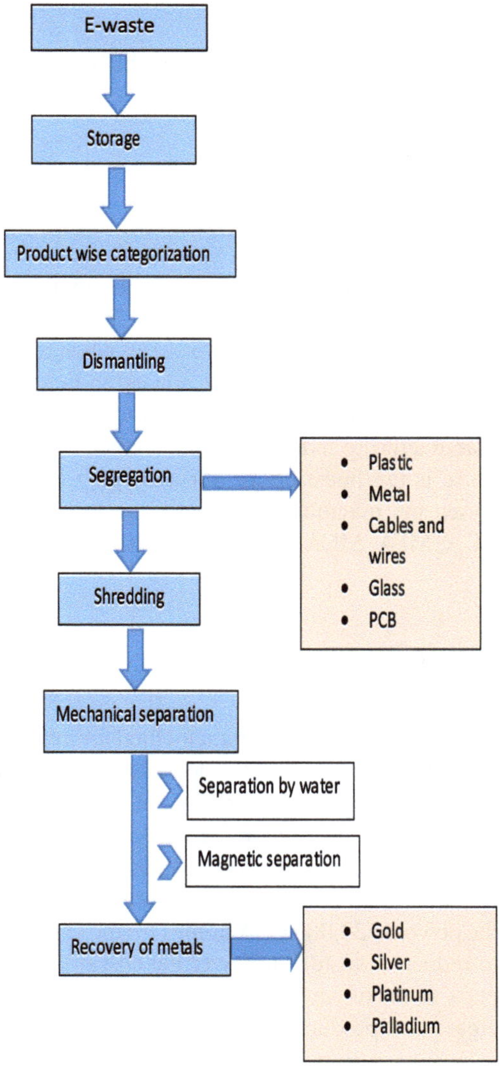

Table 2 Rate of recycling in different countries according to the latest available data, 2023

Rank	Country	Recycling rate (%)
1	China	16
2	USA	15
3	India	1
4	Japan	22
5	Brazil	0
6	Germany	52
7	France	56

Source Author's own creation

soil acidification; that's how landfills lose money. The recycling process is very important.

4.2 Reuse

Reuse is another way to manage the increase in e-waste. It works after using the equipment or for secondhand use. When e-products become e-waste, they are not useful, but some products are useful when reused. Computer cell phones, etc., can be reused. For donations in the education field, old working computers can contribute (Gaidajis et al., 2010). Many e-products are repaired and can be returned to the manufacturing industries. This method makes the generation of e-waste. Reducing reuse is the finest option for managing e-waste generation. This can significantly lessen the magnitude of e-waste generation (Kumar et al., 2019). The features of LCA, MFA, MCA, and EPR for managing e-waste are eco-friendly and important.

4.3 Reduce

Reducing is a very important method for managing e-waste; via this method, it cannot impact our environment. The construction of these devices and the use of frequent materials that go into their manufacturing stand as a great source of embodied energy. Decreasing e-waste helps preserve the benefits and reduces the energy we get from the earth (Garfì et al., 2009). Reusing the valuable metals and plastic from the old mobiles rather than producing more of them would rescue more energy, turning over the power to 24,000 USA homes for the entire year (Kumar et al., 2019). Another way to reduce e-waste is to lower the cost of new consumer products by using recycled parts to make new products. This is not an expensive proposal for constructors to create new products.

4.4 Life Cycle Assessment (LCA)

LCA is used to reduce the e-waste problem and to make the environment healthy. LCA is a high-powered tool for growing eco-design products such as computers, washing machines, heating and air conditioning devices, printers, and so on, to recognise possible environmental impacts on eco-design products (Herat et al., 2012). LCA is widely used to manage e-waste. In Europe, many researchers have chosen to use LCA to appreciate the EoL treatment of e-waste, which impacts the environment.

LCA is widely used to manage e-waste on various continents, including Europe, Asia, Australia, and others. In Asia, LCA has been applied to evaluate the effect of e-waste and e-waste management. In Korea, LCA assigns recycling prospects to environmental and economic strands. The recycling prospect employing the environmental result reveals that the benefit was for glass and circuit boards, followed by Fe, Cu, aluminium, and plastic. Disposal comprises two frameworks: landfill and recycling (Herva & Roca, 2013; Priyashantha et al., 2022). Based on their track record, recycling is the most sensible way to get rid of it. Nakamura and Kondo adopted the LCA technique in Japan, using a life cycle cost analysis that contrasted two approaches: landfilling and recycling for the disposal of e-waste. Although they prove that landfill disposal is more cost-effective than recycling, the environmental cost of landfill disposal is high. The LCA method is used in India to regulate the management of computer trash (Kidde et al., 2013).

4.5 Material Flow Analysis (MFA)

Large amounts of e-waste from industrialized nations were sent to developing nations, namely China, South Africa, and India, for recycling and reuse before the Basel Agreements' full implementation. The MFA tool analyzes the flow of materials in space and time and asks questions about how e-waste material enters recycling or disposal facilities. Tools for material flow analysis aid in waste and environmental management conclusions (Dhir et al., 2021). This tool can be relevant to progressing suitable e-waste management. This comprises a consideration of the circulation of e-waste and its evaluation employing environmental, productive, and collective value (Reuter & Van Schaik, 2016).

4.6 Multi-criteria Analysis (MCA)

According to Garfi et al. (2009), MCA is the best tool for thinking through difficult choices and resolving complex problems involving several criteria, including important aspects of the issue. In order to permit optional e-waste management programs, the MCA tool has attempted to address environmental issues, including e-waste management. MCA is often utilized for solid waste management, despite the fact that it is not widely employed for e-waste management (Herva & Roca, 2013; Vego et al., 2008). According to Rucevska et al. (2015), MCA has been proposed as a community response to e-waste management and is thus a useful tool in the combination with other e-waste management methods.

4.7 Extended Producer Responsibility (EPR)

The OECD states that EPR is an environmental policy approach that multiplication responsibility constructors to reclaim after use and is found in the polluter-pays principle (OECD, 2001; Sivaramanan, 2013). The principles of the EPR project for e-waste management are developed nations, such as the European Union (EU), Switzerland, Japan, and some states of the United States and Canada. The OECD has preferred an environmentally friendly project and announced a guidance manual for the government (OECD, 2001). Legislators are progressively adopting EPR policies to organise various kinds of wastes, like discarded cars, electrical and electronic appliances, and batteries, which require unique handling and treatment (Ashfaq & Khatoon, 2014; OECD, 2001).

5 Categorization of E-Waste

E-waste has been categorised differently depending on its composition. According to several reviews and research, there are many types of E-waste (He et al., 2024). The main types of e-waste include large household appliances, IT and telecommunications equipment, small household appliances, toys and sports equipment, and medical devices. These types contain various E-products, which are as follows.

5.1 Large Household Appliances

From our houses, various waste is produced. E-waste produces large amounts, represented as refrigerators, washing machines, TV monitors, microwaves, air conditioners, mobile phones, mixture machines, fans, pumps, dishwashers, and many more (Vats & Singh, 2014). Every year, more than 98% of household e-waste is generated by large appliances and thrown in landfills. This careless dumping impacts our environment and the loss of precious materials.

5.2 IT and Telecommunications Equipment

Personal computers, monitors, laptops, printers, keyboards, networking equipment, audio amplifiers, CDs, DVDs, and video cameras are considered IT and telecommunications equipment. It is very easy to obtain, and we get precious metals from them (Kang et al., 2020).

5.3 Small Household Appliances

Developed countries, like most towns in America, offer convenient recycling options that overlook these appliances and maintain the environment's health. These services are free of charge. These are electric shavers, hair dryers, watches, vacuum cleaners, fryers, and toasters. One of the specific reasons that these small appliances produce a mass of e-waste issues is that they are not responsible (Dixit et al., 2023).

5.4 Toys and Sports Equipment

These categories generate some of the fastest-growing e-waste bulk. Kids are quickly bored with their toys, which are too much to use and throw everywhere. Every day brings up-to-date gaming software and hardware, forcing consumers to buy new gadgets to enjoy the most recent updates and features. These are video games, electric trains, car racing sets, and any sorting equipment with electronic components (Kang et al., 2020).

5.5 Medical Devices

This category has a broad range of big and small equipment. Medical electric and electronic devices contain components that become annual e-waste data. Here, some equipment is freezers, dialysis equipment, ventilators, scanners, lab equipment, diagnostic equipment, and more. The irresponsibility and carelessness of healthcare e-waste create critical issues in e-waste. We must reduce recycling or donation for better environmental quality, which can result in severe data loss (He et al., 2024).

These are the five major categorizations, but many common components are classified in this category. These components are glass, LCD, rubber, wiring/electrical, transformer, magnetron, circuit boards, fluorescent lamps, batteries, external electrical cables, refractory ceramic fibre, and radioactive substances extracted from building blocks. E-waste consists of different categories, which vary and differ. It contains more than 1000 various components, some hazardous or non-hazardous. It contains ferrous and nonferrous metals, plastic, glass, wood and plywood, printed circuit boards, concrete and ceramic boards, and rubber (Kumar et al., 2019).

6 Why is E-Waste a Significant Problem

Electric equipment is useful in our daily lives, but becomes e-waste at the end of its useful life. Various E-waste equipment found in our surroundings can be recycled, such as computers, printers, mobile devices, laptops, and other equipment (He et al., 2024). However, e-waste affects our environment, human health, and the lives of animals and plants. Some reasons why e-waste is a significant problem are-

6.1 E-Waste Expanding

Nowadays, people are dependent on many gadgets or electronic devices which have a short lifespan, so they become faster to e-waste. For example, circuit boards usually have a life of 1 to 2 years. In 2023, over 347 MT of e-waste was generated globally (Forti et al., 2020).

6.2 Effects on the Environment

When e-waste is thrown in landfills, toxic chemicals are liberated from the products into the water, soil, air, and groundwater, affecting the land, sea animals, and plants. Thus, the harmful materials are absorbed into groundwater, and toxic chemicals are liberated into the atmosphere and soil (Stevels et al., 2023).

6.3 Health Implications of E-Waste

Refrigerators and other electronic products involve harmful chemicals and metals like Hg, Pb, Ni, Cr, PBR, and so many others; these materials affect the environment and human kidneys, blood, skin, and central and peripheral systems. It causes many diseases among the residents of e-waste recycling areas (Jaunich et al., 2020). It may even cause children, especially pregnant ladies, to suffer from IQ deficiency and developmental abnormalities. In developing countries, the impact of e-waste is damaging to people's health.

6.4 *Shipping Tons of E-Waste Overseas*

Various countries—China, India, Vietnam, Pakistan, and the Philippines—have dumped polluted e-waste that handles about 50–80% of the world's e-waste landfill sites. That's why, all over the world, shipping E-waste overseas and dumping it in developing countries, this impact is dangerous to the environment and health of the people in the country (Al-Salem et al., 2022).

7 Case Study of E-Waste

To manage e-waste, many developing countries use different parts of e-waste, such as those extracted by recycling gold, silver, platinum, copper, and others, to produce various valuable items that we can use and be eco-friendly (Grandhi et al., 2024). Such a case study is found from different countries, which are followed:

7.1 *Case Study in Japan*

Each year, the global quantity of e-waste production is 53 mt. In England, a company authorized 'Mount Recyclmore', which organized a completely revealed e-waste to emphasize observing the president attending the G7 meeting. The country of destination in the Olympics, Japan, is distinct to highlight the platform issue to do something about this (Sharma et al., 2023a, 2023b). From the tons of e-waste to making the Olympic medals in 2020. The 'Tokyo 2020 Medal Project' began in 2017, over two years, Japan's Olympic organizing committee collected e-waste such as computer laptops, mobile phones, and other diagnostic items in the country (Fig. 2). University students, even in the country's public sector, contribute electronic equipment, whether small or large, that is no longer used. Private companies are also interested in managing this organization and establishing collection centers to agree with those interested in depositing regularly. The e-waste collection was closed in 2019, and deposits are sustained from 1621 municipalities and villages, towns, cities, and wards, 90% of which are in Japan. About 6.21 mt of mobiles were used, and approximately 78,985 tons of E-waste were collected (da Silva Monteiro et al., 2025).

Extracted metals were melted and refined by the recycling process. And lastly, 70 lbs. (32 kg) of gold, 4850 lbs. (2250 kg) of bronze, and 7716 lbs. (3500 kg) of silver were regained from the collected e-waste, which would be used to make 2500 medals for the Olympic and Paralympic games. Aside from medals, Japan also created podiums with recycled tons of plastic for the Olympic ceremony. Japan is a creative country that innovates new gadgets (Das et al., 2023).

Fig. 2 Olympic *medal* made from E-waste

7.2 Case Study in India

In the government of India, a section of information technology is in the Ministry of Communication, and information technology is the central ministry for the electronic industry. In numerical order to dissertation the emergency requirement of cost-effective and eco-friendly technology, DIT is associated with encouraging R&D to progress technological solutions for e-waste management to approach environmentally friendly (Sengupta et al., 2022). The focus is on progressive recycling technology for various types of e-waste, which is most significant due to very little landfilling and zero emissions to air and water. For a successful business to recover precious metals and reuse plastic components, it must focus on making recycling eco-friendly. The R&D project is at the national level in India. Some projects are:

(a) The program mentioned, "Development of processing technology for recycling and reuse of E-waste", has been flourishingly executed at "The National Metallurgical Laboratory", Jamshedpur, India, an 'R&D' research center under the "Council of Scientific and Industrial Research" (CSIR) in 2011. In this program, national innovation has been used to extract mineral volume from the e-waste, including an extraction value of 90% (Chatterjee et al., 2016). This process is free from toxic gases and dangerous emissions. This advanced technology has successfully recycled up to almost 1 Metric Ton of e-waste. This is the focus on accepting its probable industrial implementation (Sengupta et al., 2022).

(b) A diverse project on "Environmentally sound methods" for extracting minerals from PCBs is occurring at a conveyance approved recycler, M/s. E-parisara Pvt. Ltd., Bangalore. These programs aim to evolve environmentally sound methods for desolate, segregation, and materials and reconstruction techniques for minerals of desolate PCBs (Chatterjee et al., 2016).

7.3 Case Study in the USA

Apple is the world's most precious technology company, with over 2 trillion USD market financing. Naturally, that means Apple plays a huge role in the manufacturing and management of E-waste. Almost all other companies compete for an acceptable presentation to greenwash their aspect to ensure disposition (The E-waste Problem, 2021).

Apple is one of the most capable ecologically sound organizations in the commercial enterprise, and it has become carbon neutral by 2030. This means that every product of this company will have a neutral effect on the environment as part of the recycling industry. In the industry, Apple acquired the target of the first position; the first achieved that 'holistic recycling program', which is recycling tin, cobalt, and carbon-free aluminium. Second, Apple is exploring analysis and recovery to assist in recycling; the consequence is Daisy, a robot resembling old smartphones and tablets, will remove its materials in 2018 (The E-waste Problem, 2021). Daisy crumbles the iPhone at a rate of 200 per hour. The existence of Daisy means that it is more of a PR stunt and greenwashing capability than an active and extensive solution. Thirdly, Apple's global solution, 2018, is 100% powered by renewable energy sources, an extremely magnificent achievement which is undoubtedly an illustration for more institutions to pursue (Apple, 2018).

8 Conclusion and Future Prospects

E-waste is mentioned as the rapidly increasing solid waste stream globally, about 20–50 mt each year have grown in the last 2000 years. We use electronic products in our daily lives that are becoming e-waste, which is increasing in large amounts. In this way, if we use electronic products for less recycling, the quantity of e-waste will double by 2050, reaching nearly 111 mt each year. However, we should manage e-waste recycling, reuse, and reduction to maintain our environment and human health. Every resident of their own country should know the impacts of e-waste on the environment and human health, so that legislation for using electronic equipment in every country. The foundation of e-waste collection, replacement, and recycling centers should be motivated and accomplished in partnership with private companies. In order to achieve long-term sustainable environment and resource efficiency, the electronics industry must adopt a circular economy. Organizations may take action to incorporate circularity into their operations by comprehending how end-of-use ITs and electronics management and e-waste recycling contribute to this shift. E-waste recycling equipment needs a lot of financing, but some countries grant budgets for e-waste management. In many developing countries, e-waste management is careless and unrecognized. In numerous countries, people are unaware of e-waste, applicable strategies, legislation, and management. Accomplishment of a statutory organization for EPR is frequently required and abandoned in developing countries.

An acceptable development system associated with e-waste management can play a critical responsibility in conference acceptable development ambition in 2030 globally; some management strategies for e-waste from the toxic substances and improvement of our environment and health include (1) safety management of working environment especially women and children, (2) depletion of environmental damage from the soil erosion of various toxic chemicals, air pollution from open burning, water pollution from acid and chemical leaching, (3) appreciative manage utilizing and manufacturing of resources, (4) make greater industrialism skills and strengthening the capacity of application to secure job and approach to financial growth.

Acknowledgements Mamun Mandal acknowledges the University Grants Commission (UGC, New Delhi, Govt. of India) as the author receives fellowship in the form of Senior Research Fellowship (CSIR-UGC SRF; Award No. 16-6(DEC. 2018)/2019(NET/CSIR); UGC-Ref. No.: 900/(CSIR-UGC NET DEC. 2018); dated: 02.04.2019). Anamika Roy acknowledges the Council of Scientific Research (CSIR), New Delhi, for providing fellowship in the form of Junior Research Fellowship from GoI. All authors also acknowledge the infrastructural support in the form of the DBT-BOOST program, Department of Science & Technology and Biotechnology, GoWB, GoI (vide Ref. No. 1089/BT(Estt)/1P-07/2018; dated: 24.01.2019) to the Department of Botany, University of Gour Banga, Malda, West Bengal, India.

Author Contribution All the authors have contributed to the article. Adori Khatun: Writing (Original draft preparation); Mamun Mandal: Conceptualization, resources, investigation and Writing (Review and editing); Anamika Roy: Writing (Review and editing); Abhijit Sarkar: Conceptualization, supervision, and writing (Review and editing). All the authors have read and approved the final draft of the manuscript.

Data Availability All data analysed during the preparation of this review are included in this article.

Declarations
Ethics approval: Not applicable.

Consent to participate: Not applicable.

Consent to publish: Not applicable.

Competing Interests The authors declare no competing interests.

References

Al-Salem, S. M., Leeke, G. A., El-Eskandarany, M. S., Van Haute, M., Constantinou, A., Dewil, R., & Baeyens, J. (2022). On the implementation of the circular economy route for E-waste management: A critical review and an analysis for the case of the state of Kuwait. *Journal of Environmental Management, 323*, 116181.

Apple. (2018). *Apple now globally powered by 100 percent renewable energy*. https://www.apple.com/in/newsroom/2018/04/apple-now-globally-powered-by-100-percent-renewable-energy/

Ashfaq, A., & Khatoon, A. (2014). Environmental impacts and assessment of electronic waste management. *International Journal of Current Microbiology and Applied Sciences, 3*(7), 772–779.

Baldé, C. P., Forti, V., Gray, V., Kuehr, R., & Stegmann, P. (2017). *The global e-waste monitor 2017: Quantities, flows and resources*. United Nations University, international telecommunication union, and international solid waste association.

Chatterjee, S. (2012). Sustainable electronic waste management and recycling process. *American Journal of Environmental Engineering, 2*(1), 23–33.

Chatterjee, S., Kumari, A., & Jha, M. K. (2016). Sustainable recycling technology for electronic waste. In *Sustainability in the mineral and energy sectors* (pp. 189–204). CRC Press.

Chatterjee, S., Mandal, M., Kakkar, M., Basak, G., Khan, N. B., Chakraborty, R., Popek, R., Sarkar, A., & Barman, C. (2024). An insight to strategical responses of particulate pollution in plants: From phenome to genome. *Plant Stress*, 100671.

Cheah, C. G., Chia, W. Y., Lai, S. F., Chew, K. W., Chia, S. R., & Show, P. L. (2022). Innovation designs of industry 4.0 based solid waste management: Machinery and digital circular economy. *Environmental Research, 213*, 113619.

da Silva Monteiro, M., de Oliveira, J. D., da Silva, M. M., Silva, T. T. S., de Oliveira Neto, J. F., & Santos, S. M. (2025). Strategies to improve the management of end-of-life electrical and electronic equipment: A case study in Brazil. *Journal of Hazardous Materials Advances, 18*, 100721.

Das, S., Gupta, B., & Sarkar, A. (2023). Diverse technological initiatives for e-waste management and its impact on ecosystem. In: Jawaid, M., & Khan, A. (Eds.), *Conversion of electronic waste in to sustainable products*. Sustainable Materials and Technology. Springer, Singapore.

Dhir, A., Koshta, N., Goyal, R. K., Sakashita, M., & Almotairi, M. (2021). Behavioral reasoning theory (BRT) perspectives on E-waste recycling and management. *Journal of Cleaner Production, 280*, 124269.

Dixit, R., Kumar, S., & Pandey, G. (2023). Biological approaches for E-waste management: A green-go to boost circular economy. *Chemosphere, 336*, 139177.

Forti, V., Baldé, C. P., Kuehr, R., & Bel, G. (2020). The global e-waste monitor 2020: Quantities, flows and the circular economy potential. United Nations University (UNU)/United Nations Institute for Training and Research (UNITAR)—Co-hosted SCYCLE Programme, International Telecommunication Union (ITU) & International Solid Waste Association (ISWA), Bonn/Geneva/Rotterdam.

Gaidajis, G., Angelakoglou, K., & Aktsoglou, D. (2010). E-waste: Environmental problems and current management. *Journal of Engineering Science and Technology Review, 3*(1), 193–199.

Garfì, M., Tondelli, S., & Bonoli, A. (2009). Multi-criteria decision analysis for waste management in Saharawi refugee camps. *Waste Management, 29*, 2729–2739.

Grandhi, S. P., Dagwar, P. P., & Dutta, D. (2024). Policy pathways to sustainable E-waste management: A global review. *Journal of Hazardous Materials Advances*, 100473.

He, Y., Kiehbadroudinezhad, M., Hosseinzadeh-Bandbafha, H., Gupta, V. K., Peng, W., Lam, S. S., Tabatabaei, M., & Aghbashlo, M. (2024). Driving sustainable circular economy in electronics: A comprehensive review on environmental life cycle assessment of e-waste recycling. *Environmental Pollution, 342*, 123081.

Heeks, R., Subramanian, L., & Jones, C. (2015). Understanding e-waste management in developing countries: Strategies, determinants, and policy implications in the Indian ICT sector. *Information Technology for Development, 21*(4), 653–667.

Herat, S., & Agamuthu, P. (2012). E-waste: A problem or an opportunity? Review of issues, challenges and solutions in Asian countries. *Waste Management & Research, 30*(11), 1113–1129.

Herva, M., & Roca, E. (2013). Ranking municipal solid waste treatment alternatives based on ecological footprint and multi-criteria analysis. *Ecological Indicators, 25*, 77–84.

Hossain, M. S., Al-Hamadani, S. M., & Rahman, M. T. (2015). E-waste: A challenge for sustainable development. *Journal of Health and Pollution, 5*(9), 3–11.

Ilankoon, I. M. S. K., Ghorbani, Y., Chong, M. N., Herath, G., Moyo, T., & Petersen, J. (2018). E-waste in the international context–A review of trade flows, regulations, hazards, waste management strategies and technologies for value recovery. *Waste Management, 82*, 258–275.

Islam, A., Ahmed, T., Awual, M. R., Rahman, A., Sultana, M., Abd Aziz, A., Monir, M. U., & Hasan, M. (2020). Advances in sustainable approaches to recover metals from e-waste—A review. *Journal of Cleaner Production, 244*, 118815.

Jaunich, M. K., DeCarolis, J., Handfield, R., Kemahlioglu-Ziya, E., Ranjithan, S. R., & Moheb-Alizadeh, H. (2020). Life-cycle modeling framework for electronic waste recovery and recycling processes. *Resources, Conservation and Recycling, 161*, 104841.

Kang, K. D., Kang, H., Ilankoon, I. M. S. K., & Chong, C. Y. (2020). Electronic waste collection systems using Internet of Things (IoT): Household electronic waste management in Malaysia. *Journal of Cleaner Production, 252*, 119801.

Kiddee, P., Naidu, R., & Wong, M. H. (2013). Electronic waste management approaches: An overview. *Waste Management, 33*(5), 1237–1250.

Kumar, A. M., Prakash, D., Tech, M., & Student, M. E. (2019). Risk assessment environmental impact of e-waste management. *International Journal of Engineering Science and Computing, 9*(5), 22193–22202.

Liu, K., Tan, Q., Yu, J., & Wang, M. (2023). A global perspective on e-waste recycling. *Circular Economy*, 100028.

Mandal, M., Chatterjee, S., Majumdar, S. (2022b). Outside the cell surface: encoding the role of exopolysaccharide producing rhizobacteria to boost the drought tolerance in plants. In Roy, S., Mathur, P., Chakraborty, A. P., & Saha, S. P. (eds) *Plant stress: Challenges and management in the new decade. Advances in science, technology & innovation.* Cham: Springer.

Mandal, M., Roy, A., Binha, S. K., Popek, R., Przybysz, A., Koczoń, P., Prasad, D., & Sarkar, A. (2025). Waste dumps as microplastic hotspots: A comparative investigation at urban, suburban, and rural areas of Eastern India and associated risk assessment. *Environmental Toxicology and Chemistry, 2025*, vgaf086.

Mandal, M., Roy, A., Das, S., Rakwal, R., Agrawal, G. K., Singh, P., Awasthi, A., & Sarkar, A. (2024b). Food waste-based bio-fertilizers production by bio-based fermenters and their potential impact on the environment. *Chemosphere*, 141539.

Mandal, M., Roy, A., Mitra, D., & Sarkar, A. (2024b). Possibilities and prospects of bioplastics production from agri-waste using bacterial communities: finding a silver-lining in waste management. *Current Research in Microbial Sciences, 7*, 100274.

Mandal, M., Sarkar, M., Khan, A., Biswas, M., Masi, A., Rakwal, R., Agrawal, G. K., Srivastava, A., & Sarkar, A. (2022b). Reactive oxygen species (ROS) and reactive nitrogen species (RNS) in Plants–maintenance of structural individuality and functional blend. *Advances in Redox Research, 5*, 100039.

OECD. (2001). *Extended producer responsibility: A guidance manual for governments.* OECD.

Oteng-Ababio, M. (2010). E-waste: an emerging challenge to solid waste management in Ghana. *International Development Planning Review, 32*(2).

Patel, B., Jinal, H. N., Chavan, S. M., Paul, D., & Amaresan, N. (2023). Bacteria isolated from e-waste soil enhance plant growth and mobilize trace metals in e-waste-amended soils. *International Journal of Phytoremediation, 25*(7), 900–906.

Priyashantha, A. K. H., Pratheesh, N., & Pretheeba, P. (2022). E-waste scenario in South-Asia: An emerging risk to environment and public health. *Environmental Analysis, Health and Toxicology, 37*(3), e2022022.

Quinto, S., Law, N., Fletcher, C., Le, J., Antony Jose, S., & Menezes, P. L. (2025). Exploring the e-waste crisis: Strategies for sustainable recycling and circular economy integration. *Recycling, 10*(2), 72.

Reuter, M., & Van Schaik, A. (2016). Gold–a key enabler of a circular economy: Recycling of waste electric and electronic equipment. In *Gold ore processing* (pp. 937–958). Elsevier.

Rucevska I., Nellemann C., Isarin N., Yang W., Liu N., Yu K., Sandnæs S., Olley K., McCann H., Devia L., Bisschop L., Soesilo D., Schoolmeester T., Henriksen, R., & Nilsen, R. (2015). *Waste crime—waste risks: Gaps in meeting the global waste challenge. A UNEP rapid response assessment*. United Nations Environment Programme and GRID-Arendal, Nairobi and Arendal. http://www.grida.no. ISBN

Sengupta, D., Ilankoon, I. M. S. K., Kang, K. D., & Chong, M. N. (2022). Circular economy and household e-waste management in India: Integration of formal and informal sectors. *Minerals Engineering, 184*, 107661.

Sharma, D., Nema, A., Prasad, R., Sweta, K., Sonaviya, D. R., & Karmakar, S. (2023). Global E-waste management: consolidated information showcasing best available practices. In *Global e-waste management strategies and future implications* (pp. 289–314). Elsevier.

Sharma, M., Agarwal, S., Agarwal Malik, R., Kumar, G., Pal, D. B., Mandal, M., Sarkar, A., Bantun, F., Haque, S., Singh, P., & Gupta, V. K. (2023). Recent advances in microbial engineering approaches for wastewater treatment: a review. *Bioengineered, 14*(1), 2184518. 1.

Sivaramanan, S. (2013). E-waste management, disposal and its impacts on the environment. *Universal Journal of Environmental Research & Technology, 3*(5).

Song, Q., & Li, J. (2014). Environmental effects of heavy metals derived from the e-waste recycling activities in China: A systematic review. *Waste Management, 34*(12), 2587–2594.

Stevels, A. (2023). The challenge of introducing design for the circular economy in the electronics industry: A proposal for metrics. *Circular Economy, 2*(3), Article 100051.

Suja, F., Abdul Rahman, R., Yusof, A., & Masdar, M. S. (2014). E-waste management scenarios in Malaysia. *Journal of Waste Management, 2014*.

The E-waste Problem: A Case Study of Apple. (2021). https://globuswarwick.com/2021/01/21/the-e-waste-problem-a-case-study-of-apple/

Vats, M. C., & Singh, S. K. (2014). E-waste characteristic and its disposal. *International Journal of Ecological Science and Environmental Engineering, 1*(2), 49–61.

Vego, G., Kucˇar-Dragicˇevic´, S., & Koprivanac, N. (2008). Application of multi-criteria decision-making on strategic municipal solid waste management in Dalmatia, Croatia. *Waste Management, 28*, 2192–2201

Wang, S., Su, D., & Wu, Y. (2022). Environmental and social life cycle assessments of an industrial LED lighting product. *Environmental Impact Assessment Review, 95*, 106804.

Yong, Y. S., Lim, Y. A., & Ilankoon, I. M. S. K. (2019). An analysis of electronic waste management strategies and recycling operations in Malaysia: Challenges and future prospects. *Journal of Cleaner Production, 224*, 151–166.

Zeng, X., Yang, C., Chiang, J. F., & Li, J. (2017). Innovating e-waste management: From macroscopic to microscopic scales. *Science of the Total Environment, 575*, 1–5.

Unveiling the Circularity: Challenges and Barriers in Transitioning to the Circular Practices

Kanhaiya Shah, Chandranshu Tiwari, and Anupam Sunny

Abstract In the current age of global economy, the transition from a traditional the linear economy to the circular economy is significantly important for promoting environmental sustainability and as well as economic growth. Studies have recognized that practice of the circular economy in economic sector is a multifaceted approach which needs a comprehensive understanding. The current study was carried out to understand the present scenario of traditional the linear economy and the potential use of the circular practices for resource use and waste minimization through bibliometric analysis on Scopus database. For this purpose, this review aims to highlight various barriers and challenges that creates the hinderance for widespread adoption of the circular economy practices in various economy sectors at global level. The study outlines various key challenges such as economic and financial barriers, Innovation and technological barriers, infrastructural and operational challenges, stakeholder engagement challenges, and regulatory and policy challenges with the help of content analysis of available research articles. The review also talks about various examples of innovative business models and technology-driven solutions through national and international case studies which can help in overcoming these challenges and barriers. The study concludes that an integrated approach is needed to overcome these challenges. Business sector and industries should focus on innovative business models and stakeholder collaboration, while government and policymakers should create coherent regulating frameworks and financial incentives and subsidies. There is a need for promoting the academic research for knowledge generation and technological advancement if the circular economy principles. The study emphasizes that transitioning to the circular economy is crucial for addressing the environmental and economic issues generated by the linear economy model.

K. Shah · A. Sunny (✉)
Department of Environmental Studies, Satyawati College, University of Delhi, New Delhi, India
e-mail: anupamsunny@satyawati.du.ac.in

C. Tiwari
Department of Environmental Studies, Shyama Prasad Mukherji College for Women, University of Delhi, Delhi, India

P. Singh et al. (eds.), *Circular Economy and Sustainable Innovation*,
https://doi.org/10.1007/978-3-032-00437-6_13

Keywords The circular practices · The linear economy · Barriers and challenges · Stakeholder engagement · Innovative business models · Technology-driven solutions

1 Introduction

The traditional the linear economy is based on the concept of "take-make-dispose", which is an unsustainable approach and hinders economic resiliency. The circular economy model replaces this old "take-make-dispose" method with a sustainable system that reduces resource use, waste, emissions, and energy loss and aims for a stronger and more sustainable future (Marino & Pariso, 2016; Sehnem et al., 2019; Zemanová, 2023). While, the linear economy is running out of resources and creating a lot of environmental waste, the circular economy focuses on the concept of "reduce-reuse-recycle-regenerate". Thus, the circular economy keeps environmental resources and products in use as long as possible (Sørensen, 2018). So, transitioning to the circular economy helps in achieving the sustainable economic growth. The urgent need for this transition to the circular practices is driven by several environmental issues like resource scarcity, climate change, and environmental pollution (Heshmati, 2017; Sehnem et al., 2019; Sterev, 2019; Zemanová, 2023).

With a growing global population and rising resource consumption, the linear model is becoming unsustainable for environment as well as economy (Singh and Mondal & Palit, 2022; Singh, 2016; Wang & Azam, 2024). The current widely established conventional economic practices are among the primary causes of environmental degradation (Maja & Ayano, 2021; Pörtner et al., 2023). The shift to the circular economy has started in the last few decades, and nearly every economic sector needs to start implementing a novel resource consumption approach. Among other things, the most pressing issue that is advocated by leading global environmental organizations is to include the concepts of resource efficiency and waste management (Elroi et al., 2023; Heshmati, 2017; Van Ewijk & Stegemann, 2008).

During the initial transition of the circular economy from conventional economic practices, responsible production and consumption, waste management, and product stewardship management principles were the primary means of addressing sustainability. However, various barriers and challenges, such as economic and financial barriers, innovation and technological barriers, infrastructural and operational challenges, stakeholder engagement challenges, and regulatory and policy challenges impede the widespread and global adoption of the circular economy practices (Debrah et al., 2022; Guldmann & Huulgaard, 2020; Heshmati, 2017; Tan et al., 2022). It is evident that the conventional approach of resource use overexploits and consequently depletes the ecosystems' capacity for regeneration, regardless of whether carbon usage is focused on the need for housing, food, or economic systems (Chu & Karr, 2016; Duguma et al., 2020).

This approach gives rise to an exponential increase in the global carbon footprint. The scarcity of available natural resources has led to a resolution to adopt the circular economy practices in all economic sectors. For the business sector, any long-term approach to sustainability aids companies in assessing potential risk, fostering innovation, growing by seizing new market opportunities, and utilizing cost–benefit analysis. To enforce and expedite the transition to the circular economy, the government must simultaneously provide strong incentives and the appropriate regulatory enforcement (Chenavaz & Dimitrov, 2024; Domenech & Bahn-Walkowiak, 2019). For instance, government policies were implemented to either define sustainability in 'National eco-innovation strategies' or simply highlight the ecological perspective of the circular economy (Chaparro-Banegas et al., 2023). The international scientific community and environmental organizations also have a responsibility to limit the negative environmental impacts of basic research.

Despite growing recognition of its importance, the implementation of the circular economy practices faces numerous interrelated challenges and various barriers (Chenavaz & Dimitrov, 2024; Velter et al., 2022). Key barriers and challenges include knowledge gaps, financial constraints, inadequate infrastructure, policy incoherence, and limited awareness among stakeholders. The challenges and barriers are more pronounced for developing nations' economies, where adoption efforts are made more difficult by competing priorities and a lack of resources (Faye et al., 2004; Johnson, 2021; Suzanne et al., 2020). With this understanding, the current study reviews the current global scenario of traditional the linear economy and potential role of resource efficiency and waste minimization in achieving the successful transition to the circular economy through bibliometric analysis on the Scopus database. Additionally, the study aims to identify and broadly discuss various barriers and challenges that hinder this transition with the help of content analysis of the available research literature. The study also highlights best the circular practices and case studies of innovative business models and technology-driven solutions that may help in achieving the transition to the circular economy.

2 Resource Efficiency and Waste Minimization: An Integral Aspect of the Circular Economy

2.1 Bibliometric Analysis on Available Research Articles

It is widely recognized that the traditional the linear economy leads to excessive resource wastage and environmental harm due to unsustainable use of raw materials in economic sectors and industries (Di Maio et al., 2017; Wilts et al., 2016). The circular economy model addresses the shortcomings of the linear economy by maximizing the use of raw materials during product design and also promoting reuse, recycling, and remanufacturing after product manufacturing. This is done by applying the concept of 'resource efficiency and waste minimization' in the manufacturing process of the circular economy model (Elroi et al., 2023). The concept of 'resource efficiency and waste minimization' helps to protect ecosystems, reduce pollution, and lessen climate change effects, thus ultimately supporting environmental sustainability.

To understand the current global trend depicting the transition to the circular practices by adopting the concept of 'resource efficiency and waste minimization' in the circular economy model, a bibliometric analysis was conducted on the Scopus database. Bibliometric analysis is a quantitative method that is used to analyze research publications (books, research articles, and other publications) available on a particular research field or study domain (Donthu et al., 2021; Shah & Sharma, 2022a). It uses statistical tools and mathematical techniques to analyze the structure, patterns, trends, and evolution of a research field in due course of time (Darko & Chan, 2021; Shah & Sharma, 2022b). For bibliometric analysis, two search keywords ('The circular Economy' AND 'The circular Practices') were used on Scopus database to search for the number of research publications on a global scale (Scopus, 2025). Results revealed that 298 research articles were published during the period (2016–2024), with maximum number of research articles (126) published in 2024 (Fig. 1). There was an exponential increase in the number of research publications after the year 2021.

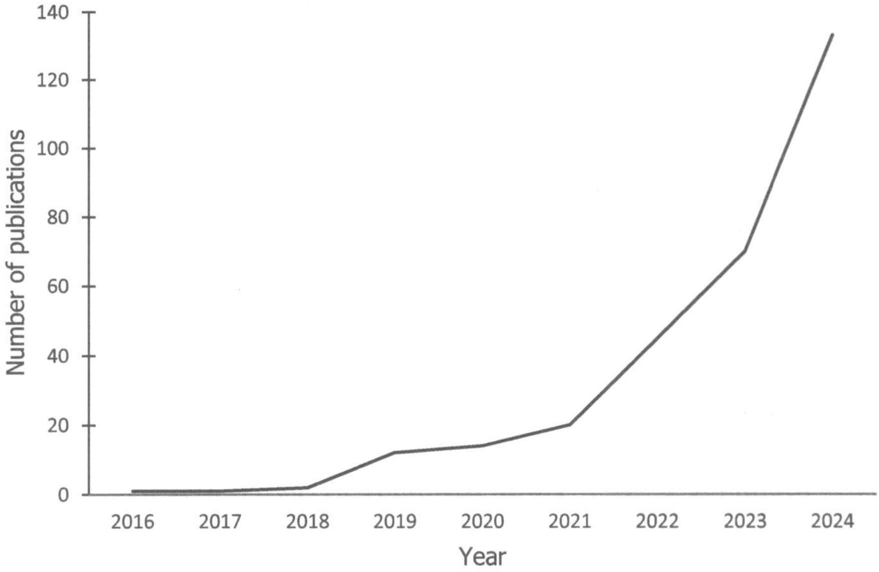

Fig. 1 Number of research articles published on the topic of 'the circular economy practices' globally during the period (2016–2024) (data retrieved from Scopus, 2025)

2.2 Content Analysis on Available Research Articles

Figure 1 provides an overview of research trends that help us in understanding the global scenario of transition to the circular economy practices from the conventional the linear economy model. This transition is crucial for promoting economic resilience and reducing environmental impact of the linear economy. Content analysis of research articles revealed that one of the most important ways to achieve this transition is designing products and related services by following the concept of 'resource efficiency and waste minimization'. This means producing environment friendly products that are durable as well as recyclable. Through this, manufacturers can make sure that the circular products will have a longer lifespan and durability. These circular products then can easily be reintroduced into production cycles after use at the end of their lifecycle.

One such example is the scheme of 'extended producer responsibility' (EPR). EPR held producers accountable for the entire lifecycle of their products that also includes take-back programs and recycling initiatives (Gupt & Sahay, 2015; Maitre-Ekern, 2021). Another example that is being used in industrial sector is the use of recycled or secondary raw materials. Primarily, this reduces the dependency on raw environmental resources. This is achieved through the efficient manufacturing process that minimizes wastage and energy consumption. Our review indicated that waste minimization and resource recovery are prioritized in the circular economy model (Fig. 2). Thus, instead of focusing solely on waste management after disposal, the emphasis should also be given to prevent waste generation in the first place.

An important principle of the circular economy, i.e., turning waste (outputs) from one process into inputs for another, can significantly reduce resource wastage and

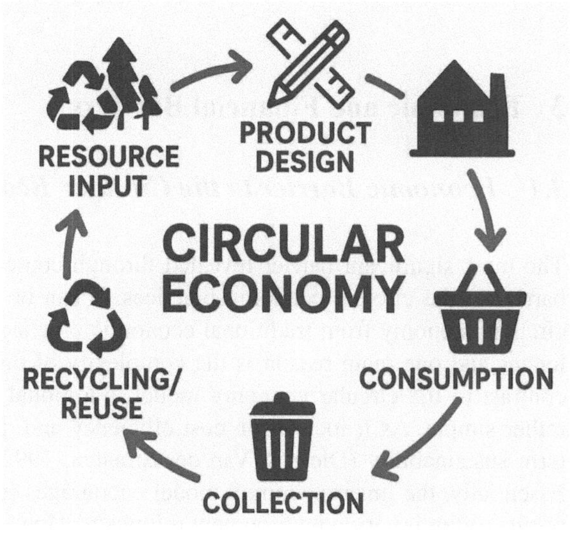

Fig. 2 An illustration depicting of an efficient manufacturing process based on one of the important principles of the circular economy, i.e., resource efficiency and waste minimization. Illustrative figure created by author with Canva

environmental pollution (Bockholt et al., 2020; Huisman & Stevels, 2006). Industries are adopting this principle through better process management, reuse, refurbishment, and remanufacturing of products (Bockholt et al., 2020; Jensen, 2025). Various initiatives such as take-back programs, easy-return policies, and recycling initiatives can help extend product life and cut down on the need for new materials (Kant & Pedersen, 2019). Research studies reported that successful transition to the circular economy has several benefits. These benefits are the reduced demand for raw materials, lower production costs, and reduced greenhouse gas emissions (Suzanne et al., 2020; Wijkman & Skånberg, 2015). Additionally, it also leads to better supply chains and creates new jobs in recycling, repair, and remanufacturing programs and initiatives. However, there are challenges such as difficulties in recycling some materials, the need for supportive policies and infrastructure, and the risk that efficiency gains could increase consumption (Herring & Roy, 2007; Zink & Geyer, 2017). Thus, research and technological development and government policies are important for helping this transition.

Government policies should encourage the use of the circular practices by offering monetary incentives, e.g., taxes on raw materials and subsidies for recycled goods (Wilts et al., 2016). Other than this, content analysis revealed few other important examples of initiatives and programs. As support from regulations and national policies is essential for industries for the successful transition to the circular economy practices, so collaboration among businesses, governments, and civil society is crucial for this. Sharing knowledge and innovative technologies helps create markets for recycled materials and supports new business models that optimize resource use (Acquier et al., 2019). Based on this, it can be said that a comprehensive approach addressing policy, infrastructure, technology, and collaboration is necessary to improve resource efficiency and waste minimization. This will enable a sustainable shift from traditional to the circular economy practices.

3 Economic and Financial Barriers

3.1 Economic Barrier to the Circular Economy Practices

The most significant barrier revealed through content analysis was the economic barrier to the circular economy practices. It can be said that the transition to the circular economy from traditional economic practices faces major economic challenges and one main reason is the complexity of the circular economy model. In contrast to the circular economy model, traditional the linear economy model is rather simple. As it focuses on cost efficiency and quick profits, rather than long-term sustainability (Dietz & Van der Straaten, 1992; Huppes & Ishikawa, 2009). Essentially, the linear economy model encourages producers to cut costs to boost profits, often ignoring environmental impacts. However, this focus on profit of the linear economy model can clash with the goals of the circular economy, especially in managing byproducts and reducing waste.

Other than this inherent complexity of the circular economy model, other economic and financial barriers are the high costs for resources, poor financial setups within companies, and lack of support from banks and other financial institutions (Roelich et al., 2015). Another major economic barrier is the lack of adequate infrastructure and resources, especially in underdeveloped and developing countries (Srivastava & Srivastava, 2006). This serves as a significant hindrance for the widespread global adoption of the circular practices. In developing and underdeveloped countries, the businesses struggle to obtain the large financial investments needed for product redesign and develop new manufacturing units which are essential for moving toward the circular economy (Ghisellin et al., 2016; Hopkinson et al., 2018; Suzanne et al., 2020).

3.2 Economic Feasibility of the Circular Economy Practices

Another issue that creates an economic barrier to transition to the circular economy is the lack of investment in infrastructure, like waste collection trucks and sorting stations. High disposal costs for the waste collection and storage also restrict funding for necessary systems. Additionally, financial institutions often do not have the right financing options to support the circular economy practices needed for recycling and remanufacturing processes. For example, Luttenberger (2020) reported that transitioning to the circular economy requires significant public and private sector investment in waste collection trucks, sorting equipment and stations, and environmentally viable waste management techniques in municipal corporations and private solid waste management companies. This financial disadvantage has created a significant hindrance for municipal corporations and private solid waste management companies that earlier were considering the transition to more sustainable the circular economic models.

Thus, it can be said that the economic feasibility of the circular economy practices is a key concern for businesses that are planning to transit to the circular economy practices (Corvellec et al., 2022). This was also influenced by absence of government subsidies and policies that favor traditional the linear models. So, although the circular practices offer long-term benefits, these issues make the circular economy model less economically viable for businesses. It was found that traditional economic systems generally favor the linear economy model approach, such as heavy reliability on raw resources. For example, companies following the linear economy model may find it more cost-effective to dispose of waste rather than invest in the installation of recycling units and development of recycling processes (Aiguobarueghian et al., 2024; Lewandowski, 2016). Aiguobarueghian et al. (2024) also reported that the circular economic models focus more on non-tangible business aspects like service delivery and product durability, often undervaluing traditional the linear economic methods that emphasize tangible resources. So as a result, businesses working under the linear economy approach may find it cheaper to throw away waste instead of recycling.

Additionally, the absence of government support and financial tax incentives and subsidies to support the circular economy practices has prevented businesses from making this transition (Asgari & Asgari, 2021). Consequently, this lack of funding complicates implementation of the circular practices in manufacturing processes, even when businesses acknowledge the environmental and social benefits of the circular economy models. Tan et al. (2022) urges that only by promoting government policies that address both environmental costs and societal well-being, these financial challenges can be addressed.

4 Innovation and Technological Barriers

4.1 Technological Barriers to the Circular Economy Practices

Research studies advocate that the circular economy can provide a sustainable solution to current environmental crisis (Grafström & Aasma, 2021; Rizos et al., 2016; Trevisan et al., 2023). However, current the circular practices do not focus on economic growth, social impacts, or reducing resource consumption. This serves as one of the significant shortcomings of proposed the circular economy model. Grafström and Aasma (2021) discussed key technological barriers such as lack of technological expertise, absence of technical knowledge, and inadequate supporting infrastructure. For example, many companies struggle with the technical knowledge required to change their business models and redesign products according to the circular economy principle. This lack of knowledge is a big barrier to implementing the circular economy practices, which involve more than simple recycling principles—reduce, reuse, and recycle (Blomsma et al., 2019; Moreno et al., 2016).

Thirumal et al. (2024) reported that as the circular economy is a multifaceted concept, so the technical knowledge of the circular economy also includes areas like business strategy, sourcing materials, product design, and managing product flows in manufacturing. Since, implementation of the circular practices is severely hindered by the lack of technological expertise, technical knowledge, and adequate supporting infrastructure. So, this also prevents companies from making necessary changes for a successful transition to the circular economy practices (Blomsma et al., 2019; Bocken et al., 2016; Kandasamy et al., 2023). Helbig et al. (2022) reported that current recycling technologies often face challenges due to a complex of raw materials used and possible contamination that have led to downcycling or disposal of waste. This waste can be used as 'input' for another manufacturing process (Cimpan et al., 2015; Srivastava et al., 2025). For example, to recover valuable resources from landfills, innovation in processes like mechanical sorting, chemical recycling, and depolymerization is needed (Enache et al., 2024; Somoza-Tornos et al., 2020).

These techniques are being used in developed countries as developing countries like India are not able to afford them due to their high price. While, Bressanelli et al. (2021) discussed that new engineering methods and better material choices are necessary for making products easier to disassemble and reuse. Hahladakis and Iacovidou (2019) reported that effective recycling and remanufacturing procedures are hampered by the absence of the circular economy-based product design and lack of knowledge on material composition of the product being recycled. Overcoming these technological issues is vital for a successful transition to the circular economy (Niwalkar et al., 2023). Scaling up the current the circular economy technological knowledge presents a considerable challenge. It is interesting that pilot projects that are initiated to adopt the circular practices are demonstrating feasibility. However, widespread adoption on a larger scale requires overcoming barriers related to lack of technological expertise, technical knowledge, and adequate supporting infrastructure. Hazen et al. (2021) also outlines the scaling up of the current the circular economy projects and initiatives as considerable challenge.

4.2 Lack of Innovative Techniques to Support the Circular Economy Practices

Mostly businesses are working under the traditional the linear economy principle. So, there is a need to develop significant technological knowledge and innovative techniques to incorporate the circular economy principle into already existing business operations and processes. For this transition, research studies discussed five main types of innovation necessary: product design innovation, business model innovation, behavioral changes, digital innovation, and policy innovation (Allwood, 2014; Coscieme et al., 2022; Guldmann et al., 2019; Suchek et al., 2021). It is noteworthy that these areas of innovation are often found to be interconnected, which adds to the already existing complexity of the transition (Guldmann et al., 2019; Wang et al., 2022). Adding to this, another key challenge is creating innovation techniques that support efficient material recovery and processing. For this purpose, use of blockchain technology for keeping the record of the products and materials in supply chains can significantly support the transition to the circular economy (Centobelli et al., 2022; Kouhizadeh et al., 2020; Santana & Ribeiro, 2022).

While it is more than just a digital log book of a product. It actually creates secure and long-lasting digital records of transactions and product data at various stages that can be shared among the stakeholders easily (Francisco et al., 2021; Shojaei et al., 2021; Schmidt et al., 2024). Blockchain can also implement 'tokenized reward systems' encouraging recycling and repairs instead of replacements (Kouhizadeh et al., 2020; Rejeb et al., 2022). However, like the pilot the circular projects, the scalability issues of blockchain technology must also be resolved. This should be done while working with large data volumes and different stakeholders like producers, consumers, and recyclers for the entire supply chain. Given that the technological knowledge and innovative techniques are very much important for the circular economy transition, there is a need to assess how they fit into current business models of the circular companies (Herring & Roy, 2007; Vivanco et al., 2016; Zink & Geyer, 2017).

5 Infrastructural and Operational Challenges

5.1 *Infrastructural and Operational Challenges: A Producer's Perspective*

Based on the content analysis of available research articles, various infrastructural and operational challenges that hinder the transition to the circular economy are listed in Table 1. Table 1 indicated that these challenges are particularly noticeable in sectors like construction, energy, transport, and waste management. Content analysis also revealed that most prominent issue is the reliance of industries on old infrastructure designed for the linear economy. It is also known as 'legacy industrial systems' which means that existing industries are optimized for the linear economy model and thus are costly to adapt in current scenario (Jayasinghe et al., 2023). Essentially, existing manufacturing and waste systems do not support the circular economy practices as they are optimized for only one-way flow of resources/materials (i.e., the linear flow of economic production) (Beamon & Fernandes, 2004; Nguyen et al., 2014). Thus, transitioning to these new models of the circular economy will require significant monetary investment and technological innovation (Kara et al., 2022; Muller et al., 2013; Neto et al., 2009).

Another major infrastructural issue is the lack of proper facilities for collecting, sorting, and recycling materials (Debrah et al., 2022). While, in developed countries, there is a lack of the support needed to effectively recover and process recycled materials in industrial plants. The infrastructure for effective recycling and material recovery is often very limited or absent (Kerdlap et al., 2019). The problem of difficulty in tracking material flows across global supply suppliers can be solved by use of blockchain technology in product supply chains (Hazen et al., 2021; Kouhizadeh et al., 2020). Moreover, adopting the infrastructural facility that support the circular principles faces challenges because it does not receive as much investment as traditional infrastructure projects.

Table 1 A list of various infrastructural and operational challenges that hinder the transition to the circular economy with their description

Infrastructural and operational challenges	Description
Lack of recycling facilities	Insufficient infrastructure to collect, sort, and process recyclable materials[a]
Poor waste segregation infrastructure	Inadequate systems for separating recyclable, hazardous, and organic waste[b]
Urban planning constraints	Cities may not be designed to accommodate material loops or decentralized reuse centres[c]
Legacy industrial systems	Existing industries and plants are optimized for the linear models and costly to adapt[d]
Insufficient data infrastructure	Poor data systems to track product lifecycle and enable the circular business models[e]
Skills gaps	Workers may lack expertise in repair, remanufacturing, or the circular product design[f]
Complex supply chains	Difficulty in tracking material flows across global suppliers[g]
Landfill dependency	Overreliance on existing landfill infrastructure discourages investment in the circular solutions[h]

[a] Debrah et al. (2022), [b] Kerdlap et al. (2019), [c] Remøy et al. (2019), [d] Jayasinghe et al. (2023), [e] Jabbour et al. (2019), [f] Sumter et al. (2021), [g] Hazen et al. (2021), [h] Aiguobarueghian et al. (2024)

5.2 Infrastructural and Operational Challenges: A Consumer's Perspective

For the consumer perspective, it is recognized that consumer resistance to refurbished or repaired products further complicates the operational challenges to the circular practices. As consumer involvement is essential for the circular practices, which rely on consumer behaviors like returning used items and choosing refurbished products (Borrello et al., 2020). To promote these behaviors, educational engagement and awareness campaigns are needed. Other than consumer resistance to buy recycled products, there are also difficulties in balancing the supply and demand for secondary materials from manufactures' side (Solomon et al., 2024). Manufacturers might be reluctant to use recycled materials because of concerns about their quality and durability, while consumers may prefer new products due to perceptions of low quality in recycled and refurbished items (Guldmann & Huulgaard, 2020; Kumar et al., 2019; Shashi et al., 2024). Although, many e-commerce websites have started to sell refurbished products at cheaper price. The business of refurbished products is one of the growing areas in e-commerce. The mismatch between supply of the circular operations and demand of recycled materials can weaken the business operations for the circular practices (Calvo-Porral & Lévy-Mangin, 2020).

It is now being recognized that transitioning to the circular practices needs new business models that should focus on reverse logistics. Reverse logistics processes such as collecting used products, transporting them for recycling or refurbishment, and integrating recovered materials into new production processes at the end. Setting up these systems is often complex and costlier compared to forward logistics (a core aspect of traditional the linear economy) (Oyedele et al., 2014; Purchase et al., 2021), especially when dealing with scattered sources of waste (Guldmann & Huulgaard, 2020).

5.3 Infrastructural and Operational Challenges: A Policy Issue

It is interesting to note that there is also a policy perspective on the infrastructural and operational challenges to the circular economy. Transitioning to the circular production needs major changes in product design and tracking materials (Olipp et al., 2024; Psarommatis et al., 2025). Products should be made to last, be easily repaired, and taken apart, which may require new production rules and policies and also teamwork across different industries. To track materials during the product's life, blockchain technology is important, but many organizations lack the necessary digital systems and data management skills which are needed for product information sharing between industries or businesses across supply chains (Francisco et al., 2021; Schmidt et al., 2024; Shojaei et al., 2021). This lack of collaboration across different industries weakens the operational and manufacturing transitions. Therefore, as costs are high, monetary benefits may take time to appear for the circular businesses (Sikander, 2024).

The uncertainty of information sharing between industries or business across supply chains affects investors and businesses due to weak policy and regulatory frameworks, especially in underdeveloped and developing countries (Mhatre et al., 2023). The existing frameworks in developed countries are often inconsistent and incomprehensive, which discourages investment and slows down the adoption of the circular business practices (Wysokińska et al., 2020). Additionally, there is a lack of government-sponsored incentives like tax subsidies or recycling quotas for businesses that have adopted the circular practices. So, these businesses have little motivation to invest in the circular operations and infrastructure in future initiatives and projects. The transition to the circular system is expensive from infrastructural and operational perspective. It needs upfront investments in new technology, improved infrastructure, and skill training for workers, along with comprehensive production policies and guidelines.

6 Stakeholder Engagement Challenges

6.1 Lack of Awareness of the Circular Practices Among Stakeholders

A significant challenge to the transition to the circular economy practices is being recognized as the widespread lack of awareness and understanding among key stakeholders. These stakeholders have different interests, so businesses must balance their needs and expectations to ensure a successful transition to the circular economy practices (Cosa, 2024; O'Higgins, 2010). This is known as the process of stakeholder engagement. It is recognized that extended producer responsibility (EPR) policies have been successfully implemented and have increased awareness of recycling. However, many small businesses have not embraced broader the circular concepts such as resourceful usage of waste, extending product lifespans, and implementing return policies (Brown et al., 2023; Gupt & Sahay, 2015; Maitre-Ekern, 2021).

In general, policymakers lack a comprehensive understanding of currently evolving the circular economy concepts, which impedes the development of effective supportive policy guidelines and regulations (Milios, 2018; Waddell, 2017; Yu et al., 2022). Without this much-needed conceptual foundation, creating policy frameworks that may aid in implementing and incentivizing the circular practices such as repair, reuse, and recycling becomes exceedingly difficult (Harris et al., 2016). Kazancoglu et al. (2021) proposed that businesses and governments can effectively reduce consumer resistance and accelerate the transition to the circular economy by combining education, trust-building measures, financial incentives, and collaborative efforts. Considering that the consumer demand for items and goods produced through the circular practices appears limited, management committees often dismiss the circular practices as a high-cost initiative with uncertain returns (Schumacher & Green, 2021; Tukker, 2015).

6.2 Consumer Resistance Towards the Circular Products

Several research studies reported that consumer resistance towards the circular products poses a major challenge to adopting the circular economy practices (Harris et al., 2016; Singh & Giacosa, 2019). Negative views on repair, reuse, and second-hand products harm the demand for the circular options. This situation makes it even harder for businesses to embrace these practices. As consumer resistance creates low demand for the circular goods. Luthra et al. (2022) reported that partnerships in supply chains can help move towards the circular practices by tackling issues like high return rates and supply shortages that are arising due to consumer resistance. When companies work together, they can build trust among consumers and share benefits. It has been discussed earlier also that governments can support this shift by offering subsidies for sustainable designs or tax breaks for the circular products (Balke et al., 2017; Chenavaz & Dimitrov, 2024). Moreover, there should also be campaigns from governments and educational institutions to inform consumers about the benefits of the circular economy. This will encourage consumers to buy more the circular products. By promoting education and collaboration among stakeholders, businesses can overcome the stakeholder engagement challenges (Geme et al., 2023). To overcome this resistance, it's essential to tackle the psychological, social, and economic barriers while promoting awareness, trust, and convenience (Camacho-Otero et al., 2020; Singh & Giacosa, 2019).

6.3 Overcome Consumer Resistance—Effective Business Strategies

To overcome the consumer resistance for recycled products, these products, such as refurbished or durable goods, should be marketed for their high quality, reliability, and environmental friendliness with proper certifications to counter negative biases (Bhagat, 2024; Choudhury et al., 2024; Nygaard, 2023). Business organizations need to design the circular products and services that are attractive, accessible, and affordable to consumers while also addressing concerns about quality and current fashion trends (see Table 2). Table 2 lists various case studies of companies that have successfully overcome consumer resistance to the circular economy practices and products. Pesce et al. (2020) reported that insufficient efforts by governments, educational institutions, and nonprofit organizations (NGOs) to create education among individuals and businesses about the circular economy practices hinder widespread adoption of the circular practices. Sales et al. (2021) reported that educating consumers about the overexploitation of resources in the linear economy model and its connection to high consumption of raw resources will reduce consumer resistance (Salas et al., 2021).

6.4 Overcome Consumer Resistance- Awareness and Engagement

Educating and promoting the circular practices as part of a sustainable lifestyle through social campaigns that advertise recycling, reusing, and buying refurbished goods can drive behavioral change among consumers (Esposito et al., 2024). Government policies that subsidize the circular products and/or implement taxes on the linear economy products can shift consumer preferences toward sustainable options (Wasserbaur et al., 2022). Offering discounts or rewards for returning used items or purchasing the circular products, such as trade-in programs for electronics or clothing, can motivate consumers to engage with the circular economy businesses (Hazen et al., 2017; Yamaguchi, 2021). Grassroots-level community engagement by

Table 2 Case studies of companies overcoming consumer resistance to the circular economy practices and their impacts

Company	Challenge	Solution	Impacts
Patagonia	Consumer stigma around second-hand or repaired clothing	– Repair services for damaged clothing – Trade-in program with store credits – Resale of refurbished items	– Extended product lifecycle – Strengthened customer loyalty – Promoted environmental responsibility[a]
IKEA	Consumer reluctance to reuse furniture is due to inconvenience and low awareness	– Take-Back Program for used furniture – Furniture rentals and refurbished sales – Products designed for disassembly and recycling	– Reduced waste – Made the circular options more accessible – Positioned IKEA as a sustainability leader in home solutions[b]
H&M	Consumers reluctance to buy recycled clothes	– In-store collection of used clothes – Incentives like discounts or vouchers	– Reduced textile waste – Boosted consumer participation in the circular practices[c]
Fairphone	Modular smartphones for easy repair and upgrades	– Modular smartphones for easy repair and upgrades – Phones designed with replaceable parts	– Extends product lifespan – Reduces electronic waste, promotes repair and reuse[d]
Interface	Convincing customers of the value of recycled carpets	– ReEntry Program: Old tiles are recycled into new ones – Mission Zero is to eliminate negative environmental impact	– Reduced virgin material use – Attracted eco-conscious consumers[e]
Zara	Limited consumer knowledge and participation in textile recycling	– Home and in-store collection of used garments – Sorted for donation, recycling, or repurposing	– Increased consumer convenience – Strengthened Zara's sustainability positioning[f]

[a] Dzombak et al. (2020), [b] Ström et al. (2020), [c] Watson (2021), [d] Peters (2022), [e] Cline et al. (2015), [f] Oestereich (2024)

collaborating with local organizations to create awareness campaigns and infrastructure for recycling or repair services will aid in consumer trust-building for the circular products (Ugwu et al., 2025). By combining education, trust-building measures, convenient systems, financial incentives, and collaborative efforts, governments and businesses can effectively reduce consumer resistance and accelerate the transition to the circular economy.

7 Regulatory and Policy Challenges

7.1 Inefficient Regulatory Frameworks

From a regulatory and policy perspective on the circular economy transition, the main challenge in moving to the circular economy practices is the lack of clear frameworks to help businesses in the sustainable transition (Table 3). While various individual organizations have implemented frameworks, but theses regulatory frameworks on waste, recycling, and product design are often spread across different sectors and fragmented. Mubarik et al. (2024) discussed the need of a widely accepted regulatory framework that would be easy to adapt globally across all economic sectors. Saidani et al. (2017) also reported that a comprehensive framework would address key issues like product design, raw material selection, and waste management for the circular businesses. Moreover, after implementing such a framework, businesses could use metrics and tools to measure their success and find areas for improvement in the post-implementation phase.

Recently, research studies reported implementation of few important urban development guidelines to promote the circular economy practices (Cavaleiro de Ferreira & Fuso-Nerini, 2019; Levoso et al., 2020; Petit-Boix & Leipold, 2018).

Table 3 A list of various policy and regulatory challenges that hinder the transition to the circular economy with their description and impacts

Policy and Regulatory Challenge	Description	Impact on the Circular Economy Transition
Fragmented regulations	Regulations on waste, recycling, and product design	Creates confusion and inefficiencies; discourages holistic the circular solutions[a]
Outdated waste legislation	Existing laws often focus on disposal and incineration rather than reuse, repair, and recycling	Encourages the linear practices like landfilling over the circular alternatives[b]
Inadequate incentives	Lack of tax benefits or subsidies for the circular practices such as eco-design, remanufacturing, etc.	Makes the circular options less economically viable compared to traditional approaches[c]
Subsidies for the linear practices	Continued subsidies for raw material extraction, fossil fuels, and waste incineration	Distorts the market against the circular products and services[d]
Standardization and certification gaps	Lack of consistent standards for recycled materials, secondary products, and repair services	Reduces trust and uptake among consumers and industries[e]
Trade barriers	Restrictive rules around the import/export of waste and recycled materials	Hinders international the circular business models and supply chains[f]

[a] Mubarik et al. (2024), [b] Chioatto and Sospiro (2023), [c] Asgari and Asgari (2021), [d] Kumar et al. (2020) [e] Sharma et al. (2021), [f] Yamaguchi (2021)

One example is the trend of using the circular economy concepts in urban planning to create 'smart and sustainable cities' (Chourabi et al., 2012). For a smart city, innovative approaches driven by technology are essential for improving resource use, reducing waste, and enhancing the quality of life for its residents (Mutha & Pokharel, 2009). Without clear frameworks, policymakers and urban developers struggle to implement the circular economy practices effectively for sustainable urban development (Balke et al., 2017; Chenavaz & Dimitrov, 2024). High initial investment costs and uncertain returns discourage many urban development companies from adopting the circular practices for designing the urban smart cities (Hegeman, 2021; Gonçalves & Carvalho, 2022).

7.2 *Lack of Corporate and Industrial Guidelines*

The lack of corporate and industrial guidelines for the circular economy transition poses challenges for businesses, particularly small and medium enterprises (SMEs). Unlike larger industries that have dedicated sustainability teams with sufficient manpower and resources, SMEs often do not have the knowledge or funding needed for complex the circular production systems (Rizos et al., 2016; Sharma et al., 2021). Sharma et al. (2021) also reported that this issue is made worsened by the absence of an international body to assess and certify the circular economy practices. This situation makes it tough for SMEs to compete against larger companies or even demonstrate their brand value to investors and consumers for future growth and investment. This gives rise to the situation where SMEs might hesitate to adopt the circular practices due to uncertainty about the benefits.

It is anticipated that to ensure a successful transition to the circular economy practice, both national and international governing authorities must come together to actively set clear targets and create supportive policy (Domenech & Bahn-Walkowiak, 2019; Heshmati, 2017; Saidani et al., 2017). The return on initial investments for the businesses adopting the circular practices may take longer than in the traditional the linear economy. This situation leads to uncertainty for businesses and investors. The absence of any established markets for refurbished or recycled products also adds to this uncertainty. It is crucial to develop a reliable investor market that reduces risks and promotes long-term financial sustainability for the circular businesses (Bressanelli et al., 2019; Hopkinson et al., 2018). It can be said that the policy and regulatory support is essential for moving the circular economy from being a 'niche approach' to a 'common practice'.

7.3 Inadequate Governmental Policies

It has been discussed earlier that ineffective government policies and regulations make it hard for industries to adopt the circular economy practices globally. Moreover, research studies also indicated that a lack of government support, like funding for research, waste management infrastructure, and public education, slows down the use of the circular strategies (Chenavaz & Dimitrov, 2024; Salmenperä et al., 2021; Salvador et al., 2022). While, the European Union (EU) has taken important steps with its 'the circular economy action plan' (CEAP), which includes rules like recycling targets, eco-design needs, and producer responsibility policies in developed countries (Brown et al., 2023; Chioatto & Sospiro, 2023). However, there are few laws requiring companies to manage their products after use in most underdeveloped and developing countries. Without clear policy or regulatory frameworks, businesses often stick to old ways of making and using products (Hu et al., 2021; Park et al., 2018; Toffel, 2003). For example, poor waste collection systems and limited access to the circular technologies create significant challenges to adopting these practices (Debrah et al., 2022).

8 Overcoming Challenges and Barriers: Business Models and Case Studies

8.1 Innovative Business Models

It has been earlier discussed that a successful the circular economy model specifically addresses the shortcomings and problems of the linear "take-make-dispose" economic model (Hopkinson et al., 2018; Rashid & Malik, 2023). Considering that the successful transition to the circular economy is dependent on technological compliance and consumer approval, adaptation is dependent on a variety of factors, including consumer willingness to cooperate and the ability to expand operations and processes smoothly (De Mattos & De Albuquerque, 2018). The transition to the circular economy requires businesses to reconsider old the linear business models, in which goods are manufactured, utilized, and then discarded, and instead implement the circular practices that keep resources in use for as long as possible (Awan & Sroufe, 2022; van Loon & Van Wassenhove, 2020). Several creative business models are emerging to enable this transformation, with an emphasis on product lifecycle extension, waste reduction, and resource reuse (see Table 4). Table 4 lists various case studies of companies that have successfully incorporated the principles of the circular economy into their innovative business models.

For example, companies that rely on the circular supply chain make an effort to use renewable, recyclable, or biodegradable resources in their goods and continuously recirculate them into the manufacturing process (van Loon & Van Wassenhove, 2020). This requires a much-needed cooperation with suppliers and manufacturers

Table 4 Case studies of companies incorporating the principles of the circular economy into their innovative business models

Company	The circular economy model	Key principles	Impact
Loop	The circular shopping platform (Reusable packaging system)	Partners with brands to provide products in reusable packaging, which is collected, cleaned, and reused	Reduces single-use plastic, promotes reusable packaging, and reduces waste[a]
Tesla	Battery recycling and second-life applications	Recycles old batteries and repurposes materials for new products	Reduces waste, promotes resource reuse, and minimizes environmental impact[b]
Dell Technologies	Closed-loop recycling for electronics	Recycles used electronics and uses recycled plastics in new products	Reduces electronic waste, promotes sustainable manufacturing, and minimizes resource extraction[c]
Airbnb	Sharing platform for unused spaces	Leveraging underutilized resources (spare rooms)	Maximizes resource use, reduces demand for new construction, and minimizes waste[d]
Uber	Sharing platform for unused cars	Car-sharing service utilizing underused vehicles	Reduces manufacturing demand, maximizesresource use, and minimizes waste[e]
Philips	Product-as-a-service model to promote recycling of service-based lighting products	Through the lighting-as-a-Service model, Philips retains ownership and ensures reuse/recycling	Lowers upfront costs for customers and enables end-of-life material recovery[f]
Unilever	Sustainable packaging and product life extension	Develops recyclable and biodegradable packaging, and encourages product refills	Reduces plastic waste, encourages product reuse, and supports sustainability[g]
TerraCycle	Closed-loop recycling for hard-to-recycle materials	Partnerships for recycling difficult materials like plastic packaging into new products	Reduces reliance on virgin materials, lowers waste, and promotes recycling[h]

[a] Long et al. (2020), [b] Cui et al. (2023), [c] Bakker et al. (2014), [d] Quattrone et al. (2016), [e] Roblek et al. (2021), [f] Schoemaker et al. (2020), [g] van den Berg and Bekker (2015), [h] Szaky (2019)

who value sustainable sourcing and material reuse. A closed-loop system is an innovative business model concept in which a corporation is responsible for its products' complete life cycle, from manufacture to disposal at the end of the product's life cycle (Camilleri, 2019; Winkler, 2011). The goal is to design products so that when they reach the end of their lifecycle, they may be returned, disassembled, and remanufactured into new products, resulting in a 'closed loop'. Upcycling adds more value through introducing innovative designs and processes that help in reducing the environmental impact of producing entirely new goods (Galila et al., 2024; Sung, 2023).

These innovative business models contribute to speeding up the transition to the circular economy by providing incentives to minimize waste, conserve resources, and produce long-term value of products (Asgari & Asgari, 2021; Bocken et al., 2016). By incorporating these approaches, the companies not only contribute to a more sustainable world but also position themselves to lead in a market that is constantly evolving, where customers are now attaching a higher value to sustainability and sustainable products. Businesses can help drive the transition to the circular economy while simultaneously benefiting from much greater customer loyalty, lower material costs, and improved innovation (Vivanco et al., 2016).

8.2 Technology-Driven Solutions

Technology-driven solutions facilitate the transition to the circular economy through improved resource management, reducing waste generation, and extending the product's lifecycle. Examples of such technology-driven solutions are compiled in Table 3. For instance, 'artificial intelligence' (AI) and 'machine learning' (ML), although they are still in a developing stage, are used to optimize product supply chains and identify the future product maintenance requirements, which leads to longer product lifecycles and less waste generation (Younis et al., 2022). Additionally, the use of blockchain technology in the supply chain that supports transparency, accountability, and traceability throughout the products' life cycle (Kadio, 2023; Schmidt et al., 2024; Shojaei et al., 2021).

'Internet of Things' (IoT) devices provide real-time product tracking, monitoring, and assessment, thus allowing companies to provide "product-as-a-service" models (Sallam et al., 2023). For example, Philips offers light products such as incandescent, fluorescent, and LED lights as a service rather than selling bulbs (Schoemaker et al., 2020; also see Table 3). Interestingly, in contrast to traditional mechanical recycling, advanced technology-driven processes such as chemical recycling degrade the plastic into the original monomers, which enables unlimited recycling loops for future use of the product (Sanchez et al., 2020; Seyi-Lande et al., 2024). The EV battery recycling through technology-driven innovation facilitates an efficient and sustainable method for recovering valuable materials from end-of-life batteries in current times, where the number of electric vehicles has significantly increased (Silvestri et al., 2024).

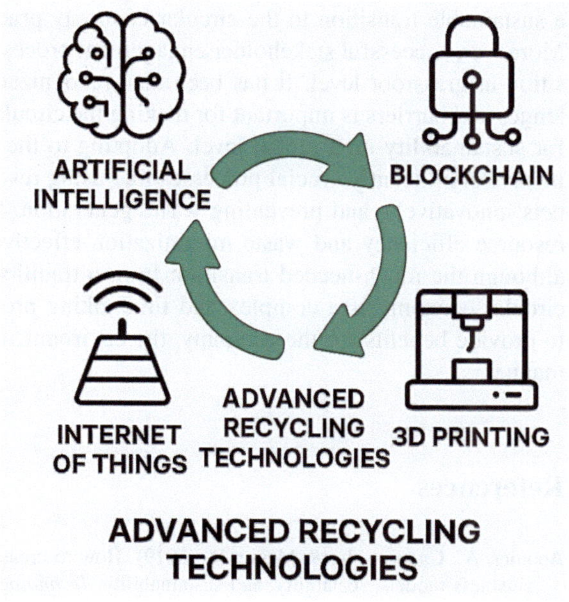

Fig. 3 Illustration of key advanced recycling technology-driven solutions that can help in transitioning to sustainable the circular practices. Illustrative figure created by authors with Canva

Adidas' technological innovation, Futurecraft 4D sneakers, used additive manufacturing through '3D printing' (Nekin Joshua et al., 2023), which enables on-demand manufacture of the product based on consumer preferences and needs (Priyadarshini et al., 2022; Sauerwein et al., 2019; Tavares et al., 2023). It reduces surplus inventory of the product in warehouses and encourages localized manufacturing. The Refactory, an initiative of Renault (a French multinational automobile manufacturer), uses both AI and IoT technologies to increase its manufacturing efficiency while refurbishing automobiles and recycling mechanical parts (Istriteanu et al., 2024). This new business model also includes EV battery recycling to facilitate its second-life applications. ECONYL, a regenerated nylon derived from ocean and landfill waste, such as fishing nets and old carpets, is produced by the Aquafil company. Aquafil uses the chemical recycling (CR) method to produce ECONYL, which is then utilized in the manufacturing of luxury clothes, carpets, and swimwear (Dahlbom & Martvall, 2025). When applied with a sustainable approach, these technology-driven solutions are changing industries to more regenerative, sustainable economy models (Fig. 3).

9 Conclusion

The current review concludes that the transition from a traditional the linear economy to the circular economy is challenging task. It requires a collaborative effort of all the stakeholders. Identifying and addressing the barriers and challenges while also proposing creative business models and technology solutions, is essential for a sustainable transition to the circular economy practices in all economic sectors. Moreover, successful stakeholder engagement process is much needed for this transition at grassroot level. It has been also recognized that overcoming these challenges and barriers is important for making the circular economy practices effective for sustainability on a global level. Adopting to the circular practices should also involve significantly crucial processes like using resources wisely, designing products innovatively, and preventing waste generation. It might help in achieving the resource efficiency and waste minimization effectively. The study advocates that although the much-needed transition from a traditional the linear economy to the circular economy is a complex and time-taking process. It also has the potential to provide benefits for the economy, the environment, and society in a sustainable manner.

References

Acquier, A., Carbone, V., & Massé, D. (2019). How to create value(s) in the sharing economy: Business models, scalability, and sustainability. *Technology Innovation Management Review, 9*(2).

Aiguobarueghian, I., Adanma, U. M., Ogunbiyi, E. O., & Solomon, N. O. (2024). Waste management and the circular economy: A review of sustainable practices and economic benefits. *World Journal of Advanced Research and Reviews, 22*(2), 1708–1719.

Allwood, J. M. (2014). Squaring the circular economy: the role of recycling within a hierarchy of material management strategies. In *Handbook of recycling* (pp. 445–477). Elsevier.

Asgari, A., & Asgari, R. (2021). How the circular economy transforms business models in a transition towards the circular ecosystem: The barriers and incentives. *Sustainable Production and Consumption, 28*, 566–579.

Awan, U., & Sroufe, R. (2022). Sustainability in the circular economy: Insights and dynamics of designing the circular business models. *Applied Sciences, 12*(3), 1521.

Bakker, C., Wang, F., Huisman, J., & Den Hollander, M. (2014). Products that go round: Exploring product life extension through design. *Journal of Cleaner Production, 69*, 10–16.

Balke, V., Evans, S., Rabbiosi, L., & Monnery, S. A. (2017). Promoting the circular economies. In *Green industrial policy* (pp. 121–13). UN Environment Report.

Beamon, B. M., & Fernandes, C. (2004). Supply-chain network configuration for product recovery. *Production Planning & Control, 15*(3), 270–281.

Bhagat, C. (2024). Greenwashing vs. genuine efforts: The role of transparent communication in building corporate credibility. *Journal of Law for Future Security, 1*(2), 51–62.

Blomsma, F., Pieroni, M., Kravchenko, M., Pigosso, D. C., Hildenbrand, J., Kristinsdottir, A. R., Kristoffersen, E., Shahbazi, S., Nielsen, K. D., Jönbrink, A. K., & Li, J. (2019). Developing the circular strategies framework for manufacturing companies to support the circular economy-oriented innovation. *Journal of Cleaner Production, 241*, 118271.

Bocken, N. M., De Pauw, I., Bakker, C., & Van Der Grinten, B. (2016). Product design and business model strategies for the circular economy. *Journal of Industrial and Production Engineering, 33*(5), 308–320.

Bockholt, M. T., Kristensen, J. H., Colli, M., Jensen, P. M., & Wæhrens, B. V. (2020). Exploring factors affecting the financial performance of end-of-life take-back program in a discrete manufacturing context. *Journal of Cleaner Production, 258*, 120916.

Borrello, M., Pascucci, S., Caracciolo, F., Lombardi, A., & Cembalo, L. (2020). Consumers are willing to participate in the circular business models: A practice theory perspective to food provisioning. *Journal of Cleaner Production, 259*, 121013.

Bressanelli, G., Perona, M., & Saccani, N. (2019). Challenges in supply chain redesign for the circular economy: A literature review and a multiple case study. *International Journal of Production Research, 57*(23), 7395–7422.

Bressanelli, G., Pigosso, D. C., Saccani, N., & Perona, M. (2021). Enablers, levers and benefits of the circular economy in the electrical and electronic equipment supply chain: A literature review. *Journal of Cleaner Production, 298*, 126819.

Brown, A., Laubinger, F., & Börkey, P. (2023). New Aspects of EPR: Extending producer responsibility to additional product groups and challenges throughout the product lifecycle. *OECD Environment Working Papers.*

Calvo-Porral, C., & Lévy-Mangin, J. P. (2020). The circular economy business model: Examining consumers' acceptance of recycled goods. *Administrative Sciences, 10*(2), 28.

Camacho-Otero, J., Tunn, V. S., Chamberlin, L., & Boks, C. (2020). Consumers in the circular economy. In *Handbook of the circular economy* (pp. 74–87). Edward Elgar Publishing.

Camilleri, M. A. (2019). The circular economy's closed loop and product service systems for sustainable development: A review and appraisal. *Sustainable Development, 27*(3), 530–536.

Cavaleiro de Ferreira, A., & Fuso-Nerini, F. (2019). A framework for implementing and tracking the circular economy in cities: The case of Porto. *Sustainability, 11*(6), 1813.

Centobelli, P., Cerchione, R., Del Vecchio, P., Oropallo, E., & Secundo, G. (2022). Blockchain technology for bridging trust, traceability and transparency in the circular supply chain. *Information & Management, 59*(7), 103508.

Chaparro-Banegas, N., Mas-Tur, A., Park, H. W., & Roig-Tierno, N. (2023). Factors driving national eco-innovation: New routes to sustainable development. *Sustainable Development, 31*(4), 2711–2725.

Chenavaz, R. Y., & Dimitrov, S. (2024). From waste to wealth: Policies to promote the circular economy. *Journal of Cleaner Production, 443*, 141086.

Chioatto, E., & Sospiro, P. (2023). Transition from waste management to the circular economy: The European Union roadmap. *Environment, Development and Sustainability, 25*(1), 249–276.

Choudhury, R. R., Islam, A. F., & Sujauddin, M. (2024). More than just a business ploy? Greenwashing as a barrier to the circular economy and sustainable development: A case study-based critical review. *The Circular Economy and Sustainability, 4*(1), 233–266.

Chourabi, H., Nam, T., Walker, S., Gil-Garcia, J. R., Mellouli, S., Nahon, K., Pardo, T. A., & Scholl, H. J. (2012, January). Understanding smart cities: An integrative framework. In *2012 45th Hawaii international conference on system sciences* (pp. 2289–2297). Institute of Electrical and Electronics Engineers (IEEE).

Chu, E. W., & Karr, J. R. (2016). Environmental impact: Concept, consequences, measurement. *Reference Module in Life Sciences*, B978–0.

Cimpan, C., Maul, A., Jansen, M., Pretz, T., & Wenzel, H. (2015). Central sorting and recovery of MSW recyclable materials: A review of technological state-of-the-art, cases, practice and implications for materials recycling. *Journal of Environmental Management, 156*, 181–199.

Cline, A., LeMay, S., & Helms, M. M. (2015). A framework for reverse logistics: The case of post-consumer carpet in the US. *International Journal of Commerce and Management, 25*(4), 466–489.

Corvellec, H., Stowell, A. F., & Johansson, N. (2022). Critiques of the circular economy. *Journal of Industrial Ecology, 26*(2), 421–432.

Cosa, M. (2024). Equity integration in corporate social responsibility: Analyzing stakeholder, legitimacy, and social activism dynamics. In *CSR, Governance and value* (pp. 133–154). Singapore: Springer Nature Singapore.

Coscieme, L., Manshoven, S., Gillabel, J., Grossi, F., & Mortensen, L. F. (2022). A framework of the circular business models for fashion and textiles: the role of business-model, technical, and social innovation. *Sustainability: Science, Practice and Policy, 18*(1), 451–462.

Cui, J., Tan, Q., Liu, L., & Li, J. (2023). Environmental benefit assessment of second-life use of electric vehicle lithium-ion batteries in multiple scenarios considering performance degradation and economic value. *Environmental Science & Technology, 57*(23), 8559–8567.

Dahlbom, M., & Martvall, A. (2025). Maturity of fibre-to-fibre recycling in Europe-Assessment of recycling companies in Europe. In *Assessment of recycling companies in Europe*, Report C874.

Darko, A., & Chan, A. P. C. (2021) Applying science mapping in built environment research. In *Secondary research methods in the built environment* (pp. 117– 132). Routledge.

De Mattos, C. A., & De Albuquerque, T. L. M. (2018). Enabling factors and strategies for the transition toward the circular economy (CE). *Sustainability, 10*(12), 4628.

Debrah, J. K., Teye, G. K., & Dinis, M. A. P. (2022). Barriers and challenges to waste management hindering the circular economy in Sub-Saharan Africa. *Urban Science, 6*(3), 57.

Di Maio, F., Rem, P. C., Baldé, K., & Polder, M. (2017). Measuring resource efficiency and the circular economy: A market value approach. *Resources, Conservation and Recycling, 122*, 163–171.

Dietz, F. J., & Van der Straaten, J. (1992). Rethinking environmental economics: Missing links between economic theory and environmental policy. *Journal of Economic Issues, 26*(1), 27–51.

Domenech, T., & Bahn-Walkowiak, B. (2019). Transition towards a resource efficient the circular economy in Europe: Policy lessons from the EU and the member states. *Ecological Economics, 155*, 7–19.

Donthu, N., Kumar, S., Mukherjee, D., Pandey, N., & Lim, W. M. (2021). How to conduct a bibliometric analysis: An overview and guidelines. *Journal of Business Research, 133*, 285–296.

Duguma, L., Kamwilu, E., Minang, P. A., Nzyoka, J., & Muthee, K. (2020). Ecosystem-based approaches to bioenergy and the need for regenerative supply options for Africa. *Sustainability, 12*(20), 8588.

Dzombak, R., Rosen, C., Pendleton-Knoll, S., O'Neill, K., Chesbrough, H., Strand, R., & Critcher, C. (2020). *Waste not, want not: Eliminating Patagonia's pre-and post-consumer textile waste.* The Berkeley-Haas Case Series. University of California, Berkeley. Haas School of Business.

Elroi, H., Zbigniew, G., & Agnieszka, W. C. (2023). Enhancing waste resource efficiency: The circular economy for sustainability and energy conversion. *Frontiers in Environmental Science, 11*, 1303792.

Enache, A. C., Grecu, I., & Samoila, P. (2024). Polyethylene terephthalate (PET) recycled by catalytic glycolysis: A bridge toward the circular economy principles. *Materials, 17*(12), 2991.

Esposito, B., Sica, D., Malandrino, O., & Supino, S. (2024). Social media on the route to the circular economy transition from a dialogic perspective: Evidence from the agri-food industry. *British Food Journal, 126*(1), 64–79.

Faye, M. L., McArthur, J. W., Sachs, J. D., & Snow, T. (2004). The challenges facing landlocked developing countries. *Journal of Human Development, 5*(1), 31–68.

Francisco Luis, B. M., Pedro Víctor, N. C. U., Valentín, M. M., & Esteban, R. F. (2021). Blockchain as a service: A holistic approach to traceability in the circular economy. In *Blockchain technologies for sustainability* (pp. 119–133). Singapore: Springer Singapore.

Galila, A., Kaseireldeil, S., Abdalkhaliq, N., Farouk, E., Eichner, M., & Sarhan, Y. (2024). Enhancing sustainability and resource efficiency through upcycling: A comprehensive review and analytical–based framework for evaluating building upcycled products. *Innovative Infrastructure Solutions, 9*(7), 270.

Geme, T., Nijman, E., Ntawuhiganayo, E. B., & Negesa, D. (2023). Stakeholder knowledge and perceptions of the circular economy in Ugandan cities. *Frontiers in Sustainability, 4*, 1117814.

Ghisellini, P., Cialani, C., & Ulgiati, S. (2016). A review on the circular economy: The expected transition to a balanced interplay of environmental and economic systems. *Journal of Cleaner Production, 114*, 11–32.

Gonçalves, B. D. S. M., Carvalho, F. L. D., & Fiorini, P. D. C. (2022). The circular economy and financial aspects: A systematic review of the literature. *Sustainability, 14*(5), 3023.

Grafström, J., & Aasma, S. (2021). Breaking the circular economy barriers. *Journal of Cleaner Production, 292*, 126002.

Guldmann, E., & Huulgaard, R. D. (2020). Barriers to the circular business model innovation: A multiple-case study. *Journal of Cleaner Production, 243*, 118160.
Guldmann, E., Bocken, N. M., & Brezet, H. (2019). A design thinking framework for the circular business model innovation. *Journal of Business Models, 7*(1), 39–70.
Gupt, Y., & Sahay, S. (2015). Review of extended producer responsibility: A case study approach. *Waste Management & Research, 33*(7), 595–611.
Hahladakis, J. N., & Iacovidou, E. (2019). An overview of the challenges and trade-offs in closing the loop of post-consumer plastic waste (PCPW): Focus on recycling. *Journal of Hazardous Materials, 380*, 120887.
Harris, F., Roby, H., & Dibb, S. (2016). Sustainable clothing: Challenges, barriers and interventions for encouraging more sustainable consumer behaviour. *International Journal of Consumer Studies, 40*(3), 309–318.
Hazen, B. T., Mollenkopf, D. A., & Wang, Y. (2017). Remanufacturing for the circular economy: An examination of consumer switching behavior. *Business Strategy and the Environment, 26*(4), 451–464.
Hazen, B. T., Russo, I., Confente, I., & Pellathy, D. (2021). Supply chain management for the circular economy: Conceptual framework and research agenda. *The International Journal of Logistics Management, 32*(2), 510–537.
Hegeman, P. D. (2021). Fund model innovations for the circular economy investing. In *Research handbook of innovation for the circular economy* (pp. 295–306). Edward Elgar Publishing.
Helbig, C., Huether, J., Joachimsthaler, C., Lehmann, C., Raatz, S., Thorenz, A., Faulstich, M., & Tuma, A. (2022). A terminology for downcycling. *Journal of Industrial Ecology, 26*(4), 1164–1174.
Herring, H., & Roy, R. (2007). Technological innovation, energy efficient design and the rebound effect. *Technovation, 27*(4), 194–203.
Heshmati, A. (2017). A review of the circular economy and its implementation. *International Journal of Green Economics, 11*(3–4), 251–288.
Hopkinson, P., Zils, M., Hawkins, P., & Roper, S. (2018). Managing a complex global the circular economy business model: Opportunities and challenges. *California Management Review, 60*(3), 71–94.
Hu, H., Liu, J., Zhu, Q., & Chen, Q. (2021). An integrated supply chain management system for end-of-life tires in China and its promotion barriers: A stakeholder perspective. *Resources, Conservation and Recycling, 164*, 105214.
Huisman, J., & Stevels, L. N. (2006). Eco-efficiency of take-back and recycling, a comprehensive approach. *IEEE Transactions on Electronics Packaging Manufacturing, 29*(2), 83–90.
Huppes, G., & Ishikawa, M. (2009). Eco-efficiency guiding micro-level actions towards sustainability: Ten basic steps for analysis. *Ecological Economics, 68*(6), 1687–1700.
Istriteanu, S., Băjenaru, V., & Badea, F. (2024). The automotive industry's transition to the circular economy through digital transformation. *International Journal of Mechatronics and Applied Mechanics, 15*, 101–112.
Jabbour, C. J. C., de Sousa Jabbour, A. B. L., Sarkis, J., & Godinho Filho, M. (2019). Unlocking the circular economy through new business models based on large-scale data: An integrative framework and research agenda. *Technological Forecasting and Social Change, 144*, 546–552.
Jayasinghe, P. A., Derrible, S., & Kattan, L. (2023). Interdependencies between urban transport, water, and solid waste infrastructure systems. *Infrastructures, 8*(4), 76.
Jensen, H. H. (2025). 6th R–Recycle: The circular Closure. In *The circular economy opportunities and pathways for manufacturers: Manufacturing renewed* (pp. 201–237). Cham: Springer Nature Switzerland.
Johnson, H. G. (2021). *Economic policies towards less developed countries*. Routledge.
Kadio, A. K. C. (2023). The benefits of the circular economy for companies' the circular supply chains in a French port area. *European Journal of Logistics, Purchasing and Supply Chain Management, 11*, 31–47.

Kandasamy, J., Venkat, V., & Mani, R. S. (2023). Barriers to the adoption of digital technologies in a functional the circular economy network. *Operations Management Research, 16*(3), 1541–1561.

Kant Hvass, K., & Pedersen, E. R. G. (2019). Toward the circular economy of fashion: Experiences from a brand's product take-back initiative. *Journal of Fashion Marketing and Management: An International Journal, 23*(3), 345–365.

Kara, S., Hauschild, M., Sutherland, J., & McAloone, T. (2022). Closed-loop systems to the circular economy: A pathway to environmental sustainability? *CIRP Annals, 71*(2), 505–528.

Kazancoglu, I., Sagnak, M., Kumar Mangla, S., & Kazancoglu, Y. (2021). The circular economy and the policy: A framework for improving the corporate environmental management in supply chains. *Business Strategy and the Environment, 30*(1), 590–608.

Kerdlap, P., Low, J. S. C., & Ramakrishna, S. (2019). Zero waste manufacturing: A framework and review of technology, research, and implementation barriers for enabling the circular economy transition in Singapore. *Resources, Conservation and Recycling, 151*, 104438.

Kouhizadeh, M., Zhu, Q., & Sarkis, J. (2020). Blockchain and the circular economy: Potential tensions and critical reflections from practice. *Production Planning & Control, 31*(11–12), 950–966.

Kumar, V., Sezersan, I., Garza-Reyes, J. A., Gonzalez, E. D., & Al-Shboul, M. D. A. (2019). The circular economy in the manufacturing sector: Benefits, opportunities and barriers. *Management Decision, 57*(4), 1067–1086.

Levoso, A. S., Gasol, C. M., Martínez-Blanco, J., Durany, X. G., Lehmann, M., & Gaya, R. F. (2020). Methodological framework for the implementation of the circular economy in urban systems. *Journal of Cleaner Production, 248*, 119227.

Lewandowski, M. (2016). Designing the business models for the circular economy—Towards the conceptual framework. *Sustainability, 8*(1), 43.

Long, Y., Ceschin, F., Mansour, N., & Harrison, D. (2020). Product–service systems applied to reusable packaging systems: A strategic design tool. *Design Management Journal, 15*(1), 15–32.

Luthra, S., Sharma, M., Kumar, A., Joshi, S., Collins, E., & Mangla, S. (2022). Overcoming barriers to cross-sector collaboration in the circular supply chain management: A multi-method approach. *Transportation Research Part e: Logistics and Transportation Review, 157*, 102582.

Luttenberger, L. R. (2020). Waste management challenges in transition to the circular economy–case of Croatia. *Journal of Cleaner Production, 256*, 120495.

Maitre-Ekern, E. (2021). Re-thinking producer responsibility for a sustainable the circular economy from extended producer responsibility to pre-market producer responsibility. *Journal of Cleaner Production, 286*, Article 125454.

Maja, M. M., & Ayano, S. F. (2021). The impact of population growth on natural resources and farmers' capacity to adapt to climate change in low-income countries. *Earth Systems and Environment, 5*(2), 271–283.

Marino, A., & Pariso, P. (2016). From the linear economy to the circular economy: Research agenda. *International Journal of Research in Economics and Social Sciences, 6225*(6), 2249–7382.

Mhatre, P., Gedam, V. V., Unnikrishnan, S., & Raut, R. D. (2023). The circular economy adoption barriers in built environment-a case of emerging economy. *Journal of Cleaner Production, 392*, 136201.

Milios, L. (2018). Advancing to the circular Economy: Three essential ingredients for a comprehensive policy mix. *Sustainability Science, 13*(3), 861–878.

Mondal, S., & Palit, D. (2022). Challenges in natural resource management for ecological sustainability. In *Natural resources conservation and advances for sustainability* (pp. 29–59). Elsevier.

Moreno, M., De los Rios, C., Rowe, Z., & Charnley, F. (2016). A conceptual framework for the circular design. *Sustainability, 8*(9), 937.

Mubarik, M. S., Kontoleon, A., & Shahbaz, M. (2024). Beyond the hurdles: Exploring policy obstacles in the path to the circular economy adoption. *Journal of Environmental Management, 370*, 122667.

Muller, D. B., Liu, G., Løvik, A. N., Modaresi, R., Pauliuk, S., Steinhoff, F. S., & Brattebø, H. (2013). Carbon emissions of infrastructure development. *Environmental Science & Technology, 47*(20), 11739–11746.

Mutha, A., & Pokharel, S. (2009). Strategic network design for reverse logistics and remanufacturing using new and old product modules. *Computers & Industrial Engineering, 56*(1), 334–346.

Nekin Joshua, R., Aravind Raj, S., & Sakthi Balan, G. (2023). Sustainable design in the era of additive manufacturing: A review. In *International Conference on Recent Advances in Industrial and Systems Engineering* (pp. 521–530). Singapore: Springer Nature Singapore.

Neto, J. Q. F., Walther, G., Bloemhof, J., Van Nunen, J. A. E. E., & Spengler, T. (2009). A methodology for assessing eco-efficiency in logistics networks. *European Journal of Operational Research, 193*(3), 670–682.

Nguyen, H., Stuchtey, M., & Zils, M. (2014). Remaking the industrial economy. *Mckinsey Quarterly, 1*, 46–63.

Niwalkar, A., Indorkar, T., Gupta, A., Anshul, A., Bherwani, H., Biniwale, R., & Kumar, R. (2023). The circular economy-based approach for green energy transitions and climate change benefits. *Proceedings of the Indian National Science Academy, 89*(1), 37–50.

Nygaard, A. (2023). Is sustainable certification's ability to combat greenwashing trustworthy? *Frontiers in Sustainability, 4*, 1188069.

O'Higgins, E. R. (2010). Corporations, civil society, and stakeholders: An organizational conceptualization. *Journal of Business Ethics, 94*, 157–176.

Oestereich, E. M. L. V. (2024). *The circular economy in fast fashion-the case of Zara*. Master's thesis, Universidade NOVA de Lisboa (Portugal).

Olipp, N., Woschank, M., & Hoffelner, M. (2024). Exploration of the framework conditions for measures to reduce resource consumption in the manufacturing industry with a focus on the circular economy: Systematic secondary data research. *Production & Manufacturing Research, 12*(1), 2431723.

Oyedele, L. O., Ajayi, S. O., & Kadiri, K. O. (2014). Use of recycled products in UK construction industry: An empirical investigation into critical impediments and strategies for improvement. *Resources, Conservation and Recycling, 93*, 23–31.

Park, J., Díaz-Posada, N., & Mejía-Dugand, S. (2018). Challenges in implementing the extended producer responsibility in an emerging economy: The end-of-life tire management in Colombia. *Journal of Cleaner Production, 189*, 754–762.

Pesce, M., Tamai, I., Guo, D., Critto, A., Brombal, D., Wang, X., Cheng, H., & Marcomini, A. (2020). The circular economy in China: Translating principles into practice. *Sustainability, 12*(3), 832.

Peters, J. G. (2022). *The economic and ecological impact of shifting to a modular smartphone design*. Bachelor's thesis, University of Twente.

Petit-Boix, A., & Leipold, S. (2018). The circular economy in cities: Reviewing how environmental research aligns with local practices. *Journal of Cleaner Production, 195*, 1270–1281.

Pörtner, H. O., Scholes, R. J., Arneth, A., Barnes, D. K. A., Burrows, M. T., Diamond, S. E., Duarte, C. M., Kiessling, W., Leadley, P., Managi, S., & McElwee, P., (2023). Overcoming the coupled climate and biodiversity crises and their societal impacts. *Science, 380*(6642), eabl4881.

Priyadarshini, J., Singh, R. K., Mishra, R., & Kamal, M. M. (2022). Adoption of additive manufacturing for sustainable operations in the era of the circular economy: Self-assessment framework with case illustration. *Computers & Industrial Engineering, 171*, 108514.

Psarommatis, F., May, G., & Azamfirei, V. (2025). Product reuse and repurpose in the circular manufacturing: a critical review of key challenges, shortcomings and future directions. *Journal of Remanufacturing*, 1–38.

Purchase, C. K., Al Zulayq, D. M., O'Brien, B. T., Kowalewski, M. J., Berenjian, A., Tarighaleslami, A. H., & Seifan, M. (2021). The circular economy of construction and demolition waste: A literature review on lessons, challenges, and benefits. *Materials, 15*(1), 76.

Quattrone, G., Proserpio, D., Quercia, D., Capra, L., & Musolesi, M. (2016). *Who benefits from the "sharing" economy of Airbnb*? *arXiv preprint*arXiv:1602.02238.

Rashid, S., & Malik, S. H. (2023). Transition from the linear to the circular Economy. In *Renewable Energy in The circular Economy* (pp. 1–20). Cham: Springer International Publishing.
Rejeb, A., Zailani, S., Rejeb, K., Treiblmaier, H., & Keogh, J. G. (2022). Modeling enablers for blockchain adoption in the circular economy. *Sustainable Futures, 4*, 100095.
Remøy, H., Wandl, A., Ceric, D., & Timmeren, A. V. (2019). Facilitating the circular economy in urban planning. *Urban Planning, 4*(3), 1–4.
Rizos, V., Behrens, A., Van der Gaast, W., Hofman, E., Ioannou, A., Kafyeke, T., Flamos, A., Rinaldi, R., Papadelis, S., Hirschnitz-Garbers, M., & Topi, C. (2016). Implementation of the circular economy business models by small and medium-sized enterprises (SMEs): Barriers and enablers. *Sustainability, 8*(11), 1212.
Roblek, V., Meško, M., & Podbregar, I. (2021). Impact of car sharing on urban sustainability. *Sustainability, 13*(2), 905. Tukker, A. (2015). Product services for a resource-efficient and the circular economy–A review. *Journal of Cleaner Production, 97*, 76–91.
Roelich, K., Knoeri, C., Steinberger, J. K., Varga, L., Blythe, P. T., Butler, D., Gupta, R., Harrison, G. P., Martin, C., & Purnell, P. (2015). Towards resource-efficient and service-oriented integrated infrastructure operation. *Technological Forecasting and Social Change, 92*, 40–52.
Saidani, M., Yannou, B., Leroy, Y., & Cluzel, F. (2017). How to assess product performance in the circular economy? Proposed requirements for the design of the circularity measurement framework. *Recycling, 2*(1), 6.
Salas, D. A., Criollo, P., & Ramirez, A. D. (2021). The role of higher education institutions in the implementation of the circular economy in Latin America. *Sustainability, 13*(17), 9805.
Sallam, K., Mohamed, M., & Mohamed, A. W. (2023). Internet of Things (IoT) in supply chain management: Challenges, opportunities, and best practices. *Sustainable Machine Intelligence Journal, 2*, 1–3.
Salmenperä, H., Pitkänen, K., Kautto, P., & Saikku, L. (2021). Critical factors for enhancing the circular economy in waste management. *Journal of Cleaner Production, 280*, 124339.
Salvador, R., Barros, M. V., Donner, M., Brito, P., Halog, A., & De Francisco, A. C. (2022). How to advance regional the circular bioeconomy systems? Identifying barriers, challenges, drivers, and opportunities. *Sustainable Production and Consumption, 32*, 248–269.
Sanchez, F. A. C., Boudaoud, H., Camargo, M., & Pearce, J. M. (2020). Plastic recycling in additive manufacturing: A systematic literature review and opportunities for the circular economy. *Journal of Cleaner Production, 264*, 121602.
Santana, S., & Ribeiro, A. (2022). Traceability models and traceability systems to accelerate the transition to the circular Economy: A systematic review. *Sustainability, 14*(9), 5469.
Sauerwein, M., Doubrovski, E., Balkenende, R., & Bakker, C. (2019). Exploring the potential of additive manufacturing for product design in the circular economy. *Journal of Cleaner Production, 226*, 1138–1149.
Schmidt, J. L., Sehnem, S., & Spuldaro, J. D. (2024). Blockchain and the transition to the circular economy: A literature review. *Corporate Social Responsibility and Environmental Management, 31*(3), 2010–2032.
Schoemaker, P. J., Day, G. S., & Rao, G. (2020). Converting strategic ambiguity to competitive advantage: How Philips Lighting solved the challenge of LED technology disruption. *Strategy & Leadership, 48*(2), 10–17.
Schumacher, K., & Green, M. L. (2021). The circular economy in the high-tech world workshop report. *NIST Special Publication, 10*, 1500–2204.
Scopus Database. (2025). *Elsevier B.V.* https://www.scopus.com/search/form.uri?display=basic#basic. Accessed on 20 Apr 2025.
Sehnem, S., Vazquez-Brust, D., Pereira, S. C. F., & Campos, L. M. (2019). The circular economy: Benefits, impacts and overlapping. *Supply Chain Management: An International Journal, 24*(6), 784–804.
Seyi-Lande, O. B., Layode, O., Naiho, H. N. N., Adeleke, G. S., Udeh, E. O., Labake, T. T., & Johnson, E. (2024). The circular economy and cybersecurity: Safeguarding information and

resources in sustainable business models. *Finance & Accounting Research Journal, 6*(6), 953–977.

Shah, K., & Sharma, G. P. (2022a). Nature-based Climate Solutions' (NbCS) in South Asia: emerging trends. *Harnessing Nature (IUCN–CEM), 4*(1), 10–12.

Shah, K., & Sharma, G. P. (2022b). A missing cog in the wheel: An Allee effects perspective in biological invasion paradigm. *Environmental Monitoring and Assessment, 194*(1), 1–10.

Sharma, N. K., Govindan, K., Lai, K. K., Chen, W. K., & Kumar, V. (2021). The transition from the linear economy to the circular economy for sustainability among SMEs: A study on prospects, impediments, and prerequisites. *Business Strategy and the Environment, 30*(4), 1803–1822.

Shashi, C. P., Cerchione, R., & Jhamb, D. (2024). What makes people hesitant from the circularity: An analysis of risk, marketing mix, cost and inconvenience. *Journal of Consumer Behaviour, 23*(1), 43–60.

Shojaei, A., Ketabi, R., Razkenari, M., Hakim, H., & Wang, J. (2021). Enabling the circular economy in the built environment sector through blockchain technology. *Journal of Cleaner Production, 294*, 126352.

Sikander, A. (2024). Innovative approaches to workforce training for the circular economy. *Bulletin of Green Environmental Technology, 1*(1), 47–57.

Silvestri, L., Forcina, A., Silvestri, C., Arcese, G., & Falcone, D. (2024). Exploring the environmental benefits of an open-loop the circular economy strategy for automotive batteries in industrial applications. *Energies, 17*(7), 1720.

Singh, P., & Giacosa, E. (2019). Cognitive biases of consumers as barriers in transition towards the circular economy. *Management Decision, 57*(4), 921–936.

Singh, R. L., & Singh, P. K. (2016). Global environmental problems. In *Principles and applications of environmental biotechnology for a sustainable future* (pp. 13–41). Singapore: Springer Singapore.

Solomon, N. O., Simpa, P., Adenekan, O. A., & Obasi, S. C. (2024). The circular economy principles and their integration into global supply chain strategies. *Finance & Accounting Research Journal, 6*(5), 747–762.

Somoza-Tornos, A., Gonzalez-Garay, A., Pozo, C., Graells, M., Espuña, A., & Guillén-Gosálbez, G. (2020). Realizing the potential high benefits of the circular economy in the chemical industry: Ethylene monomer recovery via polyethylene pyrolysis. *ACS Sustainable Chemistry & Engineering, 8*(9), 3561–3572.

Sørensen, P. B. (2018). From the linear economy to the circular economy: A basic model. *FinanzArchiv/Public Finance Analysis*, 71–87.

Srivastava, A., Pandey, S., Shahwal, R., & Sur, A. (2025). Recycling of waste into useful materials and their energy applications. In *Microbial Niche Nexus Sustaining Environmental Biological Wastewater and Water-Energy-Environment Nexus* (pp. 251–296). Cham: Springer Nature Switzerland.

Srivastava, S. K., & Srivastava, R. K. (2006). Managing product returns for reverse logistics. *International Journal of Physical Distribution & Logistics Management, 36*(7), 524–546.

Sterev, N. (2019). New industrial business models: From the linear to the circular economy approach. *Trakia Journal of Sciences, 17*(1), 511–523.

Ström, M., Thuvander, L., & Wallbaum, H. (2020). Understanding the circularity of resource consumption: Material loops of textiles in the circular economy perspective. *Sustainability, 12*(12), 5030.

Suchek, N., Fernandes, C. I., Kraus, S., Filser, M., & Sjögrén, H. (2021). Innovation and the circular economy: A systematic literature review. *Business Strategy and the Environment, 30*(8), 3686–3702.

Sumter, D., de Koning, J., Bakker, C., & Balkenende, R. (2021). Key competencies for design in the circular economy: Exploring gaps in design knowledge and skills for the circular economy. *Sustainability, 13*(2), 776.

Sung, K. (2023). Understanding upcycling and the circular economy and their interrelationships through literature review for design education. *Proceedings of the Design Society, 3*, 3721–3730.

Suzanne, E., Absi, N., & Borodin, V. (2020). Towards the circular economy in production planning: Challenges and opportunities. *European Journal of Operational Research, 287*(1), 168–190.
Szaky, T. (2019). *The future of packaging: From the linear to the circular.* Berrett-Koehler Publishers.
Tan, J., Tan, F. J., & Ramakrishna, S. (2022). Transitioning to the circular economy: A systematic review of its drivers and barriers. *Sustainability, 14*(3), 1757.
Tavares, T. M., Ganga, G. M. D., Godinho Filho, M., & Rodrigues, V. P. (2023). The benefits and barriers of additive manufacturing for the circular economy: A framework proposal. *Sustainable Production and Consumption, 37*, 369–388.
Thirumal, S., Udawatta, N., Karunasena, G., & Al-Ameri, R. (2024). Barriers to adopting digital technologies to implement the circular economy practices in the construction industry: A systematic literature review. *Sustainability, 16*(8), 3185.
Toffel, M. W. (2003). The growing strategic importance of end-of-life product management. *California Management Review, 45*(3), 102–129.
Trevisan, A. H., Lobo, A., Guzzo, D., de Vasconcelos Gomes, L. A., & Mascarenhas, J. (2023). Barriers to employing digital technologies for the circular economy: A multi-level perspective. *Journal of Environmental Management, 332*, 117437.
Tukker, A. (2015). Product services for a resource-efficient and the circular economy–a review. *Journal of Cleaner Production, 97*, 76–91.
Ugwu, C. N., Ugwu, O. P. C., Alum, E. U., Eze, V. H. U., Basajja, M., Ugwu, J. N., Ogenyi, F. C., Ejemot-Nwadiaro, R. I., Okon, M. B., Egba, S. I., & Uti, D. E. (2025). Sustainable development goals (SDGs) and resilient healthcare systems: Addressing medicine and public health challenges in conflict zones. *Medicine, 104*(7), 41535.
Van den Berg, H., & Bakker, C. (2015). A product design perspective on designing for the circular economy. *Proceedings of PLATE: Product Lifetimes and the Environment*, 1–7.
van Ewijk, S., & Stegemann, J. A. (2020). Recognising waste use potential to achieve the circular economy. *Waste Management, 105*, 1–7.
van Loon, P., & Van Wassenhove, L. N. (2020). Transition to the circular economy: The story of four case companies. *International Journal of Production Research, 58*(11), 3415–3422.
Velter, M. G. E., Bitzer, V., & Bocken, N. M. P. (2022). A boundary tool for multi-stakeholder sustainable business model innovation. *The Circular Economy and Sustainability, 2*(2), 401–431.
Vivanco, D. F., Kemp, R., & Van Der Voet, E. (2016). How to deal with the rebound effect? A policy-oriented approach. *Energy Policy, 94*, 114–125.
Waddell, S. (2017). *Societal learning and change: How governments, business and civil society are creating solutions to complex multi-stakeholder problems.* Routledge.
Wang, J., & Azam, W. (2024). Natural resource scarcity, fossil fuel energy consumption, and total greenhouse gas emissions in top emitting countries. *Geoscience Frontiers, 15*(2), 101757.
Wang, J. X., Burke, H., & Zhang, A. (2022). Overcoming barriers to the circular product design. *International Journal of Production Economics, 243*, 108346.
Wasserbaur, R., Sakao, T., & Milios, L. (2022). Interactions of governmental policies and business models for the circular economy: A systematic literature review. *Journal of Cleaner Production, 337*, 130329.
Watson, E. (2021). *The ugly side of fashion's take-back programmes.* Vogue Business.
Wijkman, A., & Skånberg, K. (2015). The circular economy and benefits for society. *Club of Rome*, 12–59.
Wilts, H., Von Gries, N., & Bahn-Walkowiak, B. (2016). From waste management to resource efficiency—The need for policy mixes. *Sustainability, 8*(7), 622.
Winkler, H. (2011). Closed-loop production systems—A sustainable supply chain approach. *CIRP Journal of Manufacturing Science and Technology, 4*(3), 243–246.
Wysokińska, Z. (2020). A review of transnational regulations in environmental protection and the circular economy. *Comparative Economic Research. Central and Eastern Europe, 23*(4), 149–168.

Yamaguchi, S. (2021). International trade and the circular economy-Policy alignment. *OECD Trade and Environment Working Papers, 2021*(2), 1–80.

Younis, H., Sundarakani, B., & Alsharairi, M. (2022). Applications of artificial intelligence and machine learning within supply chains: Systematic review and future research directions. *Journal of Modelling in Management, 17*(3), 916–940.

Yu, Y., Junjan, V., Yazan, D. M., & Iacob, M. E. (2022). A systematic literature review on The circular Economy implementation in the construction industry: A policy-making perspective. *Resources, Conservation and Recycling, 183*, 106359.

Zemanová, V. (2023). The circular economy implementation from the perspective of benefits and barriers. In *Liberec economic forum 2023. Presented at the Liberec economic forum 2023* (pp. 282–290). Technical University of Liberec.

Zink, T., & Geyer, R. (2017). The circular economy rebound. *Journal of Industrial Ecology, 21*(3), 593–602.

MIX
Papier aus verantwortungsvollen Quellen
Paper from responsible sources
FSC® C105338

If you have any concerns about our products,
you can contact us on
ProductSafety@springernature.com

In case Publisher is established outside the EU,
the EU authorized representative is:
Springer Nature Customer Service Center GmbH
Europaplatz 3, 69115 Heidelberg, Germany

Printed by Libri Plureos GmbH
in Hamburg, Germany